W0106361

Thrombosis and
Thrombolysis

Thrombosis and Thrombolysis

Edited by
E. I. Chazov
and
V. N. Smirnov

Institute of Experimental Cardiology
Cardiology Research Center
Academy of Medical Sciences of the USSR
Moscow, USSR

CONSULTANTS BUREAU • NEW YORK AND LONDON

Library of Congress Cataloging in Publication Data

Thrombosis and thrombolysis.

Translation from the Russian.
Includes bibliographical references and index.
1. Thrombosis. 2. Fibrinolysis. I. Chazov, E. I. II. Smirnov, V. N. (Vladimir
Nikolaevich), 1937–
RC394.T5T46   1986   616.1'35   86-20472
ISBN-13: 978-1-4684-1661-9   e-ISBN-13: 978-1-4684-1659-6
DOI: 10.1007/978-1-4684-1659-6

This translation is published under an agreement with
the Copyright Agency of the USSR (VAAP).

© 1986 Consultants Bureau, New York
Softcover reprint of the hardcover 1st edition 1986
A Division of Plenum Publishing Corporation
233 Spring Street, New York, N.Y. 10013

All rights reserved

No part of this book may be reproduced, stored in a retrieval system, or transmitted
in any form or by any means, electronic, mechanical, photocopying, microfilming,
recording, or otherwise, without written permission from the Publisher

PREFACE

Recent years have seen great strides in research on the pathogenesis of thromboses, unmatched by progress in other branches of hemostasiology. The orthodox concepts of the mechanisms of thrombus formation described by Virchow have come down to us as a "classical triad" of factors. Now, due to developments in molecular biology, pharmacology, and pathophysiology, they appear in a basically new light. The fruits of modern research, currently being tested or already implemented in clinical practice, have opened up the possibility of controlling the hemostatic process and developing effective antithrombotic drugs.

Much progress has been achieved in the past years, but much more remains to be achieved in such areas as the pathogenesis of venous and arterial thromboses, early diagnosis, therapy, and control of disorders. Many scientists in the U.S.S.R. are involved in studying these problems. Their data, from years of research carried out in leading laboratories and clinics in the U.S.S.R., are summarized in this monograph.

This work is written by experts in various fields of biology and medicine. It deals with new and original concepts on the structure and function of the fibrinolytic system, the role of nonenzymatic fibrinolysis in regulating physiological hemostasis, the heterogeneous and discrete patterns of the system regulating blood coagulation, the molecular mechanisms of fibrin polymerization, and the anticoagulating effects of fibrinogen/fibrin degradation products.

The use of immobilized enzymes and directed drug transport in treating thrombotic occlusions is emphasized. The latter technique is promising for the targetted transport of thrombolytics to zones of endothelial injury and to the site of a thrombus. The successful implementation of directed drug transport in thrombolytic therapy makes it possible to hope that, in the not too distant future, it may be used in the treatment of local injuries of various organs and tissues. The use of gravitational surgery techniques for treating

v

thromboses is another interesting topic outlined in this monograph.

The final section of this book is concerned with new
antithrombotic agents: streptodecase (the first immobilized
fibrinolytic drug developed in the U.S.S.R.), plasmakinase
(derived from cadaver blood), terrilytin and tricholysin (from
the fungi Aspergillus terricola and Trichothecium roseum),
urokinin, and platelet aggregation inhibitors.   Laboratory
and clinical data from trials of these preparations are presented, and an in vitro model for the screening of antithrombotic drugs is described.

The authors hope that this work will be of interest to
hemostasiologists and that it may provide an impetus to further
interesting studies.

CONTENTS

MOLECULAR MECHANISMS OF FIBRIN POLYMERIZATION AND ANTICOAGULANT ACTION OF FIBRINOGEN AND FIBRIN DEGRADATION PRODUCTS

V. A. Belitser and T. V. Varetskaya

A. V. Palladin Institute of Biochemistry
Academy of Sciences of the Ukrainian SSR, Kiev

ABSTRACT

The well-known concept assuming a sequence of end-to-end and side-to-side interactions of fibrin monomer molecules, which are involved in polymerization, is proven inadequate. The key role of the coupling of the central part of one monomer with the peripheral parts of two other monomers, the end-to-middle bindings, is established. Fibrinogen, and all of its degradation products bearing domains D, exhibited antipolymerization properties although only limited formation of fibrin monomer—inhibitor complex could be detected. Relatively stable complexes consisting of fibrin monomer and fragment D appeared under certain conditions provided that the peripheral (D) parts of monomer molecules had been reversibly damaged by acid (pH < 4). This modification prevents normal fibrin polymerization and allows the free fragment D to occupy the active binding sites of domain E of the modified monomer. The number of fragment D molecules thus fixed did not exceed three per mole of monomer. This valence is puzzling insofar as an odd number of bonds is not expected. On further investigation of the complex formation between active fragments and fibrin monomer in various systems, new facts emerge leading to new postulates on the mechanisms of protofibril and fibrin fiber assembly.

1

V. A. BELITSER ET AL.

INTRODUCTION

Fibrinogen is a high-molecular-weight protein (MW = 340,000). Its molecule consists of two subunits, each of which is comprised of three polypeptide chains: Aα (610 amino acid residues), Bβ (461 residues), and γ (411 residues). Fibrinogen polypeptide chains assume peculiar conformations in the molecule, making its general form a complicated one (Fig.1.1).

Three globules, morphologically and functionally independent, represent the large domains of the fibrinogen molecule. In addition, the molecule includes two smaller domains. These are loose mobile bodies which originate from the peripheral domains and stretch toward the middle of the molecule [1, 21]. They belong to the Aα chain, being its long COOH-terminal region. We refer to them as domains αC.

In microcrystals of modified fibrinogen, electron-microscopic details are clearly visible. In the fibrinogen molecule, seven separate small globules are distributed over the whole length of the three large domains [3]. The domains must be considered functional subunits of the molecule. The molecular mechanisms of fibrin(ogen) function are being analyzed mainly at the level of these subunits.

It has been possible to study the functions of domains because, during fibrinogen splitting by plasmin or trypsin, the destruction of the molecule is limited to domains αC and the bridges between large domains, although in the free fragment state these domains remain virtually intact. Thus, peripheral domains can be isolated from fibrinogen or fibrin molecules almost undamaged and preserving their functional activity.

Fibrinogen proteolysis by plasmin proceeds stage by stage. First the enzyme destroys domains αC. The first intermediate product, i.e., fragment X, is formed, retaining all of the three large domains. In the two following stages, the domains are separated (from each other), being converted to fragments D and E with the intermediate fragment Y (Fig. 1.2).

The peripheral domains are converted to fragment D, and the central one to fragment E. The peripheral domain is usually referred to as domain D, and the central domain is known as domain E. One should remember that domains D and E are similar but not identical to fragments D and E.

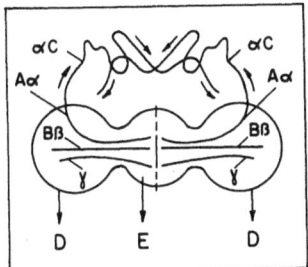

Fig. 1.1. Fibrinogen molecule. E) Central domain;
D) peripheral domains; αC) domains. Aα,
Bβ, γ) Polypeptide chains. Dashed line:
border of molecular subunits.

Important information on fibrinogen and its fragments
was obtained by thermal denaturation, i.e., the structural
transition from order to disorder caused by heat. Protein
denaturation is known to be a cooperative transition, a kind
of fusion. This transition can be revealed calorimetrically
through measurement of heat adsorption. The fibrinogen molecule was found to contain regions which denature independently of each other at different temperatures [4-6]. Fragment
D includes thermolabile and thermostable regions, and fragment
E has only thermostable ones [5, 6] (Fig. 1.3). According to
evidence accumulated by our laboratory, in cooperation with
the P. L. Privalov Laboratory at the Protein Institute, Academy of Sciences of the USSR, the fibrinogen molecule includes twelve ordered fusible structures. Each domain D contains three thermolabile regions and one thermostable, ordered region; domain E comprises two thermostable, fusible regions. The last two fusible regions are thermolabile and belong to the αC domains. With respect to the thermodynamic
parameters of fusion, the thermolabile regions of domains D
behave like protein of a globular type, while thermostable
regions of domains D and E resemble superspiral rodlike structures [6]. The thermostable component of the D fragment was
isolated after proteolysis of that fragment and was studied
in detail [7]. The degree of α-spiralization was found to
be very high, confirming its superspiral feature. It must
be a part of the rod which connects globular parts of the D
and E domains.

The structure of fibrinogen is not rigid. This protein
can change its configuration in a fundamental manner. The

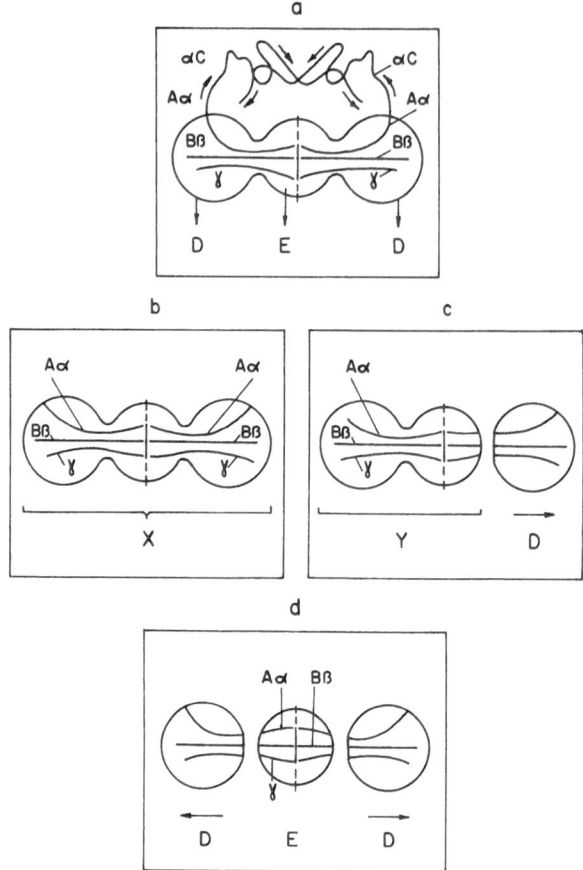

Fig.1.2. Fibrinogen molecule cleavage by plasmin. a) Intact
        molecule; b) fragment X; c) fragments Y and D; d)
        fragments D and E.

fibrinogen molecule may assume elongated as well as spherical forms. Several molecular species can often be found in
a single electron-microscopic fibrinogen specimen [2]. Probably, reversible shape transformations are taking place continuously. A single conformation out of many potential ones
is favored under certain conditions of the medium and may
acquire stability [8, 9].

    The prevailing form of fibrinogen molecules seen in
electron-microscopic microphotographs depends, among other

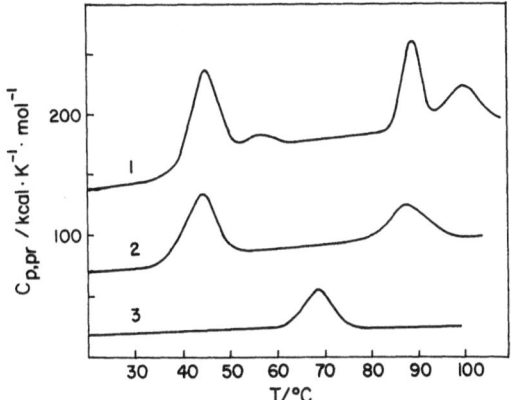

Fig. 1.3. Curves of fibrinogen melting (1) and its
proteolytic fragments D (2) and E (3) in
0.05 glycinic buffer, pH 3.5 [6].

things, on the method of staining used. Negative staining
with sodium phosphotungstate tends to make fibrinogen molecules spherical or, rather, vesicular. We have found no
spherical forms at ionic strength 0.15-0.2 using uranyl acetate as a negative contraster. However, at a very low ionic
strength (<0.05), spheres (vesicles) seemed to be the only
observable form of fibrinogen molecules [10]. If the ionic
strength of the fibrinogen solution is increased from 0.05
to 0.15, the "vesicles" disappear. The propensity to form
vesicles is evidently induced by low ionic strength. Transition to the state of vesicles, as well as some other transitions, is obviously a reversible process.

When activated by thrombin, fibrinogen is converted to
a monomeric fibrin; it then aggregates in an orderly manner
and polymerizes to form a fibrin network.

The conversion of fibrinogen to monomeric fibrin represents a typical transition of an inert precursor protein to
its active form. This is performed as a rule by proteolytic
enzymes which cleave specific peptide bonds in the polypeptide chain of the precursor. In our case, a specific enzyme,
thrombin, activates fibrinogen by cleaving small peptides (A
and B fibrinopeptides) from NH_2 terminals of polypeptide
chains Aα and Bβ; in this way, fibrinogen is converted to the
active fibrin monomer which polymerizes readily.

The term "chains Aα and Bβ" refers to the peptide sections A and B to be released as fibrinopeptides. Chains Aα and Bβ, deprived of fibrinopeptides, become polypeptide chains of fibrin: α and β. The fibrinogen chemical composition is given by the formula (Aα, Bβ, γ)$_2$, and that of fibrin (α, β, γ)$_2$.

The fibrinogen molecule is known to consist of two subunits, each of which has single Aα, Bβ, and γ chains. Upon complete activation, four fibrinopeptides are cleaved off from the fibrinogen molecule (two A and two B). Fibrinopeptides A are split off more rapidly than fibrinopeptides B. The difference is great if the thrombin concentration is low; then, and first, fibrin lacking fibrinopeptides A but retaining fibrinopeptides B (i.e., fibrin des-A) accumulates. In the case of intense thrombin action, fibrin lacking both types of fibrinopeptides (i.e., fibrin des-AB) is produced. Both forms of fibrin polymerize readily, but the molecules of fibrin des-AB form more stable polymers.

The initial product of polymerization, in any case, is a long thread two monomers thick, called protofibril (Fig. 1.4). The latter gains in length due to the addition of monomeric fibrin molecules. Having reached the critical (medium-dependent) length, the protofibrils begin to aggregate laterally, interacting through specific regions of their surfaces. Finally, thick branching fibrin fibers are constructed (at the expense of protofibrils) to form the three-dimensional fibrin network. The mechanism of protofibrin formation differs considerably from that of the final fibrin network assembly. These are two qualitatively different "polymerization" stages. In fact, the whole process can hardly be called polymerization and may be characterized rather as a biological structure assembly. But the term "fibrin polymerization" has become popular, while "fibrin assemblage" is seldom used. Although we have to go along with this usage, it should be remembered that fibrin "polymerization" is a complicated process resembling the construction of a biological structure.

FIBRIN MONOMER POLYMERIZATION

In considering the mechanism of fibrin polymerization, the first question to be answered concerns the general type of binding of monomeric molecules. For many years the idea predominated that fibrin molecules first reacted "end-to-end"

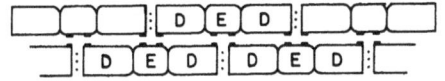

Fig. 1.4. Protofibril (see text).

and then "side-to-side." This idea has become passé. Molecules were found to interact by means of complementary polymerization sites located in the middle of one molecule and
at the end of another molecule. Therefore, in order to form
a specific monomer—monomer bond, the molecules have to react
exclusively in the end-to-middle manner. This was directly
proved electron-microscopically. If the end-to-middle interaction takes place, the dimers must be 1.5 times longer than
the monomers. The trimers must be twice as long. In the case
of end-to-end interaction, the length of the particles would
be doubled on dimerization and tripled on trimerization. Measurements performed on electron-microscopic images confirm the
end-to-middle model. The end-to-middle interaction immediately explains the formation of protofibrils with a constant
width of two monomeric units.

Deeper insight into the mechanisms involved in protofibril assembly was achieved by studies of the structures responsible for each elementary event.

As mentioned above, the plasmin hydrolysis product fragment E and the hydrophilic product of bromocyanide splitting
— the N-terminal disulfide knot (N-DSK) — are assumed to be
derivatives of domain E (the central domain) of the fibrinogen
molecule and the plasmic fragment D — a derivative of the peripheral (D) domain. This was completely borne out immunochemically. Identical antigenic determinants are demonstrated
in the domains and corresponding fragments [11] by means of
electron microscopy.

A specific affinity of certain fragments and domains is
expressed by the selective formation of complexes. The immobilization of one of the reactants on a solid carrier, usually sepharose, proved useful in the study of complexing. The
presence in fragment D of specific reactive centers essential
for polymerization was found in the following experiments.
Fragment D freely passes through the column with immobilized
fibrinogen. However, when the latter is converted to the
(immobilized) monomeric fibrin by thrombin, fragment D is re-

tained in the column. To separate it from the column, it is
necessary to change the pH drastically and to use solubilizing
agents such as urea.

In the structure of fragment D there are, evidently, specific sites with strong selective affinity for such polymerization centers which appear in the fibrinogen molecule as
a result of activation.

To simplify the discussion concerning the fibrin polymerization mechanism, we propose to use some simple designations
for the polymerization sites (centers) of fibrin(ogen) molecules. The sites of domain E may be called E_1 and E_2, of
which E_1 is a center acquired on fibrinopeptide A release,
and E_2, on fibrinopeptide B release. Domain D of fibrin,
and fibrinogen as well, contains four polymerization sites,
two of which are complementary to E_1 and two to E_2. We call
them D_1 and D_2, respectively. In domain D of fibrin des-A
there must be two E_1 sites. An intermediate with only one
E_1 site has been postulated but not proved. The two fibrinopeptides A are split off virtually in a single act.

The specific reactive sites of domain D represent parts
of its thermolabile regions. Thermal denaturation makes fragment D entirely inert. Denaturation of D_1 and D_2 prevents
fibrin monomer polymerization. As mentioned above, domain
D is thermolabile and its polymerization sites D_1 and D_2 cease
functioning after denaturation. In contrast, sites E_1 and
E_2 of the CNBr derivative of domain E function normally in
spite of the strongly denatured state of the domain as a
whole.

Domain E is a partner of fragment D in the monomer—fragment D complexing reaction. But the isolated fragment E
often does not react with fragment D. Literature data on
this topic are inconsistent, although it probably depends on
some variations in fragment D structure. Thus, this question
remains open.

The functional properties of domain E are manifested in
its CNBr derivative (N-DSK). It contains intact NH_2 terminals of all three polypeptide chains. If fibrinopeptides are
cleaved off from N-DSK by thrombin, the immobilized N-DSK
acquires the ability to bind fragment D. This simple system
demonstrates the interaction of domains D and E, which is the
fundamental factor of fibrin polymerization.

Since the specific affinity for the activated domain E is characteristic of both domain D of the intact fibrin monomer molecule and the free fragment D, the competition between them for domain E is inevitable. This competition manifests itself in the inhibition by fragment D of fibrin monomer polymerization. The value of the inhibitory effect may be regarded as a measure of fragment D activity. A more detailed description of the phenomenon of polymerization inhibition is given in the next section.

We studied the dependence of fragment D activity on the conditions under which proteolytic fragmentation of fibrinogen took place. The presence of Ca^{2+} ions was found to be of primary importance. Many previous publications noted that, during fibrinogen hydrolysis, the inhibitory activity first increases considerably and then sharply decreases to a low final value [12]. Even at low concentrations (10^{-4} M), $CaCl_2$ was found to increase the level of maximum activity and to prevent activity lowering [13] (Fig. 1.5). Fragment D isolated in the presence of Ca^{2+} maximum hydrolyzate activity is a very strong inhibitor, whereas that obtained in the absence of $CaCl_2$ after the decrease in inhibitory activity hardly affects the polymerization process. Furthermore, Ca^{2+} was found to terminate fibrinogen hydrolysis at a certain level and stop the formation of low-molecular-weight degradation products [14].

Haverkate and Timon [15] succeeded in showing that calcium ions render the COOH-terminal amino acid sequence of the γ chain (MW 13,000) resistant to hydrolysis by the enzyme. The ability of fragment D to bind calcium ions was investigated in [16]; a dissociation constant of 10^{-5} M was established.

Calcium is now known to be fixed in the last quarter of the polypeptide γ chain of fragment D.

Olexa and Budzynski [17] succeeded in localizing the polymerization center D_1 in the γ chain and isolating a peptide where the center is present. The COOH-terminal peptide of the γ chain with MW 13,000 was thoroughly hydrolyzed and, among the set of small peptides obtained, one peptide proved to carry the active center D_1. The immobilized monomeric fibrin was successfully used to isolate this peptide. Due to E_1, it specifically binds the D_1-containing peptide. The latter was identified as the COOH-terminal part of the γ chain

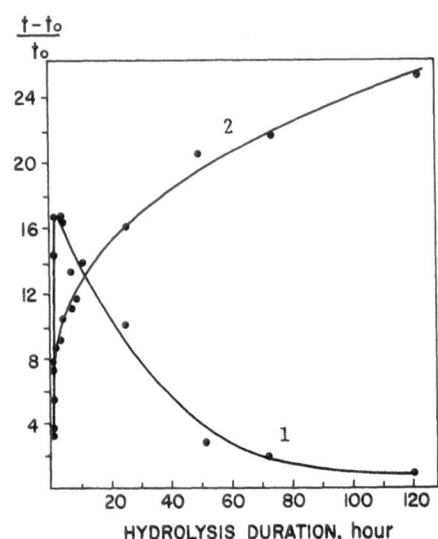

Fig. 1.5. Influence of $CaCl_2$ on inhibitory activity of fibrinogen plasmic hydrolysis products. 1) Hydrolysis
without $CaCl_2$; 2) hydrolysis in the presence of
$1 \cdot 10^{-4}\,M\,CaCl_2$. Weight ratio of fibrin monomer and
fibrinogen hydrolyzate is 1:0.5 and 1:0.37 for (1)
and (2), respectively.

which contains amino acid residues 372-411. This peptide
turned out to be the strongest inhibitor of fibrin polymerization. The experiments performed with acidic fibrin monomer
preparation [17] showed that equimolar quantities of fragment
D and peptide 372-411 exhibited about the same inhibitory effect on fibrin polymerization.

    The D_1 center, being a low-molecular-weight peptide (MW
probably even lower than the whole 372-411 sequence), cannot
be thermolabile by itself. Inactivation of D_1 due to thermal
denaturation of fragment D may be due to a displacement of
D from its exposed position on the surface of the fragment D.

    We now consider in detail the domain E polymerization
centers E_1 and E_2. Since these centers result from the cleavage of fibrinopeptides, the newly formed NH_2-terminal amino
acid sequences could be expected to function as E_1 and E_2.
This assumption has been confirmed. Peptides identical to
NH_2 terminals of the fibrin α and β chains have been synthe-

sized [18-20]. For the α chain these are Gly-Pro-Arg; for the β chain, Gly-His-Arg-Pro. The specific properties of the E_1 center are revealed in the synthetic tripeptide Gly-Pro-Arg which inhibits fibrin polymerization.

The peptide acted as a competitive inhibitor interfering with the functions of the true E_1 sites of fibrin monomer molecules. Certain derivatives of this tripeptide were found to be more active inhibitors. In these derivatives free arginine carboxyls were converted to CO—NH groups, rendering the synthetic products even more similar to the Gly-Pro-Arg sequence of the fibrin α chain. All the components of the inhibiting peptide were needed for the inhibition phenomenon; the absence of any component (for instance, Gly) or its substitution by any other amino acid led to a fully inert product.

Thus, the Gly-Pro-Arg tripeptide, as a "doubler" of polymerization center E_1, possesses a strong affinity for center D_1. When fibrin monomer is polymerizing in the presence of this peptide, there is a competition between it and E_1 of domain E for the D_1 of domain D. On the other hand, the COOH-terminal peptide of the γ chain, containing center D_1, competes with D_1 of domain D for E_1 of domain E. In either case, fibrin polymerization is retarded.

The Gly-His-Arg-Pro synthetic peptide, the doubler of E_2 polymerization sites, does not influence fibrin polymerization. This is not unexpected since the faster reaction between E and D_1 alone is sufficient to ensure protofibril formation from monomeric fibrin. The E_2–D_1 coupling contributes nothing to the rate of the polymerization process (although the strength of the resulting polymeric fibrin structure is increased by the E_2–D_1 pair). Thus, the inability of the synthetic Gly-His-Arg-Pro to retard fibrin polymerization becomes evident.

In spite of this negative result, one has to assume the existence of a specific affinity of peptide Gly-His-Arg-Pro to fibrinogen and fibrin. These proteins have normal domains D where both D_1 and D_2 must be present. Therefore, not only Gly-Arg-Pro must be bound to these proteins, but also the peptide Gly-His-Arg-Pro. In fact, by means of labeled specimens of these synthetic peptides, their binding to fibrinogen and fibrin could be readily confirmed [19].

A hereditary abnormality of fibrinogen, namely fibrinogen Detroit, should be mentioned here. Thrombin will normally remove fibrinopeptides from this fibrinogen, but fibrin
polymerization hardly takes place and, hence, severe hemorrhage occurs. Kudryk et al. [20] discovered that the structure
of the polypeptide chain of Aα of fibrinogen is changed in
this pathology due to a genetic mutation. In position 19 of
this chain, arginine is replaced by serine. In fibrin, Arg
19 occurs in position 3 and is included in the sequence Gly-
Pro-Arg of the E_1 polymerization site. In fibrinogen Detroit
[21] this sequence is transferred into Gly-Pro-Ser. The fibrinogen Detroit pathology shows that the conclusions concerning the importance of every amino acid in the NH_2-terminal
sequence Gly-Pro-Arg based on in vitro experiments are also
valid in vivo.

In general, the highly ordered molecular structures of
living organisms are built up on the basis of innumerable
noncovalent bonds (H bonds and others). These bonds provide
a level of structural order determining the formation of
strong covalent bonds at certain strategically important
positions. According to this principle, fibrin fibers are
formed through noncovalent systems and fortified by additional covalent (isopeptide) bonds. The process of protofibril
assembly is due entirely to noncovalent interactions. The
absence of covalent bonds is seen from the fact that protofibrils are dissociated into monomers by urea or other solubilizing agents. Let us briefly review noncovalent bonds
participating in the mechanism of fibrin monomer polymerization. There is strong evidence pointing to the importance
of hydrogen-bond formation as an element of the mechanism of
fibrin self-assembly [22]. It appears necessary to assume
the presence on the surface of the fibrin molecule of sets
of amino acids fit to form intermolecular hydrogen bonds.
These amino acids are histidine, a hydrogen acceptor, and
probably tyrosine, a hydrogen donor. They must be situated
in positions ensuring precise contacts to form the system of
H bonds. A theory was developed by Scheraga and Laskowski
[22] that enabled them to predict the processes of release
or binding of H^+ accompanying fibrin monomer polymerization
at different pH's of the medium. How these amino acids are
actually distributed on the fibrin molecule remains to be determined.

In 1971, data appeared showing that some nonpolar chemical groups of fibrin molecules contribute to the mechanism

of fibrin polymerization by forming hydrophobic bonds. At
a concentration not exceeding 1.0 M, NaCl inhibits fibrin
polymerization and increases the transparency of fibrin gels,
but, at a concentration of 1.1 M or higher, this salt accelerates polymerization and makes fibrin gels more turbid [23].
Similar effects could be observed with other salts, and the
effectiveness increases in parallel with the salting-out capacity according to the position of the salt in the well-known
Hoffmeister anion series. These results are in agreement
with those of a more recent investigation [24]. The polymer-
ization-accelerating salt concentrations proved much lower
than those which are needed to induce a real salting-out of
proteins. Low-molecular-weight hydrophobic substances, artificially bound to protein molecules, are known to facilitate
the salting-out process. Based on these facts, it was concluded that certain exposed nonpolar amino acid residues participate in the polymerization mechanism by creating hydrophobic bonds.

In collaboration with the Department of Biophysics at
Kiev State University, we obtained data [1] suggesting that
some hydrophobic bonds are formed with the participation of
domains αC. In these experiments the hydrophobic probe ANS
was used. Exposed hydrophobic regions were detected after
removal of αC domains from fibrinogen. In the native fibrinogen such regions are absent, obviously because of their isolation from the medium by certain hydrophobic components of
αC domains. As is known, hydrophobic amino acid residues are
accumulated in the most peripheral part of the αC domain.
This region must be responsible for hydrophobic binding. The
other part of the domain αC is hydrophilic, flexible, and
devoid of rigidity. Because of this part, the terminal hydrophobic part can be moved from one position to another.

In solution, fibrinogen molecules are mostly globular,
probably because their (peripheral) D domains are bent and
brought together due to interaction with the αC domains. On
the other hand, the αC domains probably act as "helpers" in
the polymerization process. This is evident from the fact
that fibrin polymerization is hampered with progressive degradation of the αC domain. The proposition is suggested
that, during fibrin monomer polymerization, domains αC change
over their intramolecular partner, the D domains, for intermolecular ones making bridges between fibrin molecules. It
is known that, under certain conditions, fibrinogen forms
fibrinlike structures with a characteristic cross-striation

system. In this assembly of fibrinlike aggregates of fibrinogen domains, αC may play the decisive role. Probably, αC
domains of fibrinogen shift to the intermolecular positions,
as occurs during fibrin polymerization. It must be stressed
that the characteristic feature of fibrin fibers (a cross-
striation pattern) may arise in fibrinogen polymers without
activation by thrombin. In this case, the fibrinlike structure is probably determined by domain αC intermolecular
bridges.

Another factor favoring polymerization consists of electrostatic interaction between oppositely charged loci of fibrin molecules. Anionic and cationic groups of these molecules
must be specifically distributed producing either positive
charges in the same region of the molecule or negative charges
in different regions. For instance, in the γ chain, there
is an anion region in which 7 of 17 (286-304) amino acid residues are represented by aspartic acid, and cation side chains
are absent. There are cationic regions too. Therefore, despite the negative net charge, the molecules tend to bind
with each other by electric charge attraction. This tendency
contributes to fibrin polymerization. The elimination of
this former attraction by high ionic strength leads to a
marked inhibition of polymerization.

In connection with the role of electric charges in the
fibrin molecule, it is logical to mention protamine sulfate
and the paracoagulation reaction. Protamine sulfate induces
gelatinization of soluble fibrin oligomers and early products
of fibrin hydrolysis. This process is called paracoagulation
and is a well-known diagnostic test. We interpret the paracoagulation phenomenon to be a result of binding between several molecules containing anionic sites and cationic groups
of a single protamine molecule which is a long linear poly-
cation. According to electron-microscopic studies, the polymeric structure of fibers of the clot (gel) resulting from
paracoagulation resembles that of fibrin. This indicates
that protamine does not interfere with the ordered polymerization, but rather probably binds some excess charges of
the anion loci and enhances polymerization. Protamine sulfate is known to cause a positive reaction in gel formation
only if its molecules are long enough; even mild degradation
makes paracoagulation impossible.

ANTIPOLYMERIZING EFFECT OF FIBRINOGEN
AND FIBRIN FRAGMENTS. COMPLEXES FORMED
BY FIBRIN MONOMER. PROTOFIBRIL LATERAL BINDING

We have discussed the inhibition of fibrin polymerization by substances which have a specific affinity for one
of the reactive sites responsible for polymerization. Because these processes are of great interest, they are elaborated further in this section.

In experiments concerned with the study of fibrin polymerization, monomeric fibrin is the initial substance. It
may originate from fibrinogen (+fibrin), or fibrin monomer
solutions preserved under conditions excluding polymerization
may be used. The fibrinogen—thrombin system, as well as purified monomeric fibrin solutions, is useful in studying polymerization. Two kinds of monomeric fibrin preparations are
presently being used. They are obtained by dissolving fibrin
clots either in 1 M NaBr at pH 5.3 [25] or in weak acetic
acid at low temperature [26-28]. In a number of cases both
of these preparations appeared to be equally useful [29].
However, their preparations are not identical and one has
to decide in each case which kind of fibrin monomer is to be
selected. Acidic monomer is especially sensitive to specific
competitive inhibitors such as fragment D and, therefore, is
preferable in competition experiments (the cause of the higher
sensitivity of acidic monomer is described below).

Because of this peculiarity of acidic monomer, as well
as its high stability toward acids, reproducibility, and simplicity of preparative steps, we decided to use it for the
diagnostic estimation of fibrinogen degradation products (FDP).

The clotting time of the acidic fibrin monomer preparation was measured under standard conditions in a sample containing the material of interest and in a control sample.
Thus, the inhibitory effect of an agent present in the sample
can be estimated quantitatively. Since the inhibitory effect
is a function of inhibitor concentration, its quantity in the
sample can be inferred. This method was applied successfully
in the nephrologic clinic and especially for early diagnosis
of a kidney transplant rejection [30]. The application of
this method to estimating FDP in blood plasma (no previous
removal of fibrinogen from plasma is made) is most interesting. But an exact determination of the original fibrinogen
level in plasma is necessary. To achieve this, the technique

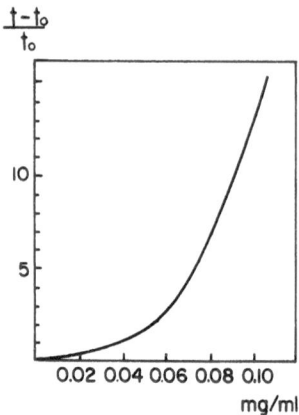

Fig. 1.6. Dependence of the inhibitory effect [(t −
t_0)/t_0] on inhibitor concentration. t_0, t)
Coagulation time in the absence and presence of inhibitor, respectively.

suggested by Belitser et al. [31] can be used. Fibrinogen
contribution to inhibition is revealed in control samples
with standard plasma. This method allows detection of early
fibrinogen degradation products that are not present in blood
serum. The use of this method for diagnostic purposes gave
encouraging results, although the data obtained must be regarded as preliminary.

Among the inhibitors of polymerization, the best known
are FDP fragments X, Y, and D. The principle of their action
is clear. Having active domains D, but lacking active domains
E (E_1 and E_2), the fragments compete with domains D of the intact fibrin monomer molecules for domain E offering it an alternative "useless" interaction. Intact fibrinogen, by the
state of its domains, corresponds well to these inhibitors.
In fact, it was found [32] to be a specific inhibitor of fibrin polymerization. Fibrinogen with fragments X, Y, and D
forms a group of inhibitors with similar properties. A re-
markably peculiar concentration dependence of the inhibitory
effect was found in experiments with this group of inhibitors
[33]. At the lowest inhibitor concentrations, the weak re-
tarding effect rises linearly as the concentration increases. At certain higher concentrations, the linearity
vanishes and the slope of the curve becomes progressively

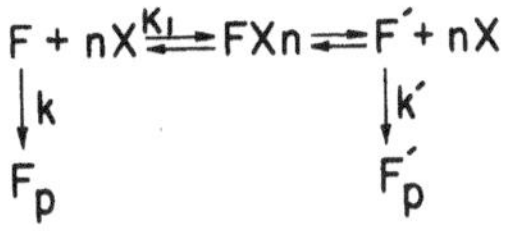

$$F + nX \underset{K_L}{\rightleftharpoons} FXn \rightleftharpoons F' + nX$$

$$\downarrow k \qquad\qquad\qquad \downarrow k'$$

$$F_p \qquad\qquad\qquad F'_p$$

Fig. 1.7. Reaction occurring in the fibrin-inhibi-
tor system. From left to right: probable
catalytic transition (conversion) of the
native fibrin monomer (F) to a modified
form (F'). The complex including n inhibitor molecules (X) is formed at the
intermediate stage of this conversion.
Vertical arrows k and k' express the polymerization (p) of F and F'.

steeper. A maximum steepness is then reached and linearity
is reestablished (Fig. 1.6).

Such a transition from a weak to a much more efficient
inhibitor at the middle concentrations is by no means a general rule. Usually, the effectiveness (the effect per inhibitor weight unit) decreases with an increase in concentration because of the gradual approach toward a saturation
state. A dramatic gain in effectiveness of an agent in a certain concentration range may be due to a cooperative transformation of the object induced by that agent. In our case,
it must be assumed that, at sufficient inhibitor concentrations, the fibrin monomer is modified, becoming especially
susceptible [34]. Electron-microscopic observations clearly
revealed that, in the presence of FDP, the fibrin monomer
undergoes damage and produces defective clots [35, 36]. Even
slightly damaged fibrin monomer molecules probably react with
the inhibitor relatively strongly.

If added at advanced stages of fibrin polymerization,
FDP neither retard the further stages of polymerization nor
produce morphological damages. It seems probable that fibrin
monomer is the only target for the initial FDP attack.

Fibrin fibers, and even soluble polymers formed in the
presence of FDP, are free from the latter. Evidently, as
the polymerization proceeds, all of the inhibitors are with-
drawn from the oligomeric intermediate. Thus, specific inhibitors act not as firmly sticking ligand molecules, but
rather catalytically, like enzymes.

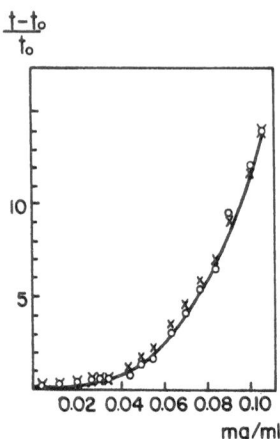

Fig. 1.8. Theoretical and experimental curve of
           fibrin monomer polymerization in the
           presence of inhibitor. See Fig. 1.6 for
           designation. o) Experimental data; x)
           theoretical values of inhibitory ef-
           fect, as calculated in [36].

In several former studies on the influence of FDP (and
fragment D, particularly) on fibrin polymerization, only a
linear dependence could be observed of the inhibitory effect
on inhibitor concentration. In these studies, inhibitor
preparations of low activity, obtained with no calcium added,
were used. As judged by the values of the inhibitory effects,
the region of cooperative increase in inhibitor efficiency
was not covered in these studies. Thus, the general conclusion about a constant proportionality of the inhibitory
strength and the concentration of the inhibitor appears er-
roneous.

Based on the suggestion outlined here, the three parts
of the curve presented in Fig. 1.6 may be interpreted as follows. The first part corresponds to the weak efficiency of
the inhibitor interacting with the unmodified fibrin monomer.
The middle part reflects the cooperative structural modification of fibrin monomer that becomes much more susceptible
(Fig. 1.7). The third part is the region of the highest inhibitor efficiency where practically all of the fibrin monomer is found in the modified form.

Using mathematical methods of kinetic analysis, we developed a quantitative model of the reactions taking place
in a system consisting of fibrin monomer polymerizing in the
presence of fragment D — a strong, specific polymerization
inhibitor. This system has been described qualitatively here.
Theoretical curves have been derived which coincide well with
the experimental ones (see Fig. 1.8).

Further analysis of the kinetics of the cooperative fibrin monomer—fragment D reaction led to the conclusion that
the cooperativity characteristic n amounts to 3 [34], i.e.,
the fibrin monomer molecule must bind three fragment D molecules to get transformed and become highly susceptible. The
calculated complex of fibrin monomer with three fragment D
molecules under usual conditions cannot be more than a transi-
ent step and its whole lifetime must be quite negligible.

To make the studies of specific inhibitors of polymerization more quantitative, appropriate activity measurements
are needed and certain activity units must be chosen. Activity units derived from the inhibition effects recorded at
equal concentrations of different inhibitors are unacceptable
because of the nonlinear character of the concentration-de-
pendence curve.

An alternative principle was used. A certain inhibitory
effect was arbitrarily chosen to serve as a standard inhibition unit; namely, a tenfold increase in clotting time of a
given solution of fibrin monomer (using the above acidic fibrin monomer preparation). For each inhibitor preparation to
be tested, the quantities were determined at which, under
strictly constant test conditions (pH, temperature, ionic
strength, etc.), they produced an inhibitory effect corresponding exactly to 1 standard activity unit. Since the amount
of the inhibitor possessing 1 activity unit is then known,
one can easily calculate the activity units per mg or μM of
the inhibitor.

Using this method we measured the activity of purified
fibrinogen, as well as plasmic fragments X, Y, and D [37]
(see Table 1.1). The values of molar activities shown in the
right column of the table are most informative. Molar activities of fragments Y and D are practically identical, while
the activity of fragment X is about twice as high. This
could be expected theoretically.

TABLE 1.1. Content of Anticoagulation Activity Units
in Fibrinogen and Its Fragments (μ/mg and
μ/mole)

	μ/mg	μ/mole
Fragment D	5.81 ± 0.06	0.58 ± 0.01
Y	3.97 ± 0.15	0.63 ± 0.02
X	4.34 ± 0.03	1.12 ± 0.01
Fibrinogen	2.02 ± 0.02	0.67 ± 0.01

The fundamental cause of the inhibitor characteristics
of fibrinogen and its fragments X, Y, and D is the presence
in these molecules of native components of the peripheral
structures (D) and the absence of the activated central
structure (E). The D_1 and D_2 sites remaining in the inhibitor molecules provide selective affinity for fibrin monomer,
but because active E_1 and E_2 are lacking, the participation
in fibrin polymerization is excluded, making those molecules
specific inhibitors of polymerization. In fragment X, the
former polymerization sites D_1 and D_2, now acting as inhibitory sites, are doubled; hence, the doubling of inhibitory
activity. The specific activity (μ/mg) is higher in fragment
D than in fragment Y simply because the molecular mass of
fragment Y is higher. The calculated difference in the specific activity between fragments D and Y is in agreement with
that shown in Table 1.1.

It is somewhat unexpected that fibrinogen, as evident
from its molar activity, is less active than fragment X. This
may be connected with the presence of domains αC which hold
the two domains D of the fibrinogen molecules close together.
This might hinder sterically the interaction between domains
E of monomeric fibrin and domains D of fibrinogen.

Now, there is every reason to regard the sites D_1 and
D_2, available either in domain or in fragment D, as the main
source of the inhibitory effects. It is interesting that in
the light species of fragment D, in which part of its γ chain
is cleaved off, the activity may be well preserved [38]. But
if another (the COOH-terminal) part of the same chain happens

to be lost on hydrolysis, a light fragment D arises with no inhibitory activity. A quite small segment of the γ chain seems to be especially responsible for the inhibitory activity.

Specific inhibitors — fibrinogen and its three fragments (X, Y, D) — seem to act by very similar, practically identical, mechanisms based on the function of D_1 and D_2 sites. In mixtures, different inhibitors behave quite additively, and the content of activity units is found experimentally to be equal to the sum of the activity units of the particular components. Inhibitors, in amounts of equal inhibitory unit content, may replace each other without changing the effect of the mixture. A minor quantity of an inhibitor added to a system in which fibrin monomer has been modified by another inhibitor acts with the highest efficiency regardless of the modifying inhibitor.

However, such an identity of interaction mechanisms does not extend to all of the specific inhibitors of fibrin polymerization. A fragment deriving from the covalently stabilized fibrin dimer DD (a well-known inhibitor of fibrin polymerization) differs from the other specific polymerization inhibitors. Its potency to compete for fibrin monomer is exceptionally high but the structure-modifying capacity is reduced considerably. The initial rate of fibrin polymerization, as recorded spectrophotometrically, is diminished by DD to a lesser extent than by the monomer D. In the region of middle inhibitor concentrations, where the effect of fragment D is strongly expressed, the effect of fragment DD remains relatively low [39] (Fig. 1.9.). These results pertain to pure fragment DD preparations. If contaminated with monomeric D, the DD preparations exhibit higher inhibitory activity. The same is true if fragment D is added to purified DD. When fragment D is present at a concentration sufficient to induce quickly the total modification of fibrin monomer, addition of a small amount of DD gives a striking inhibitory effect which is much stronger than that of an equal amount of fragment D. Thus, the dimer DD possesses the highest polymerization-retarding power but is a poor structure-modifying agent.

The above facts corroborate the concept of two independent functions of specific polymerization inhibitors: a structural modification which makes fibrin monomer molecules susceptible and the competitive inhibition itself.

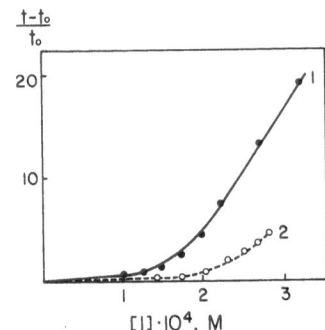

Fig. 1.9. Concentration dependence on inhibition
of fibrin monomer polymerization by
monomer D (1) and dimer DD (2).

Under special conditions we obtained relatively stable
fibrin monomer—fragment D complexes in solution. These complexes could be isolated by gel filtration [40]. In detailed
studies of these complexes we first tried to specify the important features of these complexes. In order to ascertain
that specific polymerization sites produced by thrombin are
essential for complex formation, we determined whether fibrinogen, substituted for fibrin, would be able to bind any fragment D. In fact, there was no sign of a complexing of fragment D with fibrinogen. So, specific polymerization sites
are crucial indeed. Experiments showed that, in order to
achieve fibrin monomer—fragment D complexes, a high fragment
D concentration was needed and the fragment must be present
in at least threefold excess.

Another prerequisite for a firm fibrin monomer—fragment
D complex is a slight reversible disorder of the monomer molecule. In the fibrinogen—thrombin system under physiological
conditions, no complexes arise despite the presence of fragment D. Yet all sorts of fibrin monomer (acidic, NaBr, urea
preparations) are able to form measurable amounts of monomer—
fragment complexes under our experimental conditions. All
of the fibrin monomer preparations are probably slightly dis-
ordered, and this seems important for firm fragment binding
[41].

The acidic monomer has been studied most carefully. The
stock solution of this kind of fibrin monomer is kept at mild-
ly acid pH (3.7-4.0). To induce polymerization it must be

brought to neutrality. Polymerization starts after a few
minutes at room temperature. During this period the acid-
modified monomer becomes normal, regaining its full ability
to clot. Addition of calcium ions accelerates the normalization process. If neutralization takes place in the presence
of fragment D at high concentrations, the monomer is rapidly
and quantitatively coupled with the fragment. Upon normalization of the modified monomer, its ability to bind fragment
D vanishes. Fibrin monomer of the NaBr or urea kinds, on returning to nondenaturating conditions in the presence of fragment D, becomes partially complexed and partially polymerized. In these cases the transformation of disturbed proteins
to normal is evidently so rapid that a considerable percentage of molecules manage to join each other and form the normal polymer, avoiding the formation of a complex with the
fragment. In the fibrinogen—thrombin mixture the newly formed
fibrin molecules must be free of any defects, and no stable
fibrin monomer—fragment D complex formation is feasible. The
competition with fragment D manifests itself only in some effects of inhibition of fibrin polymerization.

The modification responsible for the fibrin monomer's
loss of ability to clot and its readiness to bind fragment
D evidently concerns labile domain D; the much more stable
domain E remains active in the modified state. This situation somewhat resembles that characteristic of the immobilized fibrin monomer. Here, intermolecular reactions between
D and E domains are impossible, and thus domains E react
freely with the dissolved fragment D without competition with
domains D of fibrin.

The inferiority of fragment D in competition with the
domain of the fibrin monomer is not observed if the fragment
D is substituted by its dimer. The latter has a very strong
affinity to domain E and reacts with fibrin monomer even in
the system fibrinogen—thrombin [42]. The complex fibrin monomer—fragment DD is more stable than fibrin monomer—fragment
D. The main cause of this difference is probably thermodynamic. Since the two D structures are covalently bound in the
DD fragment, the tendency to dissociate is reduced due to the
entropy factor (the increase in entropy is larger when D-E-D
dissociates instead of E-DD).

We now have to consider the question concerning the
stoichiometry of the fibrin monomer—fragment complexes. Complexes obtained under various conditions have been analyzed

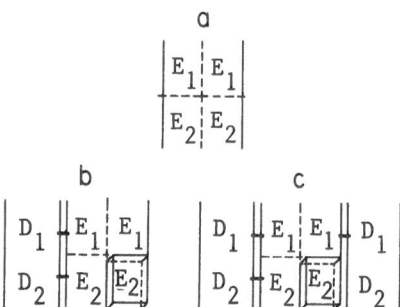

Fig. 1.10. Hypothetical change in conformation of
            domain E explaining its three valences
            for fragment D. D) Fragment D; E)
            domain E; E_1, E_2, D_1, D_2) specific poly-
            merization sites.

[43]. The main results are summarized as follows. The
amount of fragment D bound per mole of fibrin monomer is
found to increase as the concentration of fragment D increases until a final maximum of 3 moles of fragment D per mole
of fibrin monomer is reached. This saturation level was recorded in all experiments, the general result being 2.95 ±
0.28 moles of fragment per mole of fibrin monomer. Analogous
experiments with a pure DD preparation showed that complexes
with a 1:1 relationship of components were solely produced.
Interestingly, in D-DD mixtures (at high concentrations), two
forms of complexes were formed: containing three fragments
D per one fibrin monomer or one D and one DD per one monomer.
If a logical assumption is made that the DD occupies two
places for the D monomer, then the conclusion is that fibrin
monomer still appears to have three valences. When the first
two valences are occupied by a DD fragment, the remaining
one may be occupied by a D fragment but not by a DD.

The presence of three valences in the fibrin monomer
molecule is a puzzle, for one could easily explain two or
four valences since domain E is built of two identical half
domains with one E_1 and one E_2 in each half domain, for a
total of four polymerization sites. But in order to explain
the three valences something more complicated must be supposed. It seems plausible that the D_1 and D_2 sites of the
first fragment D react with both E_1 and E_2 of one of the half
domains of the monomer molecule. Then, a conformational

change must occur, partially separating the E_1 and E_2 of the other half domain. Thus, two other fragments D are needed to occupy the binding sites of that half E domain. This makes the valence equal to three (see Fig. 1.10).

Polymerization sites of domains D and E, responsible for protofibril formation (namely the E_2–D_2 pair) are also involved in lateral binding of protofibrils during the second phase of fibrin assembly [44]. The noncovalent bridges E_2–D_2 may connect protofibrils as well as fibrin monomer molecules within them. Both of these types of bridges are structurally important. Bonds between E_1 and D_1, produced by thrombin rapidly on fibrinopeptide A removal, seem to be located invariably inside the protofibrils.

Protofibril elongation due to fibrin monomer polymerization proceeds until a critical length is reached. This length depends on pH and other medium conditions. Lateral aggregation of the protofibrils follows, leading to the thickening and branching of fibrils. Finally, a continuous, three-dimensional network — the solid phase of fibrin — appears. The physical reason for the critical point, the limit of the purely longitudinal growth of protofibril, is probably the energetic advantage of the laterally associated state in comparison with the state of free protofibrils in solution, appearing when the length exceeds a certain value. The energy of the lateral association depends on the sum of the energies of interaction of the monomeric units belonging to the adjacent protofibrils. As the length of the protofibrils increases, their association energy becomes larger. At the critical point it is equal to the energy which keeps the protofibrils in the dissociated state.

The formation of a certain part of E_2–D_2 bonds must be one of the mechanisms of the lateral protofibril association. Other mechanisms are also operating. One is the electrostatic interaction of oppositely charged loci of the monomeric units. This is clearly seen from the effect of ionic strength known to reduce by screening the mutual attraction of positive and negative charges. The influence of ionic strength has been analyzed in our laboratory [45].

In this study, the "acidic" kind of fibrin monomer was used. Its solution was neutralized to start the polymerization process, which took place at ionic strength 0.15 or 0.30. The polymerizing samples were carefully watched. After a lag period, the samples suddenly began to become turbid and gelation ensued. During the lag period, when no visible changes occurred, only protofibril assembly is known to proceed. The increase in turbidity and gelation are signs of lateral protofibril association. Thus, the critical point of the transition from the stage of protofibril formation to the second stage of fiber assembly could be approximated.

Only a few of our results are mentioned here. Samples of fibrin monomer, ionic strength 0.30, were brought to ionic strength 0.15 by dilution after various periods of incubation (in the scope of the longer clotting time characteristic of the ionic strength 0.30). Since the clotting time at ionic strength 0.15 is much shorter, it had to be expected that this change of ionic strength would shorten the clotting time. The most remarkable fact is that, if the preincubation of ionic strength 0.30 lasted only a little longer than the time for clotting of samples tested at a constant ionic strength of 0.15, the clotting occurred at just the moment of dilution. This means that the degree of polymerization sufficient to clot at ionic strength 0.15 is reached at both ionic strengths almost simultaneously, i.e., the rate of polymerization is affected comparatively little by the ionic strength. But, at a higher level (0.30), this degree of polymerization is far from the clotting point and the polymerization has to continue. Since the solution remains clear, the polymerization process is limited by protofibril formation. At ionic strength 0.30, the resulting gel is fine in comparison with the coarse one formed at ionic strength 0.15. Consequently, extra polymerization leads to a longitudinal growth of protofibrils.

The overall situation appears as follows. At higher ionic strength, the electrostatic forces involved in the lateral protofibril association become weaker and longer protofibrils are required to compensate for this loss by an increased number of nonionic bonds such as E_2-D_2.

Analogous experiments in which ionic strength was changed from 0.15 to 0.30 gave results corroborating the above concept. A considerable prolongation of clotting time was observed if this change occurred at the moment when the sample was near the clotting point (at ionic strength 0.15). The critical point of polymerization at which the protofibrils begin to associate with each other is evidently shifted to the region of greater length of the protofibrils.

Fibrin fibers vary in thickness from two to three to about a hundred protofibrils. Under conditions unfavorable for lateral interactions, when the critical length of proto-fibrils is large (for instance at high ionic strength) the fibrin network consists of long thin strands and the clots are transparent and fragile (called fine clots). Under conditions favorable for protofibril lateral association (for example at low ionic strength) the network is constructed of short thick strands and the critical length of the protofibrils is small. In this case, the clots are turbid and rigid (called coarse clots).

We became interested in whether or not the type of clot (fine or coarse) could be altered by changing the conditions of the medium under which it is stored. In particular, we questioned whether a coarse clot, formed at low ionic strength, would become fine by raising the ionic strength drastically. Experiments showed that the clot does not change but remains as turbid as before. The system of lateral bonds within the network of fibrin becomes strong and irreversible although covalent stabilization is still lacking.

Polymerization of fibrin des-A prepared with snak enzymes inducing fibrinogen clotting leads to fibrin fibers showing the normal cross-link pattern. However, the resulting clots are more fine than the usual fibrin des-AB clots under the same conditions of the medium. This is just what could be expected. The absence of E_2-D_2 lateral bridges makes the critical length of protofibrils much greater and the fibrin strands correspondingly thinner. This must be the cause of that difference, since a lowering of ionic strength (or pH) proved sufficient for achieving fibrin des-A clots that were quite normal. Evidently, the increase in electrostatic interactions compensates for the lack of E_2-D_2 interprotofibril bridges.

The saturation of cross-linked fibers, as detected electon-microscopically, goes through discernible states [46]. Initially, protofibrils assemble into loose bundles with protofibril units roughly arranged. Soon, spots of condensed protein matter appear regularly spaced along the bundles. These spots increase in size to form the main bonds of the cross-striation system, and minor intermediate bands develop gradually. The distance between the main bands (i.e., the length of the axial period) has been measured and found to increase from 20 ± 1 nm to 23 ± 0.5 nm during fibrin satura-

tion. A simple explanation of this increase may be offered.
The protofibrils, constituting a bundle at the early stage
of fiber self-organization, are fastened together only by
small sites — the nascent element of the cross-link structure.
The bulk parts of the protofibrils are still not laterally
associated and, thus, allowed to be irregularly curved. As
the lateral association proceeds and the rodlike fibers are
built up, the protofibrils are stretched and, hence, the axial
period increases.

The functional aspects of the fibrinogen—fibrin problem
have been mainly considered in this chapter. Data on the
domains of these molecules and corresponding specific reactive
sites have been summarized. In order to analyze functional
mechanisms, the interactions of different active fragments
with each other and with fibrin monomer have been scrutinized.
Both fibrin monomer polymerization and protofibril lateral
association are presented as two stages in the assembly of a
fibrin network.

REFERENCES

1. V. L. Zima, E. V. Lugovskoi, L. V. Medved', G. K. Gogolinskaya, and P. L. Privalov, "Calorimetric and spectro-
   fluorimetric evidence for structural organization and
   localization of fibrinogen α-domains," Dokl. Akad. Nauk
   SSSR 256, 480-482 (1981).
2. M. W. Mosesson, J. H. J. Wall, and R. H. Hashmeyer,
   "Identification and mass analysis of human fibrinogen
   molecules and their domains by scanning transmission
   electron microscopy," J. Mol. Biol., 153, 695-718 (1981).
3. J. Weisel, G. N. Phillips, and C. Cohen, "A model from
   electron microscopy for the molecular structure of fibrinogen and fibrin," Nature, 289, 263-267 (1981).
4. J. Donovan and E. Mihalyi, "Conformation of fibrinogen:
   calorimetric evidence for a trinodular structure," Proc.
   Natl. Acad. Sci. USA, 71, 4125-4128 (1974).
5. L. V. Medved', E. I. Tiktopulo, P. L. Privalov, and
   T. V. Varetskaya, "Microcalorimetric studies of temperature transitions in fibrinogen and its proteolytic fragments," Mol. Biol., 14, 835-841 (1980).
6. P. L. Privalov and L. V. Medved', "Domains in the fibrinogen molecule," J. Mol. Biol., 159, 665-683 (1982).
7. L. V. Medved', P. L. Privalov, and T. P. Ugarova, "Isolation of thermostable structure from the fibrinogen D-
   fragment," FEBS Lett., 146, 339-349 (1982).

8. L. Tranqui-Pouit, V. J. Marder, M. Suscilon, A. Z. Budzinski, and G. Hudry-Clergeon, "Electron-microscopic
   studies of plasmic degradation product of fibrinogen.
   Implication for the disulfide structure of fibrinogen,"
   Biochim. Biophys. Acta, 400, 189-199 (1975).
9. W. E. Fowler and H. P. Erickson, "The trinodular structure of fibrinogen-conformation by both shadowing and
   negative stains from electron microscopy," J. Mol. Biol.,
   134, 241-249 (1979).
10. V. A. Belitser, T. V. Varetskaya, and V. Ph. Manyakov,
    "On the model for the fibrinogen molecule. Consecutive
    stages of fibrin polymerization," Thromb. Res., 2, 567-578
    (1973).
11. T. M. Price, D. D. Strong, M. L. Redee, and R. F.
    Doolittle, "Shadow-cast electron microscopy of fibrino-
    gen with antibody fragments bound to specific regions,"
    Proc. Natl. Acad. Sci. USA, 78, 200-204 (1981).
12. A. Z. Budzynski, M. Stahl, M. Kopic, Z. S. Latallo,
    Z. Wegrzynowicz, and E. Kowalski, "High-molecular-weight
    products of the late stage of fibrinogen proteolysis by
    plasmin and their structural relation to the fibrinogen
    molecule," Biochim. Biophys. Acta, 147, 313-322 (1967).
13. V. A. Belitser, T. V. Varetskaya, V. M. Tolstykh, L. A.
    Tsaryuk, and T. M. Pozdnyakova, "Enhanced anticlotting
    activity of fragments D formed during plasmin hydrolysis
    of fibrinogen in the presence of calcium chloride,"
    Thromb. Res., 7, 797-806 (1975).
14. V. M. Tolstykh and T. V. Varetskaya, "Anticlotting activ-
    ity of fragment D from fibrinogen and fibrin and its de-
    pendence on calcium when obtaining fragments from fib-
    rinogen," Ukr. Biokhim. Zh., 48, 116-121 (1976).
15. F. Haverkate and G. Timon, "Protective effect of calcium
    on the plasmin of fibrinogen and fibrin fragment D,"
    Thromb. Res., 10, 803-812 (1974).
16. W. Nieuwenhuizen, J. A. M. van Ruijven-Vermeer, W. J.
    Nooijen, A. Vermond, F. Haverkate, and J. Hermans, "Re-
    calculation of calcium-binding properties of human and
    rat fibrin(ogen) and their degradation products," Thromb.
    Res., 22, 653-657 (1981).
17. S. A. Olexa and A. Z. Budzynski, "Localization of fibrin
    polymerization site," J. Biol. Chem., 256, 3544-3549
    (1981).
18. A. P. Laudano and R. F. Doolittle, "Synthetic peptide
    derivatives that bind to fibrinogen and prevent the poly-
    merization of fibrin monomer," Proc. Natl. Acad. Sci.
    USA, 75, 3085-3089 (1978).

19. A. P. Laudano and R. F. Doolittle, "Studies on syn-
     thetic peptides that bind to fibrinogen and prevent fib-
     rin polymerization. Structural requirement, number of
     binding sites, and species differences," Biochemistry,
     19, 1013-1019 (1980).
20. A. P. Laudano and R. F. Doolittle, "Influence of calcium
     ion on the binding of fibrin amino-terminal peptides to
     fibrinogen," Science, 212, 457-459 (1981).
21. B. Kudryk, B. Blomback, and M. Blomback, "Fibrinogen
     Detroit — an abnormal fibrinogen with nonfunctional NH_2-
     terminal polymerization domain," Thromb. Res., 9, 25-36
     (1976).
22. H. A. Scheraga and M. Laskowski, "The fibrinogen—fibrin
     conversion," Adv. Protein Chem., 12, 2-129 (1957).
23. E. V. Lugovskoi and V. A. Belitser, "Mechanism of fib-
     rin polymerization: a study on neutral salt effects,"
     Biokhimiya, 36, 129-137 (1971).
24. E. V. Lugovskoi, G. K. Gogolinskaya, S. G. Derzskaya,
     and V. A. Belitser, "Properties of two fibrin monomer
     species which differ in the degree of thrombin activa-
     tion. Characteristics of successively appearing active
     sites," Biokhimiya, 43, 1045-1052 (1978).
25. T. N. Donnelly, M. Laskowski, J. M. Notley, and H. A. Scher-
     aga, "Equilibria in the fibrinogen—fibrin conversion.
     II. Reversibility of the polymerization steps," Arch.
     Biochem. Biophys., 56, 369-375 (1955).
26. V. A. Belitser, T. V. Varetskaya, and L. A. Tarasenko,
     "Fibrin monomer polymerization and its dependence on
     pH," Ukr. Biokhim. Zh., 37, 665-670 (1965).
27. V. A. Belitser, T. V. Varetskaya, and G. V. Malneva,
     "Fibrinogen—fibrin interaction," Biochim. Biophys.
     Acta, 154, 367-375 (1968).
28. T. M. Pozdnyakova, A. A. Musyalkovskaya, T. P. Ugarova,
     D. D. Protvin, and V. N. Kotsyuruba, "On the properties
     of fibrin monomer prepared from fibrin clot with acetic
     acid," Thromb. Res., 16, 283-288 (1979).
29. W. E. Fowler, R. R. Hantgan, J. Hermans, and H. P. Erick-
     son, "Structure of the fibrin protofibril," Proc. Natl.
     Acad. Sci. USA, 78, 4872-4876 (1981).
30. V. A. Belitser, T. V. Varetskaya, S. N. Tsynkalovskaya,
     L. A. Tsaryuk, L.I. Shevchenko, G. A. Belitskaya, and
     Ya. M. Ena, "Determination of fibrinogen and fibrin
     splitting products by their anticoagulating effect," Ukr.
     Biokhim. Zh., 48, 521-532 (1976).
31. V. A. Belitser, T. V. Varetskaya, Yu. P. Butylin, L. A.
     Tsaryjk, L. A. Svital'skaya, Ya. M. Ena, and O. A.

Bunyak, "Assay of blood plasma fibrinogen," Lab. Delo, 4, 38-42 (1983).

32. V. A. Belitser and J. L. Chodorova, "Die kinetik der polymerization des fibrin-monomeren," Acta Biol. Med. Ger., 1, 631-640 (1958).

33. V. A. Belitser, T. V. Varetskaya, and S. N. Tsinkalovskaya, "Estimation of the specific inhibitors of fibrin polymerization inhibitory units," Thromb. Res., 3, 251-264 (1973).

34. V. A. Belitser, T. V. Varetskaya, and S. A. Kosterin, "On the mechanism of inhibition of fibrin polymerization by fibrinogen and its active fragments," Biokhimiya, 45, 157-164 (1980).

35. N. U. Bang, A. P. Fletcher, N. Alkjaersic, and S. Sherry, "Pathogenesis of the coagulation defect developing during pathological plasma proteolytic (fibrinolytic) states. III. Demonstration of abnormal clot structure," J. Clin. Invest., 41, 935-942 (1962).

36. V. A. Belitser, T. V. Varetskaya, V. Ph. Manyakov, N. I. Man'ko, I. I. Degtyaryova, E. A. Smechova, and T. F. Galanova, "Susceptibility of early and late stages of fibrin self-assembly to specific inhibitors," Thromb. Res., 3, 265-279 (1973).

37. V. A. Belitser, E. V. Lugovskoi, A. A. Musjalkovskaya, and G. K. Gogolinskaya, "Quantitation of inhibitory effect of fibrinogen and its degradation products on fibrin polymerization," Thromb. Res., 27, 261-269 (1982).

38. V. A. Belitser, T. M. Pozdnyakova, and T. P. Ugarova, "Light and heavy fractions of fragment D: preparation and examination of fibrin-binding properties," Thromb. Res., 19, 807-814 (1980).

39. T. N. Platonova, A. A. Musialkovskaya, V. M. Tolstykh, and V. A. Belitser, "Inhibition of fibrin polymerization (assembly) by fragment D and its dimer derived from fibrinogen and stabilized fibrin. Evidence for the two-step type inhibition," Biokhimiya, 45, 1780-1787 (1980).

40. V. A. Belitser, T. M. Pozdnyakova, and V. M. Tolstykh, "Detection of fragment D—fibrin complex by agarose chromatography," Thromb. Res., 14, 265-272 (1972).

41. V. A. Belitser, T. M. Pozdnyakova, E. V. Vovk, and V. N. Rybachuk, "The mechanism of coupling between the fibrin monomer and the inhibitor of its polymerization-fragment D," Biokhimiya, 48, 125-131 (1983).

42. V. A. Belitser, T. N. Platonova, and T. M. Pozdnyakova, "Difference between complexes formed by monomeric fibrin with fragment D and dimer D," Ukr. Biokhim. Zh., 55, 243-249 (1983).

43. T. M. Pozdnyakova, V. N. Rybatchuk, and E. V. Vovk, "Investigation of fragment D—fibrin monomer complex formation by salting-out fractionation," Biokhimiya, 47, 971-976 (1982).

44. S. A. Olexa and A. E. Budzinski, "Evidence for four different polymerization sites involved in human fibrin formation," Proc. Natl. Acad. Sci. USA, 77, 1374-1378 (1980).

45. V. A. Belitser and T. V. Varetskaya, "Modification of fibrin assembly as method for studying mechanism of this process," Ukr. Biokhim. Zh., 47, 567-579 (1975).

46. V. F. Manyakov, T. V. Varetskaya, and V. A. Belitser, "Electron microscopic investigations on fibrin self-assembly," Mol. Biol., 11, 1182-1189 (1977).

NONENZYMATIC FIBRINOLYSIS AND ITS ROLE IN THE ORGANISM

B. A. Kudrjashov and L. A. Lyapina

M. V. Lomonosov Moscow State University, Moscow

ABSTRACT

This chapter presents experimental data which indicate that, apart from an enzymatic (plasmin) system of fibrinolysis discovered more than 150 years ago, there exists a natural nonenzymatic fibrinolysis of a noncross-linked fibrin. This type of fibrinolysis is conditioned by complex combinations of heparin with plasma proteins, amines, and other components. The molecular mechanism of the lytic action of heparin complexes on noncross-linked fibrin, leading to the dissociation of a clot into molecules of fibrin monomer, has been investigated. Heparin complexes in the organism prevent thrombi formation at its early stage, inhibit platelet aggregation, the activity of thrombin and factor XIIIa, and block fibrinogen coagulation and the activity of a number of other protein factors of the blood-clotting system. Heparin complexes are natural humoral agents whose concentration in the bloodstream increases when the anticoagulating system is excited. Experimental and clinical studies reveal the deficiency of these complexes in the organism at late stages of atherosclerosis. Hypertrophic nonenzymatic fibrinolysis was found in most of the women examined who were in labor and suffering from obstetric complications accompanied by acute bleeding.

INTRODUCTION

The present-day concept of fibrinolysis originated from a detailed investigation of the lysis of a blood clot or fibrin by a blood plasma enzyme. In 1769, Morgagni [1] described

33

the breakdown of a clot isolated from the blood vessels of
a cadaver. Sixty-nine years later Denis [2] reported spontaneous dissociation of clots formed of the blood plasma of a
healthy subject, and in 1889 Denys and Marbaix [3] showed
that this effect is conditioned by the action of a proteolytic
enzyme present in the blood serum. In 1893, Dastre [4] suggested that this phenomenon be called fibrinolysis.

In the last century, it was firmly established that a
protein precursor of the fibrinolytic enzyme called plasminogen [5], or profibrolysin [6], is contained in the plasma.

Further investigation of fibrinolysis made it possible
to formulate the concept of the plasmin system, which consists
of the proenzyme, its proactivators and activators, and the
enzyme and its inhibitors. The molecular structure of plasmin and plasminogen was described; activators and proactivators of plasminogen were isolated from tissues, blood, and
urine; and the characteristics of natural fibrinolysis inhibitors were revealed and investigated. Plasmin preparations or certain plasminogen activators came into wide clinical use for treating thromboembolic complications.

Thus, a generally accepted concept of natural fibrinolysis as an enzymatic process was formed, and all phenomena of
blood or fibrin clot dissolution have since been regarded
from this traditional point of view.

However, clinical and experimental application of different methods of determining blood fibrinolytic activity
often yielded quite opposite results which could not be satis-
factorily explained. For example, no rise in fibrinolytic
activity was revealed [7] in the same samples of plasma taken
from experimental animals after an intravenous infusion of
thrombin. In this experiment, fibrin plates or ^{131}I fibrin
clots were used. Yet, application of the euglobulin method
showed an increase of fibrinolytic activity in the blood of
experimental animals.

Analysis of the stimulation of the anticoagulating system in animals after rapid intravenous injection of threshold
doses of tissue thromboplastin or thrombin showed a considerable increase in fibrinolysis as determined by the euglobulin
method or a promptly performed Bidwell reaction [8]. The
primary results yielded by this study showed that the rise in
blood fibrinolytic activity was linked to a discharge of endo-

genic heparin and plasminogen activator into the bloodstream [9-11]. In further experiments the first of these two agents drew our attention. In the investigations that followed, it was established that endogenic heparin does not remain in a free state in the bloodstream. It enters into a complex formation with thrombogenic proteins of the blood plasma [12-19], and these complexes possess fibrin-dissolving activity.

Administration of ^{131}I fibrin into the bloodstream of experimental animals (without anesthesia) followed by intravenous injection of a limited thrombin dose which causes a typical humoral reaction of the anticoagulating system, was accompanied by a 17-20% decrease of fibrinogen concentration in the bloodstream. However, the blood temporarily lost the ability to coagulate under the action of thrombin [20]. Investigation of ^{131}I-fibrinogen behavior in experimental unanesthetized rats brought certain researchers to the conclusion that fibrinogen loses its native properties and thrombin does not react on it [20, 21].

This phenomenon was accounted for by the fact that fibrinogen combines with heparin discharged in the blood. The resulting heparin—fibrinogen complex [22] is not affected by thrombin and has anticoagulating and lytic activity toward fibrin [14, 15, 17]. It was concluded in 1970 that (a) fibrinolysis, which takes place with the help of heparin complexes, is a physiological process for nonenzymatically dissolving only nonstabilized fibrin clots [23], and (b) there are two fibrinolytic systems in the organism, one of which is enzymatic and the other of which is nonenzymatic [24-26]. Thus, the experimental data supported the concept of nonenzymatic fibrinolysis in man and animals, which has since been developed further.

MATERIALS AND METHODS

White male rats weighing 180-250 g were used. Blood was drawn with a syringe from the vena jugularis. Intravenous injections were made in the same vein.

Our technique of fractional isolation of fibrinogen—heparin (FH), epinephrine—heparin (EH), plasminogen—heparin (PGH), plasmin—heparin (PH), thyroxine—heparin (Thyr H), thrombin—heparin (TH), and other complexes from one portion of plasma is based on the FH complex isolation technique de-

veloped by Cohen et al. [25]. Fractions of PGH, PH, EH, and
Thyr H complexes were isolated as described elsewhere [26-30].
All the preparations of heparin complexes were obtained from
citrated platelet-free plasma. The euglobulin fraction of
plasma was obtained and separated from the supernatant [31].
The EH complex was then precipitated from the latter by
bringing it to the isoelectric point (pH) at 4°C with a subsequent incubation of 15-20 min and centrifugation at 3000 rpm
for 15 min. The fraction of the EH complex was dissolved in
a volume of borate buffer equal to one-sixth of the initial
plasma. Preparations of FH, PGH, TH, and Thyr H complexes
were obtained from the euglobulin fraction of plasma dissolved
in borate buffer to which a cooled 53.3% alcohol was added
into the system to a final concentration of 8%. The samples
(pH 7.2) were kept cold for 10-15 min, and the suspension was
then centrifuged at 2000 rpm for 10 min. The precipitate
containing the FH complex was dissolved in the volume of borate solution equal to the initial volume of plasma.

Cooled 53.3% alcohol was added to the supernatant obtained after the precipitation of FH complex to a final concentration of 20% and a pH of 6.5. The precipitate which
contained the Thyr H complex, obtained after centrifugation
at 1800 rpm for 15 min, was dissolved in the volume of borate
solution equal to the initial volume of plasma. The concentration of alcohol in the II-III fraction solution was then
increased to 25% at pH 7.2. The precipitate of fraction II-
III was separated by centrifugation at 4°C at 1800 rpm for
10-15 min. Plasminogen, PH, PGH, and TH complexes were present in this fraction. To separate the complex combinations,
the precipitate was dissolved in borate solution: the concentration of alcohol in the system was brought to 17% at pH
5.2, after which the mixture was centrifuged at 1800 rpm for
10 min. The PH and PGH complexes were precipitated. The
total solution of these complexes was obtained in borate buffer (the volume equal to the initial volume of plasma). The
supernatant obtained after the last centrifugation contained
the fraction of TH complex.

The basic materials for obtaining in vitro FH and TH
complexes were preparations of fibrinogen and thrombin free
from plasminogen admixture [32, 33]. Heparin preparations
of Difco USA, Spofa, Reanal, and other firms were used to
synthesize the following complexes: PGH, PH, Thyr H, EH, serotonin—heparin (SH), norepinephrine—heparin (NEH), acetylcholine—heparin (AChH), ATP—heparin (ATP-H), ADP—heparin

(ADP-H), AMP—heparin (AMP-H), urea—heparin (UH), factor XIII—
heparin (FXIII-H), prothrombin—heparin (PrH), thromboplastin—
heparin (TrH), fibrin monomer—heparin (FmH), insulin—heparin
(IH), alloxan—heparin (AlH), acetylsalicylic acid—heparin
(ASAH), DNA—heparin (DNAH). To obtain plasmin, plasminogen
was activated by streptokinase. We also utilized factor XIII
free from fibrinogen admixture [34], prothrombin free from
plasminogen admixture [35, 36], antithrombin III [37], and
preparations of thyroxine, adenosine monophosphate (AMP),
adenosine diphosphate (ADP), and adenosine triphosphate (ATP)
(Reanal), epinephrine and norepinephrine, serotonin-creatin-
ine sulfate, insulin, urea, DNA, acetylsalicylic acid, alloxan (Spofa), and other components.

Precipitation of heparin complexes with blood proteins
was carried out by using the method [38] suggested for obtaining heparin—trypsin complex. In FH [25], PrH [39], FmH [40],
FXIII-H [41], TrH [42], IH [43], Thyr H [30], antithrombin
III—heparin (ATIII-H) [44], antiplasmin—heparin [45], and AlH
complexes the ratio of heparin to the other component was
1:4-1:6. In EH, NEH [28, 29], UH [46], and AChH [47] complexes, this ratio equaled 3:1. In other complexes the ratio was
as follows: ASAH, 10:1 [48]; DNAH, 1:6-1:12 [49]; ATPH, 2:1;
ADP, 12:1; AMP, 5:1 [50]; and SH, 15:1 [51].

The formation of heparin complexes with proteins, amines,
and biologically active substances was corroborated both by
their nonenzymatic fibrinolytic activity and by cross-paper
electrophoresis [52]. Spectrophotometric and sedimentation
analyses were used for the same purpose.

The fibrinolytic characteristics of the complexes were
studied on nonstabilized thin fibrin plates [53, 54] and in
the euglobulin fraction of plasma [55] in the presence of ε-
aminocaproic acid (at a final concentration of 3%) or some
other inhibitors of plasmin fibrinolysis. To study PH and
PGH complexes, the fibrin plates [56] were heated at 85°C for
30 min for neutralization of plasminogen activity [57]. In
a number of experiments, nonenzymatic fibrinolytic activity
was also determined according to Bidwell [8], but in the presence of ε-aminocaproic acid (EACA) (a final concentration of
3%). Anticoagulating activity of the complexes was estimated
by the thrombin time according to the generally accepted technique. Antipolymerization activity of the complexes was
studied using monomeric fibrin obtained from pure fibrinogen
[58].

We also used antiplasmin [59], EACA, Trasylol (FRG), commercial preparations of serotonin, protamine sulfate (Fluka A. G., Switzerland), and ^{35}S-heparin (Amersham).

Radioactivity was measured [60-62] in a Mark-II scintillation counter (Nuclear Chicago). The data obtained were analyzed using the Fisher-Student technique.

RESULTS

Physiological Characteristics of Heparin Complexes

The studies conducted in both a relatively pure system and in plasmic medium revealed a sequence in the formation of heparin complexes. Heparin—fibrinogen and heparin—plasminogen are the first complexes to appear. If there is a relative excess of heparin and free epinephrine in the medium, then an AEH complex is formed [30].

In combination with heparin, fibrinogen, thrombin, factor XIII, and others lose their coagulating properties. Epiephrine and serotonin also do not exhibit their intrinsic physiological characteristics in heparin complexes [29, 51] and, at the same time, these complexes acquire a number of new properties.

It has been found that almost all of the heparin complexes obtained with blood plasma proteins, amines, and other biologically active substances have a rather high anticoagulating activity and antipolymerization effect on monomeric fibrin, i.e., they inhibit the formation of monomeric fibrin aggregates [30, 63, 64]. Study of platelet aggregation demonstrated that their functional activity decreased considerably under the effect of heparin-containing complexes, e.g., with epinephrine or urea [65, 66]. The ability of complex epinephrine combinations to inhibit a primary polymerization of monomeric fibrin is also due to their antagonism for factor XIII [23]. Complex combinations of heparin with coagulating system proteins (fibrinogen, prothrombin, thrombin, thromboplastin, antiplasmins, factor XIII), anticoagulating system proteins (plasminogen, antithrombin III), amines (epinephrine, norepinephrine, serotonin, and acetylcholine), biologically active substances (DNA, urea, insulin, AMP, ADP, ATP), certain pharmacological preparations (acetylsalicylic acid), and

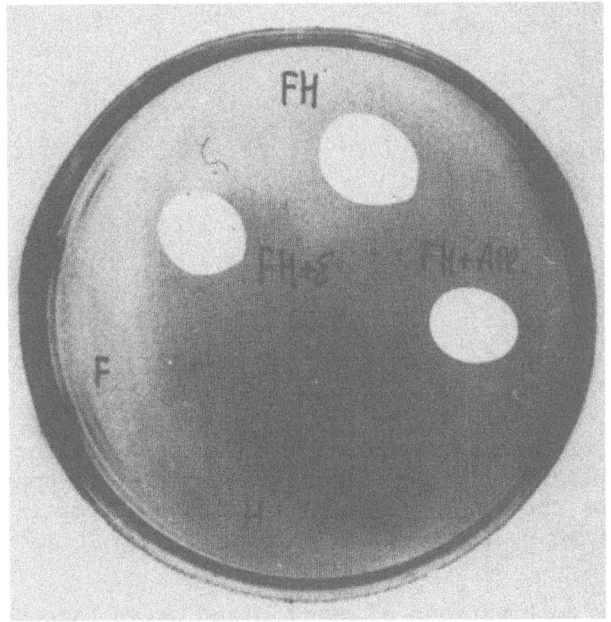

Fig. 2.1. Nonenzymatic fibrinolytic activity of fibrinogen–
heparin complex (FH). Lysis zones on noncross-
linked fibrin originating from the action of FH +
NaCl; FH + ε-aminocaproic acid; FH + antiplasmin
(AP1). F) Fibrinogen; H) heparin [30].

others perform a natural nonenzymatic fibrinolysis, i.e., all
of these heparin complexes actively dissolve both in vitro
and in vivo the fibrin clots nonstabilized by factor XIII.
However, they have no lytic effect on the cross-linked fibrin.
The parts of complexes taken separately (heparin and any second component) in doses equivalent to their content in the
complexes exert no lytic effect on noncross-linked fibrin
under the given experimental conditions.

The FH complexes isolated from blood plasma or from the
euglobulin fraction of plasma [67], as well as from the fibrinogen degradation product fraction (FDP) [68, 69], had the
same characteristics as an FH complex synthesized in vitro.
Figure 2.1 shows zones of lysis on noncross-linked fibrin in
the presence of 3% EACA resulting from the action of the FH
complex and its components taken in doses equivalent to their
content in the complex.

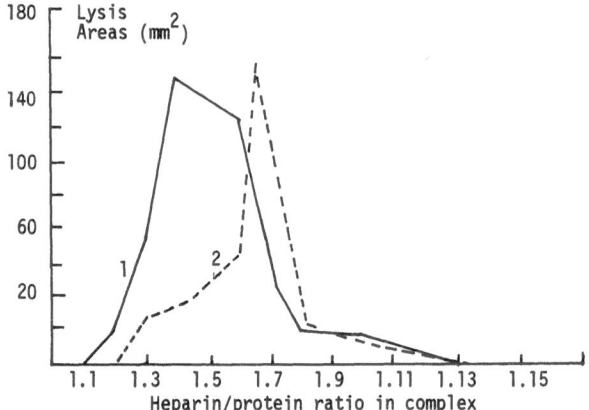

Fig. 2.2. Dependence of the lytic activity of factor XIII–
heparin (1) and fibrinogen–heparin (2) complexes
on the weight ratio of heparin and protein which
make up the complex [41].

Complex combinations of heparin and drugs (acetylsalicylic acid, insulin, etc.) have lytic properties similar to the
above-mentioned compounds. Study of the heparin complexes
showed that certain positive properties of their drug component are either intensified or changed, due to which the complexes acquire new characteristics. For example, the nonenzymatic fibrinolytic action of acetylsalicylic acid in an ASAH
complex is increased threefold [48].

All of the heparin combinations studied retain nonenzymatic fibrinolytic action on noncross-linked fibrin in the
presence of enzyme fibrinolysis inhibitors [EACA, antiplasmin,
soybean trypsin inhibitor (SBTI), and trasylol], but fail to
realize the lysis of noncross-linked fibrin in the presence
of protamine sulfate. It should be noted that heparinase completely inhibits the fibrinolytic action of FH, TH, EH complexes, etc. [30, 47].

It was found [41] that nonenzymatic fibrinolytic activity of heparin complexes is conditioned by the ratio of heparin to another component of the complex (Fig. 2.2).

Study of the fibrinolytic action of PH [26, 27] and
ocrase–heparin (OcH) complexes [70], which dissolve both cross-
linked and noncross-linked fibrin, is of special interest. In

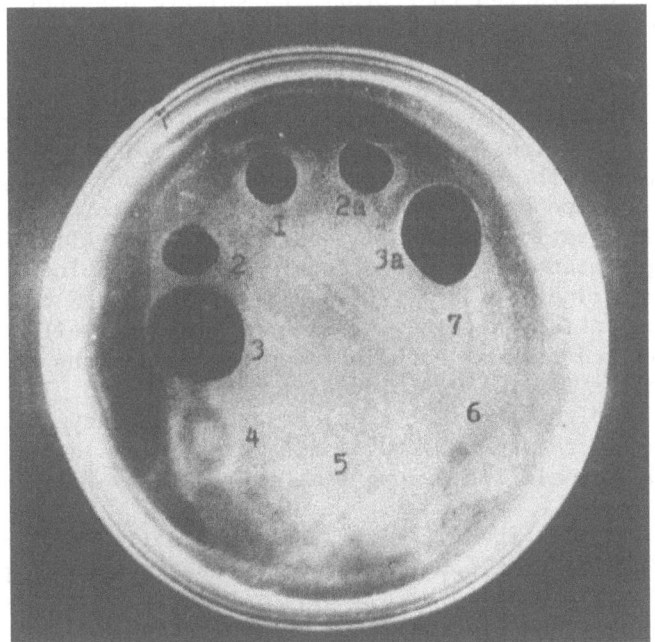

Fig. 2.3.   Zones of lysis of a noncross-linked fibrin plate
following incubation with 0.3 ml of 0.1% solution
of heparin-antithrombin III (H-ATIII) and antithrombin III—heparin—thrombin (ATIII-H-TH) complexes (2.5 h, 37°C). 1) H-ATIII (ratio 1:1); 2)
ATIII-H-TH (1:1:1); 2a) after preincubation with
3% ε-aminocaproic acid; 3) ATIII-H-TH (1:1:6); 3a)
after preincubation with 3% ε-aminocaproic acid;
4) ATIII-H-TH (7:1:6); 5) thrombin; 6) ATIII; 7)
heparin [44].

this respect, the action of PH and OcH complexes is similar
to that of the natural enzymes. However, unlike the native
plasmin and ocrase, heparin complexes retain their enzymatic
fibrinolytic activity in the presence of EACA, antiplasmin,
or SBTI. Protamine sulfate, being a heparin inhibitor, destroys the PH complex, producing a fibrinolytically inactive
protamine sulfate—heparin complex.

However, it was demonstrated that native plasmin completely loses its fibrinolytic activity in the presence of
the same concentration of EACA (3.0%), antiplasmin, or SBTI.

It was also detected that protamine sulfate does not change the fibrinolytic properties of plasmin.

Because their fibrinolytic activity is different from that of plasmin, the complexes of heparin with thrombogenic proteins and amines are called "physiological solvents" of cross-linked fibrin [23]. These agents can contain a great number of components. We have synthesized triple heparin complexes: antithrombin III—heparin—thrombin (ATIII-H-T) [44], fibrinogen—monomeric fibrin—heparin [40], and epinephrine—heparin—fibrinogen (EHF) [71]. ATIII-H-T and EHF complexes also have anticoagulating and high fibrinolytic effects on nonstabilized fibrin, both in the presence and the absence of enzymatic fibrinolysis inhibitors (Fig. 2.3). It was also demonstrated that the EHF complex retains nonenzymatic fibrinolytic activity and anticoagulating properties in the bloodstream of animals for a long time (up to 10 h) [71].

It has been found [72, 73] that blood lipoprotein lipase in a triple complex combination with heparin and lipoproteins acquires a new quality of dissolving nonstabilized fibrin clots. Consequently, this enzyme also takes part in preventing thrombi formation at an early stage.

It was shown that heparin complexes, both in vitro and in vivo, cause lysis of fresh, experimentally induced thrombi within 1.5-4 h [30]. Intravenous injection of factor XIIIa or thromboplastin into the bloodstream of rats [74], like the insertion of a thin glass needle into the jugular vein [75], leads to thrombosis.

However, this process can be effectively neutralized with heparin complexes which (due to their anticoagulating and antipolymerization activity) can prevent thrombi formation in vivo and lyse nonstabilized blood clots [76, 77]. It was found that the rate of thrombin clearance in the bloodstream directly depends on thrombin—heparin complex formation [78].

It was observed that stimulation of the anticoagulating system by an injection of a threshold dose of α-thrombin into the vena jugularis is accompanied by the accumulation of different heparin complexes in the cardiovascular system [79, 80].

An atherogenic diet leads to a dramatic fall in the formation of heparin complexes in animals due to depression of

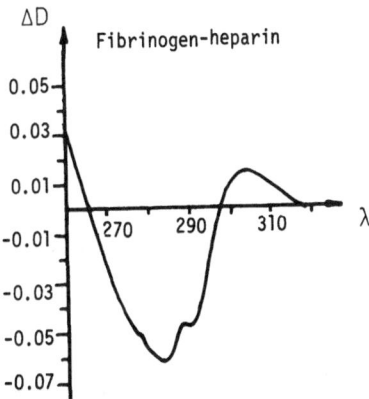

Fig. 2.4. Differential spectrum of fibrinogen-
           heparin complex [84].

the anticoagulating system [81, 82] which can be corrected by
consuming large doses of vitamin A and ethyl ethers of unsaturated fatty acids [83].

Complex Nature of Physiological Solvents of Noncross-Linked Fibrin

The existence of a heparin complex formation with the
above agents has been proved by various methods: spectrophotometry, electrophoresis, sedimentation, and radioisotopic
analyses. Sometimes during spectrophotometry in the ultraviolet range the maximum of the absorption spectrum of the complex
changed. The differential absorption spectra of the complexes
obtained in relation to their components prove the formation
and existence of these complexes in the medium (pH 7.2-7.4).
The presence of a shift in relation to fibrinogen in the differential spectrum of the complexes (e.g., FH) points to a
change in the hydrophobic surface covering of fibrinogen during the formation of FH complexes (Fig. 2.4) [84].

Infrared spectroscopy data made it possible to determine
which molecular groups of heparin and another component participate in the complex formation. Thus, NH groups in the
epinephrine molecule and OH groups of heparin interact when
the EDH complex is formed (Fig. 2.5), which is reflected in
the characteristics of the spectrum [85].

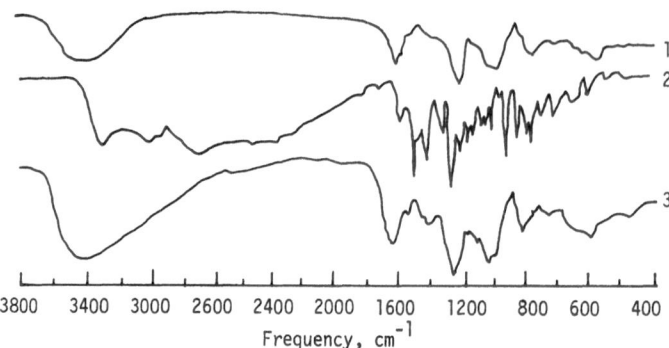

3800   3400 3000 2600 2400 2000 1600   1200   800   600   400
Frequency, cm^{-1}

Fig. 2.5.   Infrared spectra of epinephrine–heparin complex
(3), epinephrine (2), and heparin (1) [85].

Sedimentation analysis data confirm the assumption that
the heparin complex is formed [41].

Heparin complexes are thermostable. The FH complex sustains heat of up to 56-60°C for 30 min without losing its fibrinolytic properties (i.e., the ability to dissolve nonstabilized fibrin), while free fibrinogen is denatured and precipitated [25] when heated to 56°C for 3 min.

Using immunoelectrophoresis, it was found that fibrinogen in the FH-complex, obtained both in vitro and in vivo,
retains its immunologic characteristics [86].

Molecular Products of Noncross-Linked Fibrin
Dissolution by Complex Compounds of Heparin

The structure of purified bovine fibrinogen, nonstabilized fibrin, and molecular products of its dissociation by
heparin complexes has been studied by electron microscopy to
investigate the molecular transformation of noncross-linked
fibrin during nonenzymatic lysis.

A 1% solution of synthetic complexes labeled with ^{35}S-
heparin–epinephrine (EDH) or heparin–urea (UH) was used as a
solvent of noncross-linked fibrin films [60-62]. The films
were 0.7 mm thick. Drops of the solution (0.05 ml) were
poured on the film surface both in the presence and the absence of ε-aminocaproic acid (a final concentration of 3%).
Lytic activity of the heparin complexes was evaluated by the
dimensions of the lysis zones (mm^2). The area of the lysis

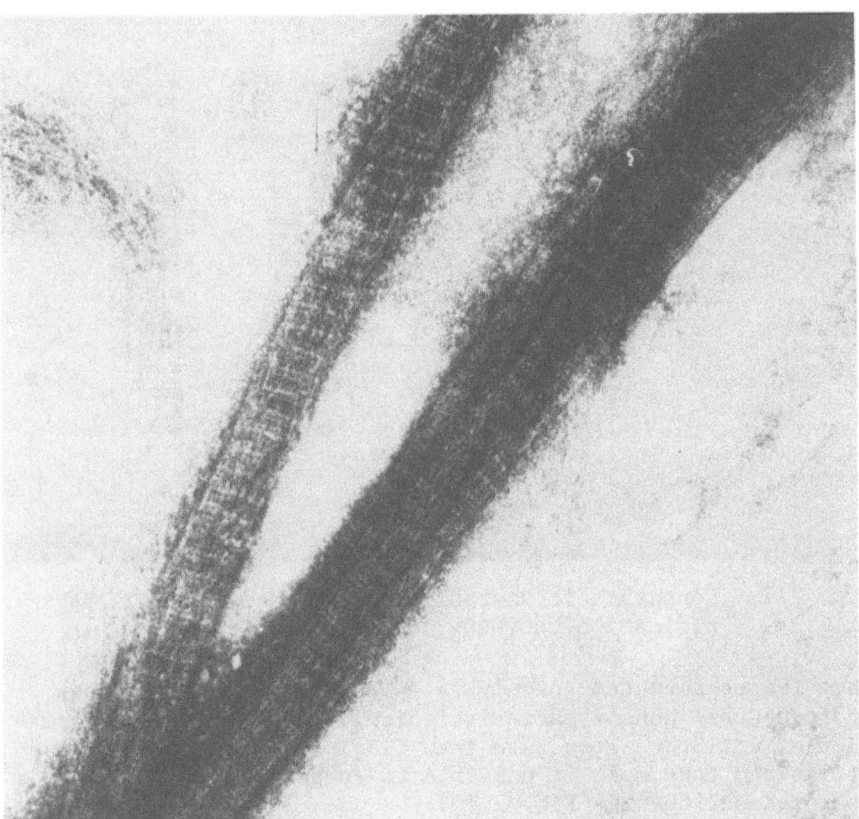

Fig. 2.6. Noncross-linked fibrin. Magnification, 120,000×.

zones produced by the heparin—urea and heparin—epinephrine
complexes averaged 223 mm² and 82 mm², respectively. In the
control, the equivalent doses of each component taken separately had no lytic effect. An electron-microscopic observation showed that a noncross-linked fibrin has the structure
of a fibrous net with cross-striation (Fig. 2.6). It was
established that the resulting fraction of nonenzymatic lysis
of noncross-linked fibrin by the heparin complexes consisted
of globular protein particles (Fig. 2.7) which were similar
to fibrinogen molecules (Fig. 2.8). About 70% of the labeled
³⁵S-heparin complexes were bound with this fraction after
precipitation of the protein by ethanol. In the control, the

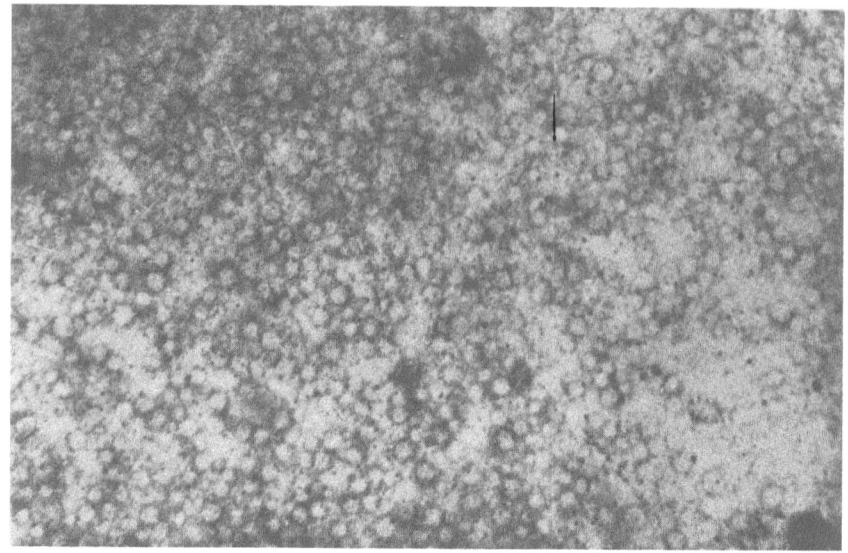

Fig. 2.7. Nonenzymatic unstabilized fibrin products. Magnification, 80,000×.

same figure amounted to only 7%. These and other findings
indicate that nonenzymatic lysis is linked to the transformation of a fibrous structure into fibrinogenlike globular particles and thus make it possible to suggest a hypothesis of
the process [60-62] (Fig. 2.10).

Hudry-Clergeon et al. [87] established that fibrinogen
particles hve the shape of noncompact globules which remain
spherical due to the presence of fibrinopeptides A and B in
the molecule. When these fibrinopeptides are separated by
the action of thrombin, fibrinogen particles open and form a
linear structure of fibrin monomer. Separated particles of
monomeric fibrin come into contact and, after primary polymerization, form a soluble fibrin.

It was found [60-62] that this noncross-linked fibrin
is dissociated from the particles of fibrin monomer by the
action of complex heparin compounds. This effect is likely
due to the formation of bonds between the particles of fibrin
monomer and ^{35}S-heparin complexes which change their molecular electrostatic potential.

It is possible that the complex compounds of heparin are
bound with fibrin monomer molecules in the loci from which

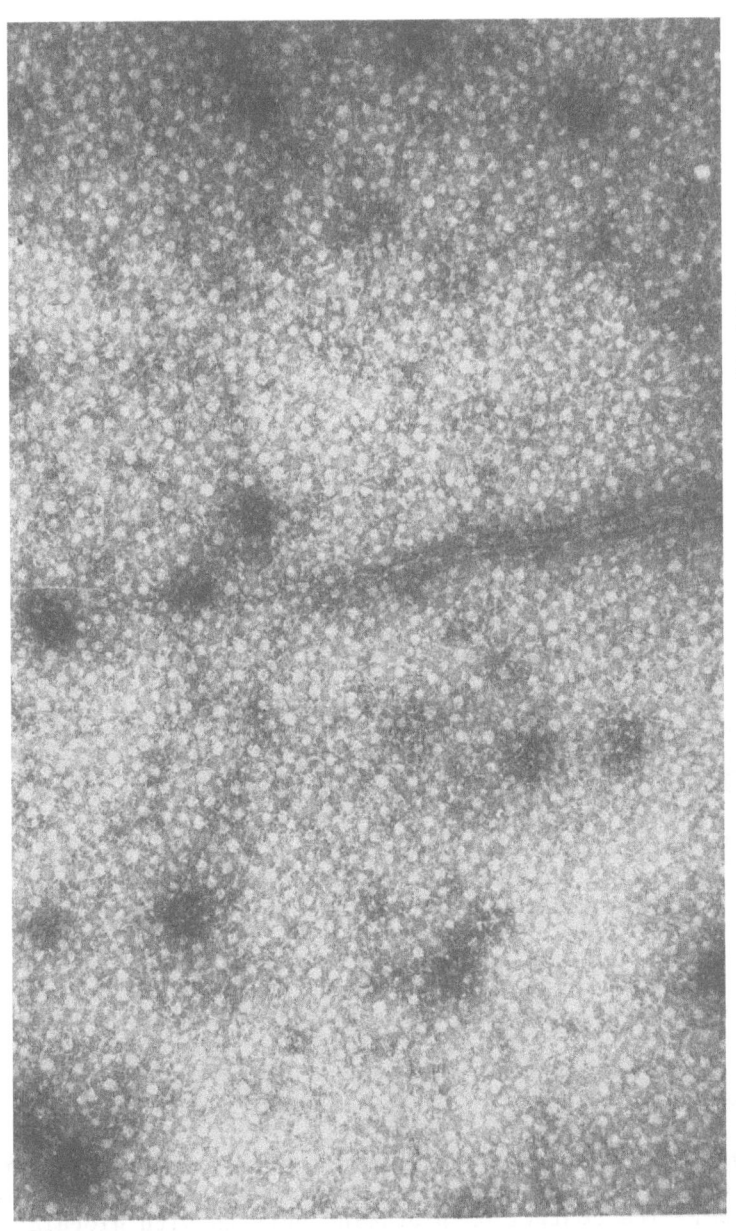

Fig. 2.8. Fibrinogen. Magnification, 80,000×.

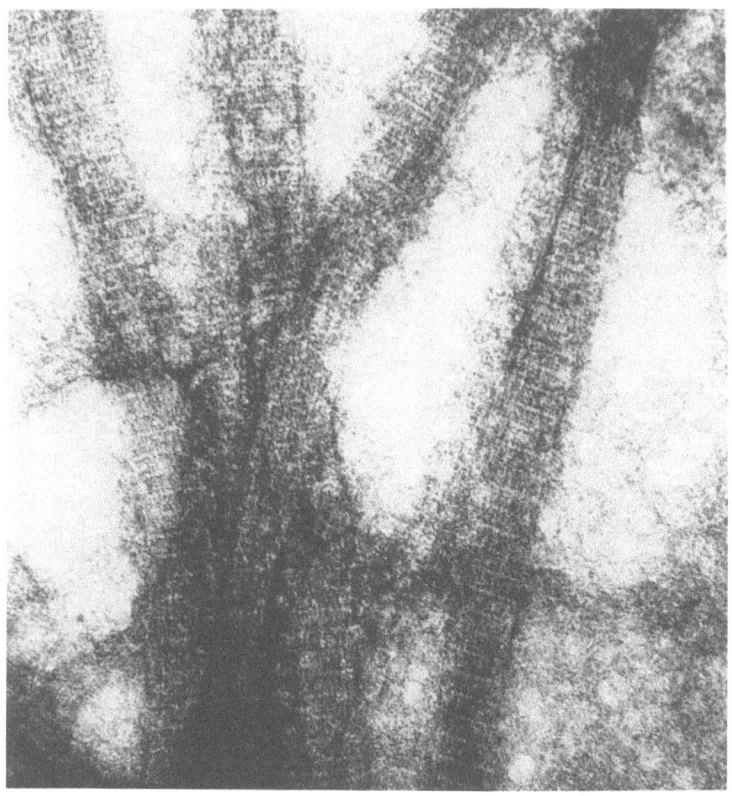

Fig. 2.9. Reconstruction of the fibrillar form of a monomer-
           ic fibrin clot of fibrinogenlike globules under
           the protamine sulfate effect. Magnification,
           120,000×.

the fibrinopeptides A and B were liberated. These structural
alterations lead to the transformation of monomeric fibrin
from a fibrous state into a dispersed globular form of the
molecule.

This hypothesis was partly confirmed by the experiments
with protamine, which is capable of dissociating heparin
complexes. The contact of protamine sulfate with globular
forms of fibrin monomer, bound with ^{15}S-heparin complexes,
leads to the reconstruction of fibrous clots of noncross-
linked fibrin (Fig. 2.9). During this process, monomeric fib-

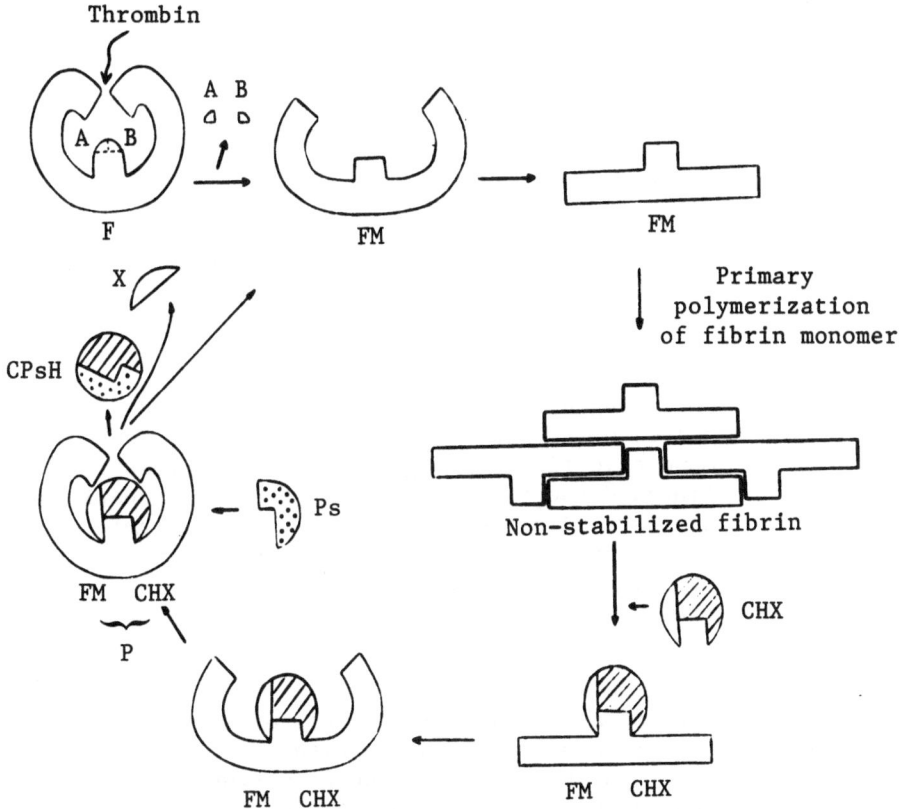

Fig. 2.10. Hypothetical scheme of heparin complex interaction with noncross-linked fibrin [60-62]. F)
Fibrinogen (globular form); A, B) A and B fibrinopeptides; FM) monomeric fibrin; P) a product of
noncross-linked fibrin dispersion with heparin
complexes; CHX) heparin complexes; Ps) protamine
sulfate; X) component of complexes (protein,
amine); CPsH) complexes of protamine sulfate with
heparin. Fibrinogen molecule model and monomeric
fibrin primary polymerization processes are given
according to Hudry-Clergeon et al. [87].

rin loses 87% of the ^{35}S-label, which is bound with protamine sulfate [60-62]. A conclusion can be drawn that the physicochemical nature of noncross-linked fibrin dissolution by the heparin complexes has nothing in common with the destruction of
fibrin to row peptides by the plasmin fibrinolytic action.

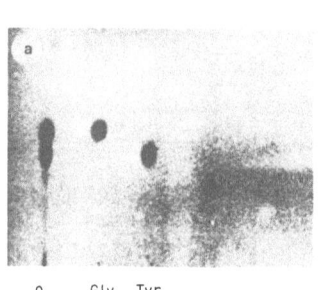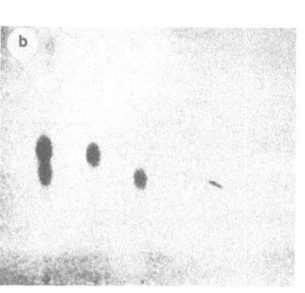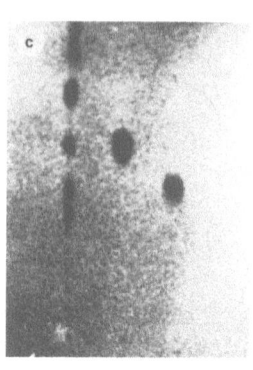

0   Gly   Tyr

Fig. 2.11. Chromatograms of phenylthiohydantoin (PTH) derivatives of NH_2-terminal amino acids. a) Control
monomeric fibrin; b) product of a noncross-linked
fibrin dispersion with the epinephrine—heparin
complex; c) product of a fibrin clot enzymatic
lysis produced by a brief plasmin action. System
of solvents: acetic acid—chloroform stabilized
with 1.5% ethanol (1:4), 0-PTH obtained by Edman
reaction [89].

The results of a biochemical analysis of the noncross-
linked fibrin lysis products coincide with those mentioned
above. A pure preparation of the monomeric fibrin, obtained
from ox fibrinogen according to the modified technique [88],
was used. Noncross-linked fibrin was obtained by adding monomeric fibrin into the Palitch borate buffer (pH 7.6) at a
final protein concentration of 0.5%. The samples were then
kept for 30 min at 37°C. The action of the EH complex on the
noncross-linked fibrin was studied according to the method described in [89]. Quantitative NH_2-terminal analysis of the
control monomeric fibrin dissolved in 0.125% acetic acid
showed that the preparation contains 3.9 moles of glycine and
2 moles of tyrosine per 1 mole of protein. That coincides
with the known content of NH_2-terminal amino acids of ox monomeric fibrin. The chromatograms given in Figure 2.11a do not
reveal any side spots, which points to the purity of the preparations. Quantitative NH_2-terminal analysis of the noncross-
linked fibrin nonenzymatic lysis products (Fig. 2.11b) also
yielded similar results (4.24 moles of glycine and 2 moles of
tyrosine per 1 mole of protein). Thus, these products do not
differ in an NH_2-terminal amino acid content from the control
monomeric fibrin. No additional NH_2-terminal amino acids were
found.

For comparison, NH_2-terminal analysis of the product of a brief plasmin action on the noncross-linked fibrin was carried out. The chromatogram given in Figure 2.11c shows a completely different composition of the NH_2-terminal amino acids.

The results obtained from the chemical analysis of NH_2-terminal amino acids indicate that the dissociation of the noncross-linked fibrin clot under the effect of an EH complex is a process of fibrin conversion from a polymeric form to the monomeric one which has nothing in common with its enzymatic (plasmin) hydrolysis [89].

Heparin Complexes and Fibrinogen Degradation Products

When the second anticoagulating system is excited, heparin complexes are found in the blood plasma. Heparin complexes (mostly FH) are present in the total fraction of fibrinogen degradation products (FDP) [68, 69]. The experiments with labeled ^{35}S-heparin showed that the FH-^{35}S complex prevails in the total FDP. Judging by the specific radioactivity, its share is not less than 56% and, according to the nonenzymatic fibrinolytic effect on noncross-linked fibrin, exceeds 90%.

It is perfect clear that the appearance and physiological action of the FH complex (which is not an FDP) and some other heparin complexes in the bloodstream is conditioned by excitation of the anticoagulating system [76, 77]. Heparin in sufficient amounts is then discharged into the bloodstream and enters into the complex formation. The data obtained demand a revision of current views on the presence and role of FDP in the bloodstream [68, 69].

It was also found that heparin-^{35}S content in the euglobulin fraction of blood plasma, obtained 7-10 min after intravenous injection of thrombin, amounted to 49.6% of its total quantity in the blood. Heparin-^{35}S contained in the FH complex accounted for 67.3%, judging by the nonenzymatic fibrinolytic effect, and for 32.4% according to its specific radioactivity. All of this indicates that about 70% of the total fibrinolytic activity in the euglobulin fraction of the plasma isolated from the blood of experimental animals (given intravenous injections of thrombin) is due to the FH-^{35}S complex [69].

DISCUSSION

It follows from the data presented in this chapter that, apart from an enzymatic (plasmin) fibrinolytic system, natural nonenzymatic fibrinolysis also exists in animals and man. It is performed by complex combinations of heparin with plasma proteins, amines, and other agents.

As known, plasmin fibrinolysis is characterized by the lysis of fibrin, both stabilized and nonstabilized, with factor XIIIa. Enzymatic fibrinolysis leads to degradation of the protein (fibrinogen/fibrin) molecule into polypeptide fragments.

It was demonstrated that nonenzymatic fibrinolysis is a dissolution of only noncross-linked fibrin via the protein depolymerization leading to its dispersion into particles of monomeric fibrin bound with heparin complexes [60-62, 89]. After the rupture of bonds with heparin complexes, monomeric fibrin particles can recover the conformation of noncross-linked fibrin fibers.

The phenomenon of natural nonenzymatic fibrinolysis has not been recorded in the scientific literature. However, there have been several investigations of "nonplasmin" fibrinolysis. Moroz and Gilmore [90] have published data on the existence of fibrinolysis mechanisms "independent of the plasminogen/plasmin system." Using a labeled fibrin, the authors found spontaneous "basal" fibrinolytic activity in normal blood plasma. The activity was retained at a certain level after the removal of plasminogen and plasmin by affinity chromatography and the addition of a number of plasmin system inhibitors to the medium. The authors assumed that certain proteinases independent of the plasmin system which condition "basal" fibrinolysis may be present in the normal plasma. Similar data on "nonplasmin" fibrinolysis and the possible presence of some proteinases in plasma have been published by Zimmerman and Lubinus [91].

These two studies, however, give no evidence supporting this viewpoint. It is quite possible that the authors who used the euglobulin fraction of plasma have, in fact, observed the phenomenon of a nonenzymatic fibrinolysis.

Chakabarti [92] showed that the incubation of citrate plasma with small doses of streptokinase or urokinase for 30 min

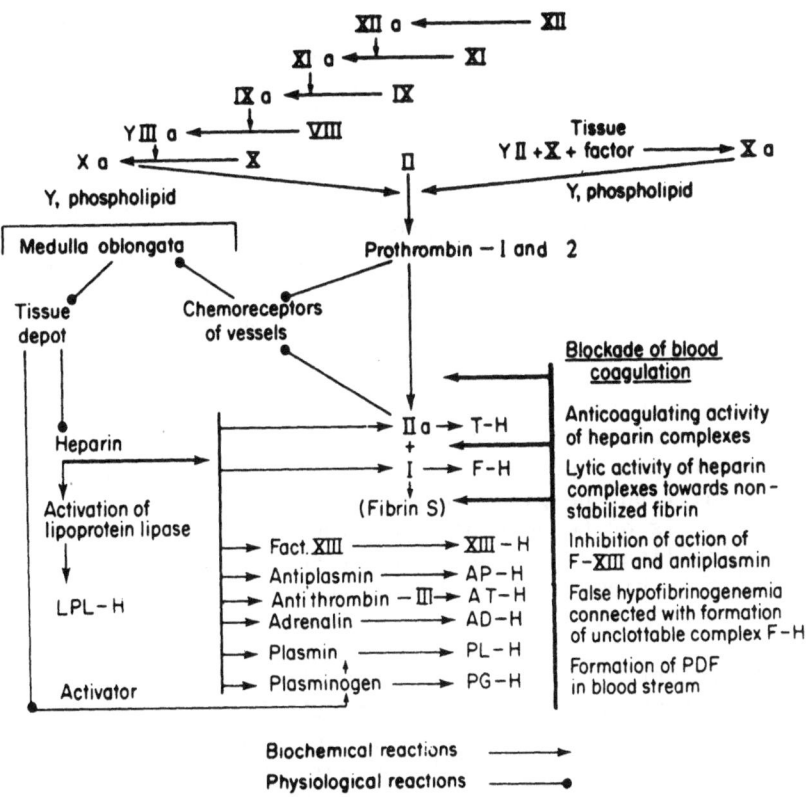

Fig. 2.12. Scheme of the interrelation between the coagulating and anticoagulating systems [77].

did not lead to the loss of fibrinogen in plasma. However, the addition of heparin prior to incubation resulted
in a fibrinogen deficiency depending on the dose of injected
heparin. This effect was observed both in vitro and after
intravenous administration of heparin (5000-7500 units). It
was also found that the addition of heparin to fibrinogen in
a pure system (before it is converted to fibrin with the help
of thrombin) in the presence of urokinase leads to the shortening of clot lysis time. On the other hand, the lysis time
under the same conditions and heparin effect was prolonged if
heparin was added after the formation of a clot with subsequent administration of urokinase. The author gave no explanation for these findings.

    The study likely involved the regularity that we de-
scribed earlier: Enzymatic components of the fibrinolytic

system form complex combinations with heparin, which have
higher activity compared with native enzymes [26, 27, 70].
A decrease in fibrinogen concentration is conditioned by its
complex formation with heparin [25], which is not clottable
by thrombin.

Interesting data have been published by Fontanyi [93].
His experience contradicts a generally accepted notion that
"heparin inhibits thrombi formation but fails to dissolve
thrombi." The author used heparin in 192 cases of acute
fresh peripheral arterial thrombosis and succeeded in 78% of
the cases (32% recanalization). In 43 cases of acute, fresh,
deep phlebothrombosis, the positive effect of massive inter-
mittent heparinization of the thrombus zone [93] amounted to
83%. The author stresses that, before his publication, there
were no clinical reports of thrombolysis caused by the use of
heparin only, but he gives no analysis of the phenomenon observed. It is quite possible that local administration of
heparin facilitated the formation of heparin complexes in the
locus of the thrombus, thus causing the acceleration of
thrombolysis.

At present, the data on nonenzymatic fibrinolysis at-
tract the attention of clinicians in the Soviet Union, but
practical aspects of the problem lie beyond the limited scope
of this chapter. We would like to point out, however, that
clinical application of the heparin complex formation and nonenzymatic fibrinolysis is in no way limited to thrombosis. It
should be noted that, during atherosclerosis, the formation
of heparin complexes in the bloodstream [81, 82] and in the
cardiovascular system [79, 8] is sharply reduced; this is supported by clinical data [94, 95]. At the same time, hyper-
function of the anticoagulating system caused by obstetric
complications and often leading to hemorrhage is linked to
the excess of various heparin complexes in the blood [96].

However, under physiological conditions, complex heparin
combinations are natural humoral agents that emerge in the
organism when the anticoagulating system is stimulated, and
they play a key role in the regulation of hemostasis [97, 98]
(Fig. 2.12).

The process of heparin complex formation leads not only
to dissolution of noncross-linked fibrin, thus preventing
thrombi formation at early stages, but also to inhibition of
factor XIII, enhanced thrombin neutralization in the blood-

stream, blocking of platelet aggregation, decreased native fibrinogen concentration, increased plasmin activity, the removal of antiplasmin, etc. Thus, heparin complexes may be regarded as an essential link in the chain of protective regulatory processes.

REFERENCES

1. J. B. Morgagni, The Seats and Causes of Diseases, Vol. III, Letter 53, Book IV, London (1769), p. 185.
2. P. S. Denis, Essai sur l'Application de la Chimie a l'Etude Physiologique du Sang de l'Homme et a l'Etude Physio-pathologique, Hygienique et Therapeutique des Maladies de Cette Humeur, Bechetjre, Paris (1838).
3. J. Denys and H. de Marbaix, "Les peptonisations provoquees par le chloroforme," Cellule, 5, 197 (1889).
4. M. A. Dastre, "Fibrinolyse dans le sang," Arch. Physiol. Pathol. (Paris), 5, 661-663 (1893).
5. L. R. Christensen and C. M. MacLeod, "A proteolytic enzyme of serum characterization, activation and reaction with inhibitors," J. Gen. Physiol., 28, 559-583 (1945).
6. E. C. Loomis, C. George, and A. Ryder, "Fibrinolysin: nomenclature, unit, assay, preparation, and properties," Arch. Biochem., 12, 1-5 (1947).
7. E. Kowalski, A. Z. Budzynski, M. Kopec, Z. S. Latallo, B. Lipinski, and Z. Wegrzynowicz, "Circulating fibrinogen degradation products (FDP) in dog blood after intravenous thrombin infusion," Thromb. Diath. Haemorrh., 13, 12-24 (1965).
8. E. Bidwell, "Fibrinolysis of human plasma," Biochem. J., 55, 497-506 (1953).
9. B. A. Kudrjashov, "Physiological anticoagulating system and its significance," Vopr. Med. Khim., 6, 3-13 (1960).
10. B. A. Kudrjashov, "Interrelations within the blood coagulating system," Nature, 184, 454-455 (1959).
11. B. A. Kudrjashov, "On the physiological anticoagulating system," Thromb. Diath. Haemorrh., 6, 371-380 (1961).
12. B. A. Kudrjashov, Regulatory Relationships between the Coagulating and Anticoagulating Blood System, X Congress of I. P. Pavlov All-Union Physiology Society, Vol. 1, Nauka, Moscow (1964), pp. 144-145.
13. B. A.Kudrjashov, T. M. Kalishevskaya, and L. A. Polyakova, "Complex heparin combinations with thrombogenic blood proteins and their physiological properties," Vopr. Med. Khim., 12, 114-116 (1966).

14. B. A. Kudrjashov, T. M. Kalishevskaya, and L. A. Lyapina, "Fibrinogen—heparin complex: a natural agent of physiological anticoagulating system," Vopr. Med. Khim., 14, 277-282 (1968).

15. B. A. Kudrjashov, Anticoagulating Blood System and Its Role in Thrombosis Prevention, Leningrad Scientific-Research Institute of Blood Transfusion, Abstracts of scientific session on fibrinolysis, June 23-26, Leningrad (1965), pp. 3-5.

16. B. A. Kudrjashov, "Regulatory relationships between the coagulating and anticoagulating blood system," 23rd International Congress of Physiologic Sciences, Abstract 203 (1965), p. 167.

17. B. A. Kudrjashov, "Zur bedeutung des plasmins im vorgang der blutgerinnungsregulationen," Z. Med., 1, 13-15 (1966).

18. B. A. Kudrjashov, "The role of some tissue and plasma factors in regulatory interrelation between the coagulating and anticoagulating system," First International Symposium on Tissue Factors in the Homeostasis of the Coagulation—Fibrinolysis System, Firence (1967), pp. 41-47.

19. B. A. Kudrjashov, "The significance of the anticoagulating system in natural prophylaxis of thrombosis (Experimental data," in: Proceedings, Union of Physiological Sciences, XXIV International Congress, Vol. VII, Abstract 744, Washington (1968), p. 248.

20. B. A. Kudrjashov, L. V. Molchanova, S. E. Shnol, and V. A. Saynikov, "Behavior of ^{131}I-fibrinogen in the blood bed of animals during activation of the physiological anticoagulating system by thrombin," Byull. Eksp. Biol. Med., No. 12, 38-42 (1965).

21. A. F. Filinkov, "Composition and formation of fibrinogen complexes in the blood bed after an intravenous injection of thrombin," Vopr. Med. Khim., 16, 59-62 (1970).

22. L. A. Lyapina, "Interaction of heparin with fibrinogen, thrombin, and trypsin," Vestn. Selsk. Nauki, No. 4, 134-137 (1968).

23. B. A. Kudrjashov, L. A. Lyapina, L. V. Molchanova, and B. A. Rustamova, "The nature of lytic action of fibrinogen-heparin and tyroxine—heparin complexes on fibrin," Vopr. Med. Khim., 14, 161-167 (1970).

24. B. A. Kudrjashov, "Physiological significance and mechanism of action of heparin complex combinations with fibrinogen, epinephrine, and norepinephrine by the regulatory reactions of the anticoagulating system," XXV International Congress of Physiological Sciences, Vol. IX, Abstract 969, Munich (1971), p. 327.

25. E. Cohen, L. Strong, N. Hughes, D. J. Mulford, J. N. Aschworth, M. Melin, and H. J. Taylor, "Preparation and properties of serum and plasma," J. Am. Chem. Soc., 68, 459-463 (1946).

26. B. A. Kudrjashov and L. A. Lyapina, "Eststehung und physiologische bedeutung des 'aktivierten' plasminogens und plasmin bei der ingangsetzung des antikoagulationssystem," Folia Haematol. (Leipzig), 98, 312-320 (1972).

27. B. A. Kudrjashov and L. A. Lyapina, "Emergence and physiological significance of 'activated' plasminogen and plasmin during excitation of the anticoagulating system," Vopr. Med. Khim., 19, 165-168 (1973).

28. B. A. Kudrjashov and L. A. Lyapina, "Der heparin—adrenalin komplex als ein humoraler faktor des antikoagulationssystem," Folia Haematol. (Leipzig), 95, 272-284 (1971).

29. B. A. Kudrjashov and L. A. Lyapina, "Heparin—epinephrine complex: a humoral agent of the anticoagulating system," Vopr. Med. Khim., 17, 46-53 (1971).

30. B. A. Kudrjashov and L. A. Lyapina, "Evolution-conditioned adaptation of hemostatic phenomena and the role of heparin complexes in maintaining the fluid state of blood in the organism," Vestn. Mosk. Univ., Ser. Biol. Pochvored., No. 4, 3-25 (1973).

31. E. Kowalski, M. Kopec, and S. Niewiarowski, "On evolution of the euglobulin method for the determination of fibrinolysis," J. Clin. Pathol., 12, 215-218 (1959).

32. R. A. Kekwick, M. E. MacKay, M.N. Nance, and B. R. Record, "The purification of human fibrinogen," Biochem. J., 60, 671-678 (1955).

33. S. Niewiarowski, Krzepniecie Krwi., Warsaw (1960), p. 125.

34. A. G. Loewy, K. Dunathan, R. Kriel, and H. L. Wolfinger, "Fibrinase. II. Some physical properties," J. Biol. Chem., 236, 134-143 (1961).

35. K. S. Stenn and E. R. Blout, "Mechanism of bovine prothrombin activation by an insoluble preparation of bovine factor X_a (thrombokinase)," Biochemistry, 11, 4502-4515 (1972).

36. S. M. Strukova, "Prothrombin activation by immobilized thrombin," Biokhimiya, 41, 643-649 (1976).

37. U. Abildgaard, "Purification of two progressive antithrombins of human plasma," Scand. J. Clin. Invest., 19, 190-199 (1967).

38. V. Mansfeld and J. Hladovec, "Studien uber reaktionen des kristallisierten trypsins. II. Beziehung zwischen der aktivitat des heparin und trypsin," Collect. Czech. Chem. Commun., 21, 1209-1213 (1956).

39. L. A. Lyapina, S. M. Strukova, and B. A. Kudrjashov,
     "Heparin—prothrombin complex formation," Vopr. Med.
     Khim., 25, 41-46 (1979).
40. B. A. Kudrjashov, L. A. Lyapina, and E. A. Grigorova,
     "Interaction of heparin with monomeric fibrin and mono-
     meric fibrin—fibrinogen complex," Vopr. Med. Khim., 26,
     318-321 (1980).
41. B. S. Kudrjashov, A. M. Ulyanov, and L. A. Lyapina, "The
     formation of complex fibrin stabilizing factor with hep-
     arin in vitro," Thromb. Res., 3, 589-602 (1973).
42. B. A. Kudrjashov and L. A. Lyapina, "Physiological char-
     acteristics of heparin—thromboplastin complex," Physiol.
     Zh. SSSR, 64, 771-776 (1978).
43. B. A. Kudrjashov, J. A. Pytel, G. M. Baskakova, and L.A.
     Lyapina, Technique for Obtaining Heparin—Insulin Complex.
     Application for invention rights No. 26022117/13,
     March 6, 1978 (granted).
44. B.A. Kudrjashov, V. E. Pastorova, and L. A. Lyapina,
     "Heparin—antithrombin II, antithrombin III—heparin—
     thrombin complexes: anticoagulating activity and lytic
     effect on noncross-linked fibrin," Biokhimiya, 46, 2024-
     2027 (1981).
45. B. A. Kudrjashov, V. E. Pastorova, and V. L. Glotova,
     "Heparin—antifibrinolysin and its significance for the
     function of the anticoagulating system," Vopr. Med.
     Khim., 13, 29-33 (1967).
46. B. A. Kudrjashov and L. A. Lyapina, "Heparin—urea com-
     plex and its physicochemical properties," Vopr. Med.
     Khim., 23, 165-168 (1975).
47. L. A. Lyapina, "Effect of proteinase inhibitors and hep-
     arin blockers on fibrinolytic activity of heparin com-
     plexes," Vestn. Mosk. Univ., Ser. Biol., No. 2, 59-63
     (1980).
48. B. A. Kudrjashov and L. A. Lyapina, "Heparin—aspirin
     complex: physicochemical characteristics," Vopr. Med.
     Khim., 23, 44-51 (1977).
49. L. A. Lyapina and B. A. Kudrjashov, "Heparin—DNA com-
     plexes: production, characteristics, and methylation
     in vitro," Biokhimiya, 45, 2189-2197 (1980).
50. L. A. Lyapina and B. A. Kudrjashov, "In vitro formation
     of AMP, ADP, and ATP complexes with heparin," Nauchn.
     Dokl. Vyssh. Shkol., Ser. Biol. Nauki, No. 9, 22-26
     (1977).
51. B. A. Kudrjashov, T. M. Kalishevskaya, and L. A. Lyapina,
     "Complex serotonin—heparin compound and its physiologi-
     cal significance," Vopr. Med. Khim., 14, 269-274 (1973).

52. S. Nakamura, K. Takeo, K. Tanaka, and T. Ueta, "Kreuzung-verfahren bei Papierelektrophorese," Hoppe-Seyler's Z. Physiol. Chem., No. 318, 115-128.

53. B. A. Kudrjashov and L. A. Lyapina, "A technique for measuring the activity of noncross-linked fibrin solvents," Lab. Delo., No. 6, 326-329 (1971).

54. B. A. Kudrjashov, L. A. Lyapina, and I. P. Baskakova, "Development of the technique for measuring nonenzymatic fibrinolytic activity of plasma and some of its fractions," Vestn. Mosk. Univ., Ser. Biol. Pochvored., No. 5, 35-39 (1974).

55. L. A. Lyapina and B. A. Kudrjashov, "Determination of heparin complexes with blood proteins in euglobulin fraction of plasma," Lab. Delo, No. 4, 246-249 (1977).

56. T. Astrup and S. Mullertz, "The fibrin plate method for estimating fibrinolytic activity," Arch. Biochem. Biophys., 40, 346-351 (1952).

57. M. Lassen, "Heat denaturation of plasminogen in the fibrin plate method," Acta Physiol. Scand., 27, 371-380 (1952).

58. K. Buluk, T. Yanuszko, and Ya. Olbromski, "Conversion of fibrin to desmofibrin," Nature, 191, 1093-1094 (1961).

59. E. C. Loomis, A. Ryder, and G. George, "Fibrinolysin and antifibrinolysin," Arch. Biochem. Biophys., 20, 444-450 (1949).

60. B. A. Kudrjashov, A. M. Ulyanov, and E. S. Zhitnikova, "Molecular forms of products of noncross-linked fibrin dissolution by complex heparin compounds: a study by scanning electron microscopy," Dokl. Akad. Nauk SSSR, 237, 965-968 (1977).

61. B. A. Kudrjashov, A. M. Ulyanov, and E. S. Zhitnikova, "Electron-microscopy study of molecular forms of products of noncross-linked fibrin dissolution by complex compounds of heparin," in: Progress in Chemical Fibrinolysis and Thrombolysis, Vol. IV, J. P. Davidson, V. Cepelak, N. M. Samama, and P. C. Desnoyers (eds.), London (1979), pp. 255-259.

62. B. A. Kudrjashov, A. M. Ulyanov, and E. S. Zhitnikova, "Electrone mikroskopische untersuchungen der molekulformen von spalproductennichtstabilisierten fibrins durch Komplexe heparin—verbindungen," Folia Haematol. (Leipzig), 107, 895-905 (1980).

63. B. A. Kudrjashov, Biological Problems in Regulating the Fluid State of Blood and Its Coagulation, Vol. 3, Meditsina, Moscow (1975), pp. 485-488.

64. L. A. Lyapina and B. S. Kudrjashov, "Nonenzymatic lysis
     of fibrin clots in the organism," VII International
     Symposium on Medical Chemistry, Spain (1980), p. 39.
65. B. A. Kudrjashov, V. E. Pastorova, and L. A. Lyapina,
     "Effect of heparin, epinephrine, and epinephrine–hep-
     arin complex on platelet aggregation and lytic activity
     of plasma in animals," Probl. Gematol. Pereliv. Krovi,
     No. 5, 28-33 (1973).
66. B. A. Kudrjashov, V. E. Pastorova, and L. A. Lyapina,
     "Effect of small doses of heparin–urea complex on plate-
     let aggregation and blood fibrinolytic activity during
     intravenous infusion and in vitro experiments," Farmakol.
     Toksikol., No. 4, 441-444 (1975).
67. B.A. Kudrjashov, L. A. Lyapina, and A. M. Ulyanov, "Sig-
     nificance of fibrinogen–heparin complex for fibrinolytic
     activity of euglobulin blood fraction after intravenous
     injection of thrombin and plasmin," Vopr. Med. Khim.,
     24, 256-260 (1978).
68. B. A. Kudrjashov, L. A. Lyapina, and A. M. Ulyanov, "The
     presence of an active fibrinogen–heparin complex in the
     total fraction of blood plasma fibrinogen degradation
     products," Biokhimiya, 42, 451-459 (1977).
69. B. A. Kudrjashov, L. A. Lyapina, and A. M. Ulyanov, "Com-
     plex fibrinogen (FH) and fibrinogen degradation products
     (FDP) in blood of rats after intravenous injection of
     thrombin," Thromb. Res., 13, 397-407 (1978).
70. B. A. Kudrjashov, L. A. Lyapina, and H. Klocking, "Uber
     fibrinolytische eigenschaften eines heparin–ocrase kom-
     plexes," Folia Haematol. (Leipzig), 103, 573-582 (1976).
71. B. A. Kudrjashov, L. A. Lyapina, E. S. Zhitnikova, and
     V. V. Obraztov, "Formation of a secondary epinephrine–
     heparin–fibrinogen complex and its characteristics,"
     Vopr. Med. Khim., 21, 65-69 (1975).
72. E. A. Malakhova, G. G. Bazajan, T. P. Levchuk, B. A.
     Kudrjashov, and V. A. Yakovlev, "On a new function of
     lipoprotein lipase in blood," Dokl. Akad. Nauk SSSR,
     231, 495-498 (1976).
73. E. A. Malakhova, G. G. Bazasian, T. P. Levchuk, and
     V. A. Yakovlev, "On a new function of lipoprotein lipase
     in blood," Thromb. Res., 12, 209-218 (1978).
74. L. A. Lyapina, A. M. Uljanov, and F. B. Shapiro, "Factor
     XIII effect of the blood plasma fibrinolytic character-
     istics during effort in animals," Vestn. Mosk. Univ.,
     Ser. Biol. Pochvoved., No. 5, 46-53 (1974).
75. L. A. Lyapina, "Simulation of thrombi formation in ani-
     mals: preventive action of heparin–urea and epinephrine–

heparin–fibrinogen complexes," Kardiologiya, No. 8, 143-144 (1978).

76. B. A. Kudrjashov, "Antithrombotic system (AS) of the animal organism, its nature, function, and pathophysiology," Abstract, VII International Congress on Thrombosis and Haemostasis, Thromb. Haemostas., 42, 459 (1979).

77. B. A. Kudrjashov, "The problems of regulation of the lipid state and coagulation of blood," in: Haemostasis and Thrombosis, Vol. 15, Proceedings of the Serono Symposia, G. G. Neri Serneri, and C. R. M. Prentice (eds.), Academic Press, London (1979), pp. 195-202.

78. B. A. Kudrjashov, L. A. Lyapina, A. M. Ulyanov, and T. N. Kovaleva, "Role of complex thrombin–heparin formation in the bloodstream and the role of liver and lungs in the absorption of heparin complexes," Thromb. Res., 8, 205-215 (1976).

79. B. A. Kudrjashov, L. A. Lyapina, and A. M. Ulyanov, "Physiological solvents of noncross-linked fibrin in the heart and blood vessels during excitation of the anticoagulating system," Kardiologiya, No. 10, 125-128 (1977).

80. B. A. Kudrjashov, G. G. Basazyan, A. M. Ulyanov, L. A. Lyapina, and G. Boneva, "The role of heparin complexes in normal heart blood circulation," J. Mol. Cell. Cardiol., 12 (Suppl. 1), 83 (1980).

81. B. A. Kudrjashov, "Depression of the function of the physiological anticoagulating system (ASC) and development of a prethrombotic state in organisms," International Society of Haematology, Third Meeting of European and African Division, Vol. 3, Abstract No. 25:04, London (1975).

82. B. A. Kudrjashov, G. G. Bazasyan, L. A. Lyapina, and N. P. Sytina, "Nonenzymatic fibrinolysis of animals fed a natural and atherogenic diet after administration of heparin–urea complex," Kardiologiya, No. 11, 72-75 (1974).

83. B. A. Kudrjashov, G. G. Basazyan, and N. P. Sytina, "Significance of vitamin A and ethyl esters of unsaturated fatty acids for the prevention of anticoagulating system depression," Kardiologiya, No. 6, 88-95 (1970).

84. K. L. Erzinkyan, M. A. Rozenfeld, A. G. Ter-Makaryan, L. A. Piruzyan, B. A. Kudrjashov, and L. A. Lyapina, "Reaction of fibrinogen–heparin complex formation: a study by spectrophotometry," Izv. Akad. Nauk SSSR, Ser. Biol., No. 6, 920-924 (1974).

85. B. A. Kudrjashov, L. A. Lyapina, A. M. Ulyanov, E. S. Zhitnikova, O. V. Podolskaya, and J. M. Ammosova, "Heparin–epinephrine complex and its physicochemical char-

acteristics," Nauchn. Dokl. Vyssh. Shkol., Ser. Biol.
Nauki, No. 1, 36-41 (1975).

86. B. A. Kudrjashov, L. A. Lyapina, E. S. Zhitnikova, and
     M. G. Krukova, "Comparative study of fibrinogen—heparin
     complex obtained in vitro and isolated from the plasma
     fraction of fibrinogen degradation products," Nauchn.
     Dokl. Vyssh. Shkol., Ser. Biol. Nauki, No. 9, 58-62
     (1979).

87. G. Hudry-Clergeon, G. Marquerie, L. Poul, and M. Suscil-
     lon, "Models proposed for the fibrinogen molecule and
     for the polymerization process," Thromb. Res., 6, 391-
     541 (1978).

88. L. A. Lyapina, A. M. Ulyanov, and E. S. Zhitnikova, "A
     procedure for obtaining monomeric fibrin," Lab. Delo,
     No. 3, 186-188 (1979).

89. E. V. Lubovski, L. A. Lyapina, G. K. Gogolinskaya, and
     S. G. Derzskaya, "NH$_2$-terminal protein amino acids ob-
     tained by dissolution of noncross-linked fibrin polymer
     by heparin—epinephrine complex," Biokhimiya, 44, 1918-
     1922 (1979).

90. Z. A. Moroz and N. J. Gilmore, "Fibrinolysis in normal
     plasma and blood. Evidence for significant mechanism
     independent of the plasminogen—plasmin system," Blood,
     48, 531-545 (1976).

91. E. Zimmerman and J. Lubinus, "Fibrinolysis in normal
     plasma independent of the plasma system," Pflügers Arch.,
     368 (Suppl.), R-9 (1977).

92. R. Chakabarti, "Effect of heparin on lysis of fibrinogen/
     fibrin," VII International Congress on Thrombosis and
     Haemostasis, Thromb. Haemostas., Abstract 1192, 42, 83
     (1979).

93. S. Fontanyi, "Theoretical, clinical, practical bases and
     technique of thrombosis with heparin," VII International
     Congress on Thrombosis and Haemostasis, Thromb. Haem-
     ostas., Abstract 1117, 42, 460 (1979).

94. V. M. Panchenko, B. A. Kudrjashov, and L. A. Lyapina,
     "On total nonenzymatic fibrinolytic activity and epi-
     nephrine—heparin complex formation in ischemic heart dis-
     ease," Klin. Med., No. 8, 37-41 (1972).

95. V. M. Panchenko and L. A. Lyapina, "Nonenzymatic fibrin-
     olysis in ischemic heart disease patients," Kardiologiya,
     No. 11, 14-18 (1974).

96. A. N. Bogarada, B. A. Kudrjashov, L. A. Lyapina, and
     A. M. Ulyanov, "Bleeding in obstetric complications and
     its dependence on nonenzymatic fibrinolysis," Akush.
     Ginekol., No. 5, 41-43 (1980).

97.   B. A. Kudrjashov, "Physiological system of regulation of the fluid state of blood and significance of its humoral components," XII Congress of the I. P. Pavlov All-Union Physiology Society, Vol. 1, Nauka, Leningrad (1975), pp. 190-191.

98.   B. A. Kudrjashov, "On the physiological principles of regulation of the liquid state of blood," XXVI International Congress of Physiological Sciences, Vol. XI, Abstract 69, New Delhi (1974), p. 23.

MECHANISMS OF ENZYMATIC FIBRINOLYSIS

G. V. Andreenko

Moscow State University, Moscow

ABSTRACT

Modern concepts of enzymatic fibrinolysis, including data
from recent years, are addressed. The basic components of
the fibrinolytic system, namely plasminogen, plasmin, activators, and inhibitors, are characterized. Plasminogen, an
inactive precursor of the fibrinolytic enzyme, exists in the
blood in mainly two forms. Plasminogen activation occurs via
both an intrinsic and extrinsic pathway. The latter, more
thoroughly studied, involves tissue activators and urokinase.
The main stages of plasminogen conversion to plasmin are described. A two-chain molecule of plasmin is formed as a result of the breakdown of two peptide links. Lysine-binding
sites of plasminogen and plasmin determine their affinity to
fibrin, antiplasmin, and certain amino acids. The possibility of nonenzymatic plasminogen activation during immobilization has been suggested. Plasminogen activators are regarded
as principal agents of fibrinolysis activation in vivo. Biosynthesis of these enzymes in cell culture was a significant
stage in the study of plasminogen activators. On the basis
of immunochemical properties, multiple forms of activators
may be divided into two main types: urokinase and tissue
activators. Streptokinase activator, which is a plasminogen
complex with small amounts of streptokinase, plays an important role in the fibrinolysis system. Activity of the complex
is enhanced by a special potentiator isolated from human plasma. Other enzymes of microbial origin are promising with respect to clinical application. Besides the previously described aspergillin, brinase, ocrase, and terrilytin, this review gives a more detailed account of new enzymes synthesized

by the lowest fungi and possessing both fibrinolytic and plasminogen-activating action. α_2-Antiplasmin is the most important inhibitor of fibrinolysis. As a result of the effect of
plasmin on fibrinogen and noncross-linked fibrin, the four
chief large-molecular fragments are formed: the early (X and
Y) and late ones (D and E). The main product formed during
lysis of cross-linked fibrin is the dimer of fragment D forming a noncovalently bound complex with fragment E.

INTRODUCTION

The enzymatic fibrinolytic system, which dissolves fibrin in man and animals, consists of the inactive proenzyme
plasminogen, plasmin, plasminogen activators, and plasmin inhibitors. Successful investigation of the chemistry, biochemistry, and physiological significance of enzymatic fibrinolysis has allowed the isolation and examination of the separate components of this system and the study of their interactions. The data obtained have made it possible to utilize
fibrinolysis in thrombolytic therapy and to evaluate its relationship to pathological conditions.

The results of long-standing research have been general-
ized in a number of monographs and reviews [1-4]. This chapter summarizes the data published in reviews over the last
five years, without giving reference to the original studies.

CHARACTERISTICS OF PLASMINOGEN AND PLASMIN,
AND PLASMINOGEN CONVERSION TO PLASMIN

Human plasminogen is a glycoprotein with a molecular
weight of approximately 92,000-94,000 and containing 2% carbohydrates [5, 6]. The primary structure of the proenzyme has
been completely deciphered; it is represented by one polypeptide chain containing 791 amino acid residues (Fig. 3.1). The
amino-terminal and carboxy-terminal amino acids are glutamic
and asparagine acids, respectively. This form is called Glu-
plasminogen or plasminogen A. During plasminogen isolation
from blood plasma and the bloodstream, the N-terminal peptide
is split off from the molecule and plasminogen is formed with
the N-terminal amino acid lysine called Lys_{77}-plasminogen or
plasminogen B [6]. Lys_{77}-plasminogen has a molecular weight
of approximately 83,000-85,000.

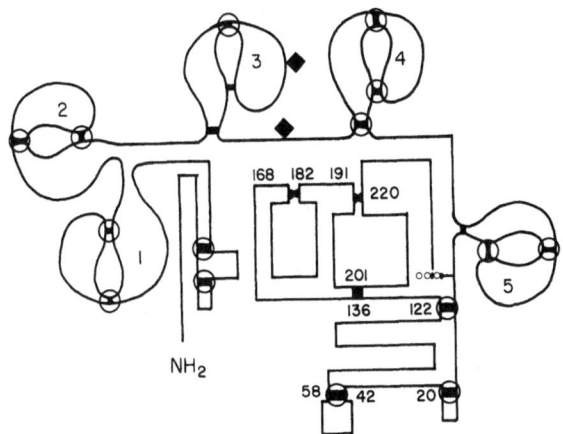

Fig. 3.1.   Hypothetical structure of human plasminogen with
            kringles. From L. Sottrup-Jensen, H. Claeys,
            M. Zajdel, T. E. Petersen, and S. Magnusson, in:
            Progress in Chemical Fibrinolysis and Thrombol-
            ysis," Raven Press, New York (1978), Vol. 3, p.
            191.

There are substantial differences in the conformation
of the two forms of plasminogen. For example, the $S_{20,w}^{0}$
value for Glu-plasminogen is 5.75, while the same value for
Lys-plasminogen is 4.79. The molecule of Lys-plasminogen is
more asymmetrical [5]. The polypeptide chain of native Glu-
plasminogen contains three parts. The N-terminal component
(MW 8800) includes amino acid residues from 1 to 76. Removal
of this component changes the plasminogen conformation. The
intermediate component (MW 57,000) includes amino acid residues from 77 to 560 and contains five kringles homologous
to the two similar structures of prothrombin [7]. In this
part of the molecule lysine binding sites are contained.
These sites form bonds with fibrin, α_2-antiplasmin, L-lysine,
and 6-aminohexanoic and tranexamic acids. The C-terminal
plasminogen component (MW 25,700) contains amino acid residues from 561 to 791 and is transformed into the light chain
of plasmin during activation [7]. There are several proofs
of human plasmin heterogeneity. Isoelectric focusing of
human Glu-plasminogen has yielded five forms of the proenzyme
with isoelectric points between 6 and 6.6. In part, such
microheterogeneity is accounted for by a different content
of sialic acids.

At least 95% of plasminogen is represented by the two forms, which differ in chemical and physical properties, during plasminogen adsorption from human plasma to columns with L-lysine, immobilized on a Sepharose matrix with subsequent elution with 6-aminohexanoic acid. The molecular weight of form F_1 is 2000-4000 daltons higher than that of F_2. F_1 also has a different degree of binding with antifibrinolytic agents and a different distribution of carbohydrates. The native F_1 form possesses two sites of glycosylation, one at Asp 288 and the other at Tyr 345; native human F_2 contains only one oligosaccharide at Trp [5].

Metabolic studies have shown that each variant of plasminogen (F_1 and F_2) is degraded and synthesized independently and is not transformed into the other.

Plasminogen synthesis takes place in the liver. During perfusion of the organ, the amount of the proenzyme is increased in the perfusate, ^{14}C-leucine is incorporated into the synthesized plasminogen, and plasminogen synthesis is inhibited by cyclohexamide [8].

Plasminogen is converted to plasmin via both an external and internal pathway. Factor XII, which participates in activation of the coagulating, kinin, and complement systems, intermediates the internal activation pathway.

External activation is effected by several physiologic activators. Plasminogen activation occurs in two phases and includes the splitting of at least two peptide bonds. During the first phase, the N-terminal part of the plasminogen molecule (preactivating peptide) is released and an altered intermediate molecule of the proenzyme is formed (Met 69 amino acid is localized on the amino-terminal part of this proenzyme). During the second stage, Arg 560—Val 561 is split under the activator effect, which leads to the formation of a plasmin molecule made up of two chains linked by disulfide residues. Peptides are not released during this phase, and disulfide bonds Cys 577—Cys 565 and Cys 547—Cys 665 stabilize the plasmin molecule [5]. Autolysis of bonds Lys 76—Lys 77, accompanied by a release of 76 amino acid residues, occurs as a consequence of plasmin formation.

A heavy plasmin chain, A, with a molecular weight of approximately 60,000 is formed from the NH-terminal plasminogen component. It contains five three-loop structures. The

B chain (or light chain) has a molecular weight of approximately 25,000 and is formed from the COOH-terminal component of the proenzyme (residues 562-791) [1-7].

Amino acid residues in the B and A chains of plasmin are linked in the following way: Met 69 or Lys 78 with Arg 561, and Val 562 with Asp 791, respectively. The carboxy-terminal part of the A plasmin chain is homologous to the three-loop structure of the prothrombin molecule. Both consists of 80 amino acid residues. Three disulfide bridges (between Cys 1 and Cys 80, Cys 22 and Cys 63, and Cys 51 and Cys 75) are responsible for their stability. Due to these bridges, a three-loop structure of the plasmin molecule is formed. The B chain of plasmin strongly resembles other serine proteases as well as the corresponding chains of prothrombin and factor Xa. His 606, Asp 646, and Ser 741 form a catalytic locus of plasmin. The A and B chains of plasmin are bonded with each other by two disulfide bonds (Cys 548-Cys 666, and Cys 558-Cys 566). The latter bridge is likely to be specific for plasmin.

Lysine-binding sites are found in the A chain of plasmin. The presence of these sites in molecules of plasminogen and plasmin accounts for their affinity to fibrin, antiplasmin, and certain amino acids with fibrin-binding properties (lysine, 6-aminohexanoic and tranexamic acids) [9]. One high-affinity binding site and four to five low-affinity binding sites for 6-aminohexanoic acids are contained in the heavy chain of plasmin. One more protein with the high-affinity binding sites has been isolated from human plasma. This protein is regarded as a physiological analog of 6-aminohexanoic acid. This protein is a glucoprotein with a molecular weight of approximately 60,000; it consists of one polypeptide chain and is extraordinarily rich in histidine. Its concentration in the plasma is 1.8 μM [9]. The nomenclature of plasmin is based on the position of the terminal amino acid of the heavy chain since the light chain of all plasmin molecules has the same residue (Val 561). Accordingly, the native plasmin is called Lys 77-plasmin. It is also possible to obtain Glu-plasmin. Conformational differences observed between Glu-plasminogen and Lys 77-plasminogen are retained in plasmin molecules as well [5].

Plasmin is a serine protease sensitive to inhibition by diisopropyl fluorophosphate and tosyllysine chloromethyl ketone. The main function of plasmin is to lyse fibrin and fibrinogen. The enzyme also affects other systems; for example,

it splits factor XII into XIIa fragments, activating several
zymogens of the complement (C1, C3, C5), and induces splitting of antihemophilic globulin, glucagon, and gamma globulin.
The enzyme hydrolyzes lysyl and arginyl bonds and has estero-
lytic activity with respect to basic amino acid esters, and
amidolytic activity toward the basic amino acid amides [5].

Besides the described way of plasminogen activation effected by specific activators, the possibility of nonenzymatic activation during plasminogen activation of DEAE cellulose has also been suggested. The conformational changes,
occurring during immobilization, induce plasminogen activation without hydrolysis of the activating peptide link [10].

Plasminogen activators which cause the conversion of
plasminogen to plasmin play the major role in the enzymatic
fibrinolysis system. Both internal and external activators
have been described. Those dependent on the Hageman factor
belong to internal activators. It has been reported that
there is a connection between complement activation and fibrinolysis. Activation of the C8 fragment of the complement
results in plasminogen conversion to plasmin. As yet, the
complement-dependent fibrinolysis has not been sufficiently
investigated [11]. As far as fibrinolytic activity emerging
during activation of the Hageman factor by a foreign surface,
this process also requires high-molecular-weight kininogen
and prekallikrein. It has not yet been established which
share of fibrinolytic activity in the organism is accounted
for by the Hageman factor [11].

The proofs that plasminogen activation by external factors is the main pathway of fibrinolysis invitation are numerous and convincing. The group of external physiologic plasminogen activators includes plasmic, vascular, and tissue
activators; urokinase; and leukocytic proteases discovered in
neutral granulocytes [1, 11]. All tissues in the organism
possess plasminogen activator activity. The highest content
of plasminogen activator has been found in vascular endothelium. Therefore, the fibrinolytic activity of tissues is linked
to their blood supply.   Epithelial cells have a lower activator activity. Plasminogen activator is localized mainly in
the lysosomal cellular fraction and, to a lesser degree, in
the microsomal fraction. The activator content in cells depends on their developmental phase and reaches maximum during
maturation and degeneration.

The tissue activator is firmly linked to the structural proteins of cells and has been comparatively recently obtained in a purified state from pig heart, ovaries, and uterus, and from rabbit kidneys [1]. The tissue activator is a serine protease with a molecular weight of 60,000 daltons and is characterized by high stability. In acidic reaction medium it sustains heating to 70°C for 30 min. Properties of the vascular activator derived from a cadaver vessel perfusate are similar to those of the tissue activator. This enzyme is a serine protease also (MW 67,000, isoelectric point 7.8-8.8) [1, 12]. In the blood plasma, plasminogen activator is labile and, in all probability, identical to the vascular activator. The plasminogen activator, isolated from the blood plasma of subjects with venous occlusion, has been shown to be similar to the vascular activator [12].

Of all the activators, urokinase, a plasminogen activator derived from human urine, has been studied most intensively. Two types of urokinase have been isolated: one with a high molecular weight (54,700) and another with a low molecular weight (31,500). The high-molecular-weight form consists of two polypeptide chains (MW 33,000 and 19,000) bonded with a disulfide link [13]. Using a radioimmunology technique, it was shown that the low-molecular-weight form of urokinase is part of the molecule of the high-molecular-weight urokinase, from which it is formed during proteolysis both in vitro and in vivo [14]. The active center is localized in the heavy chain of the high-molecular-weight urokinase. Although a specific plasminogen activator, urokinase exerts a proteolytic effect on other substrates when used in high concentrations. In the 1970s, some researchers isolated a number of plasminogen activator forms from human blood and tissues using modern biochemical preparative methods. Two activators similar to the tissue activator and differing in stability (one was acid-labile and the other was acid-stable) were isolated from normal human plasma. They are homogeneously inhibited by synthetic inhibitors, but the acid-stable activator is precipitated with plasma euglobulins while the acid-labile activator remains in the supernatant. The molecular weight of the former is 60,000 daltons, and that of the latter is 10,000 daltons [15].

A plasminogen activator with a molecular weight of 15,000-18,000 daltons and having a high esterase, fibrinolytic, and kinonogenase activity has been described [16].

Two types of plasminogen activators have been isolated from the euglobulin fraction of plasma using dextran sulfate to activate the Hageman factor and flufenamic acid to neutralize the Cl-inactivator. Type 1 is represented by the internal activators (dependent and independent of the Hageman factor) which were inhibited by the Cl-inactivator. External factors, not liable to the action of Cl-inactivators (vascular and tissue factors, and urokinase), belong to Type 2 [17].

Plasminogen activators of three types with different molecular weights and reactions to inhibitors have been isolated from uterus extracts [18]. An activator (MW 70,000) similar in kinetic characteristics to the vascular activator has been isolated from human myocardial tissue [19]. The notion of plasminogen activators has become even more complicated after these enzymes were isolated from the cell culture of a number of tissues.

Endothelial cells of bovine aorta and kidneys, and fibroblasts of human embryo lungs, form plasminogen activators whose release into the culture medium occurs during progressive cell degeneration [20]. Plasminogen activators differing in molecular weight, optimum pH, sensitivity to fibrinogen degradation product, and serum inhibitors are formed in the cell culture.

Cells of human melanocarcinoma abundantly produce plasminogen activators. The characteristics of these activators turned out to be similar to urokinase [21]. Further studies showed, however, that two forms of this activator are formed. The first consists of one polypeptide chain and is identical with the vascular or blood activator. During its proteolytic degradation, a two-chain enzyme is formed [20].

Evaluation of the data on the isolation of multiple forms of plasminogen activators from blood, cell, and tissue culture brings one to the conclusion that the two main types of activators are present in mammals. The first is similar to urokinase and the second includes tissue, vascular, and blood activators. They differ in chemical properties, specificity to fibrin, and certain synthetic substrates [22, 13]. The most significant differences between the activators lie in immunochemical characteristics. Antibodies to tissue and vascular activators do not inhibit enzymatic activity of urokinase, while antiserum to urokinase does not affect activity of the tissue activators [13].

It is of interest that the plasminogen activator extracted from tissue was reported initially to be in a low-active form. Under the effect of a special cofactor isolated from the blood, enzyme activity is considerably increased due to disclosure of its active center. At the same time, the activators formed in the cell culture have a high activator and esterase activity from the start and are insensitive to the action of plasmic cofactor [22]. In cell culture, however, a proactivator-prourokinase was found in addition to the active form when incubated with plasmin or trypsin. Only a very slow activation of this precursor takes place in the culture medium [23].

Urokinase with a low activator activity, consequently called prourokinase, has been derived from fresh urine [24]. It is also of interest that the presence of identical subunits with a molecular weight of 31,000 and an active center in molecules of urokinase and tissue activator has been reported [24]. This corroborates the previous assumption that a common precursor of urokinase and tissue activators exists which is synthesized as an inactive form and then transformed into various types of activators [25].

However, because of immunological differences between the two groups of plasminogen activators, this assumption has not been proved yet. It has also been hypothesized that only one type of plasminogen activator exists in the organism. It is this enzyme that plays an important role in the normal fibrinolytic process [13].

The most important biological properties of plasminogen activators, which provide for the physiological role played by the enzymatic fibrinolysis system, are manifested if fibrin is present in the blood. Neither vascular nor tissue activators induce plasminogen activation in the absence of fibrin. Urokinase does not have such a high specificity and can cause lysis of fibrinogen, although in the presence of fibrin it activates plasminogen more completely and rapidly [26]. Tissue and vascular activators have more than a fivefold higher affinity for fibrin than for fibrinogen [26].

In considering the problem of plasminogen activators, one cannot pass over streptokinase, an enzyme of microbial origin, as well as some other enzymes synthesized in fungi. The streptokinase activator [stoichiometric complex (1:1) of streptokinase with Glu- or Lys-plasminogen, or light B chain

of plasmin] is used widely in clinical practice [27]. Transformation of the plasminogen-streptokinase complex into a
potent activator results from intramolecular splitting [28].

An activator complex containing 86% plasminogen and
less than 2% streptokinase has been isolated by affinity
chromatography on a column of lysine-agarose with subsequent
elution with 6-aminohexanoic acid. The compound obtained has
a high activator activity and high affinity to fibrin. It
is fully devoid of antigenic properties [29]. The mechanism
of plasminogen activator by streptokinase has been studied
mainly in purified systems. In the plasma, plasmin formation
in the presence of streptokinase goes much slower and is accelerated when a fibrin clot appears [30].

A protein which substantially accelerates plasminogen
activation by streptokinase has been isolated from plasma
and purified. It is called a streptokinase potentiator of
plasminogen [31]. The potentiator increases plasminogen
activation by small amounts of streptokinase, but its presence has no effect on activation in cases of high streptokinase concentrations. The potentiator consists of beta and
gamma chains of fibrinogen, and its molecular weight varies
from 240,000 to 480,000 daltons [32].

Data have been obtained which point to the competition
of the potentiator with tranexamic acid for lysine-binding
sites of plasminogen.

The streptokinase—plasminogen—potentiator complex has a
higher activator activity than streptokinase—plasminogen or
streptokinase—plasmin [32].

Besides streptokinase synthesized by microorganisms, enzymes possessing fibrinolytic activity can be obtained from
cultures of Aspergillus fungi, Actinomyces, mycobacteria, and
putrefactive bacteria. Enzymes are formed by nonpathogenic
saprophytic fungi of the Aspergillus genus. Most of the
strains of Aspergillus oryzae are capable of synthesizing enzymes which effect fibrinolysis. Some 47 of the 50 strains of
this fungus that were studied formed fibrinolytically active
proteases [33]. Brinase and brinolase, commercial enzymic
preparations isolated from A. oryzae, are used in clinical
practice for thrombolytic therapy [2].

The fibrinolytically active enzymes terrilytin [34] and ocrase [35] have been isolated from cultures of Aspergillus terricola and A. ochraceus, respectively.

All of the three enzymes derived from Aspergillus split proteins, including fibrin and fibrinogen. They are inhibited by diisopropyl phosphate and have a molecular weight from 22,000 to 35,000 daltons. Well-known synthetic inhibitors of fibrinolysis and 6-aminohexanoic and tranexamic acids have no effect on fungi-synthesized enzymes. Brinolase activity is inhibited by α_1-antitrypsin and antiplasmin present in the blood serum; ocrase is inhibited by the agent isolated from potato [35].

Terrilytin is characterized by a higher specificity of fibrin. The initial rate of fibrin hydrolysis is ten times higher than that of fibrinogen degradation [34].

Cultures of various strains of putrefactive bacteria (Bacillus mesentericus, Bacillus subtilis, and Bacillus mycoides) grown in synthetic media form proteolytic enzymes. Biosynthesis of enzymes with two types of activity — fibrinolytic and proteolytic — takes place. Depending on the composition of the medium, either the former or the latter predominates [36]. The media have been described in which proteases with only fibrinolytic activity are synthesized.

A vast group of brown Actinomyces is capable of producing highly active fibrinolytic enzymes while performing their vital functions. The degree of fibrinolytic activity depends on the nutrient medium. The enzymes synthesized by strains A. spheroides and Actinobifida dichtomica have the highest fibrinolytic activity [37]. The enzymes listed of fungus origin have the capability to hydrolyze fibrin and fibrinogen and do not activate the plasmin system.

A new plasminogen activator called tricholysin has been isolated from the cultured fluid of a nonpathogenic saprophytic fungus Trichothecium roseum. It is the first microorganism since streptokinase which produces an enzyme capable of activating human and animal plasminogen in vitro and in vivo. Besides an activator effect, tricholysin exerts a direct fibrinolytic effect on fibrin [1].

Most of the strains (165 of the 207 examined) of the Trichothecium fungus used to synthesize the antibiotic tri-

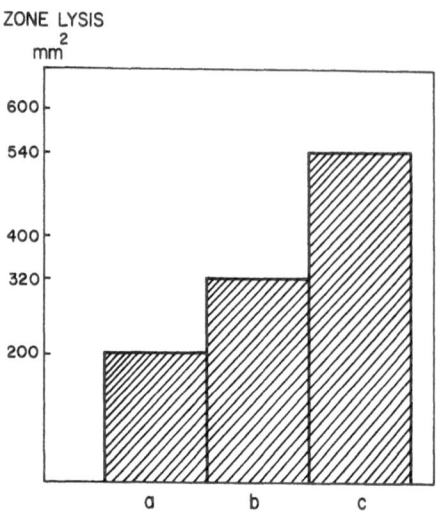

Fig. 3.2.   Plasminogen activation on fibrin plates following
            addition of tricholysin. Fibrinolytic activity
            following addition of the enzyme (0.03 ml; concen-
            tration, 1 mg/ml).   a) Heated fibrin plates; b) un-
            heated fibrin plates; c) unheated fibrin plates
            with the addition of plasminogen (4 mg/ml). See
            [40].

chothecin carry out the biosynthesis of enzymes with a high
fibrinolytic activity [38]. The maximal fibrinolytic activity in culture was observed after 96 h of growth.

After precipitation of the protein by acetone and subsequent purification by gel filtration, with fractionation on
a CM-Sephadex C-50, the enzymatic preparation with an activity
of up to 5000 arbitrary units/mg protein was obtained. Besides fibrinolytic and activator activity, tricholysin can effect hydrolysis of synthetic esters (37.0 µg TAME mg/protein)
and moderate lysis of casein (27.9 µg tyrosine/mg). The optimal conditions for fibrinolytic activity of fibrinolysin are
achieved at pH 7.0 and 37°C [39]. Using isoelectric focusing,
tricholysin was separated into four fractions with greatly
different fibrinolytic and activator properties. One of the
fractions had mostly an activator action and minimal fibrinolytic activity.

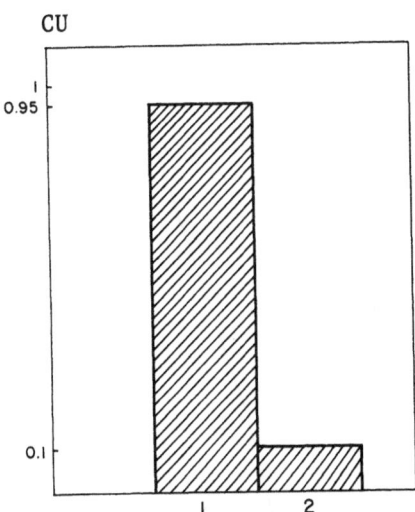

Fig. 3.3. Alteration in plasminogen concentration measured
by caseinolytic activity after intravenous injection of tricholysin to white rats. 1) Prior to
infusion; 2) after infusion. See [40].

The ability of tricholysin to activate plasminogen has
been observed in vitro. Zones of fibrin lysis on nonheated
plates or those enriched by plasminogen exceeded the lysis
zones on heated plates by 2.5- to 2.7-fold. The activator
properties of the enzyme are labile: They disappear within
2-4 days of storage, while the fibrinolytic activity of tricholysin is nearly fully retained [39].

A 10-min incubation of tricholysin with human and animal
plasma led to a 7-fold increase in fibrinolytic activity and
an 11.3-fold increase in the amount of plasminogen activator.
The time of lysis of plasmic euglobulin clots was decreased
by 2.3-fold. The concentration of plasminogen was reduced
(Fig. 3.2). The high level of plasma fibrinolytic activity
was maintained for 1.5 h. The rate of plasminogen activation
in human plasma was 2.6-fold higher than in rat plasma (Fig.
3.4).

These specific differences may be explained both by a
higher specificity of tricholysin to human plasminogen and by
the presence of an agent analogous to the streptokinase potentiator in human plasma [31].

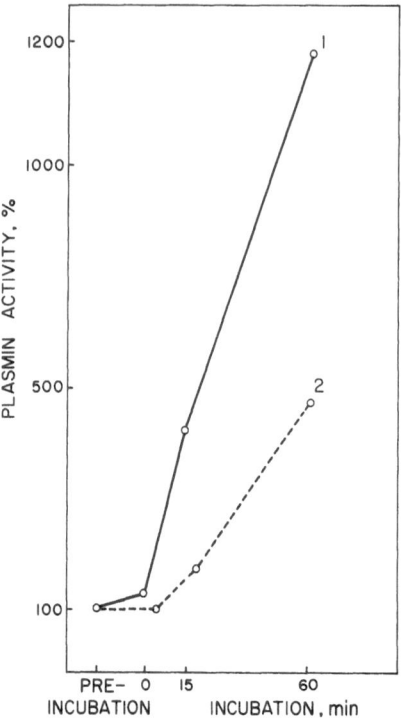

Fig. 3.4.   Activity of plasmin (in %) synthesized during the
            incubation of human (1) and rat (2) blood plasma
            with tricholysin. See [40].

     Intravenous administration of 130 µg/ml tricholysin to
animals resulted in an increase of plasmic fibrinolytic activity and a sixfold acceleration of the euglobulin clot lysis.
Ten minutes after injection of tricholysin, the amount of
plasminogen activator exceeded the initial level by 20-fold;
the raised level was retained for 2 h. The increase in amount
of plasminogen activator and the decrease of plasminogen concentration in animal blood demonstrate that tricholysin administered into the bloodstream effects the conversion of endogenous plasminogen to plasmin. This was manifested by an increase
of plasmin activity following injection of tricholysin in animals. Also, repeated injection of the preparation at the moment of highest plasmin activity did not cause noticeable
changes in the fibrinolysis indices since the first tricholysin infusion had led to conversion of plasminogen to plasmin
(Fig. 3.5) [40].

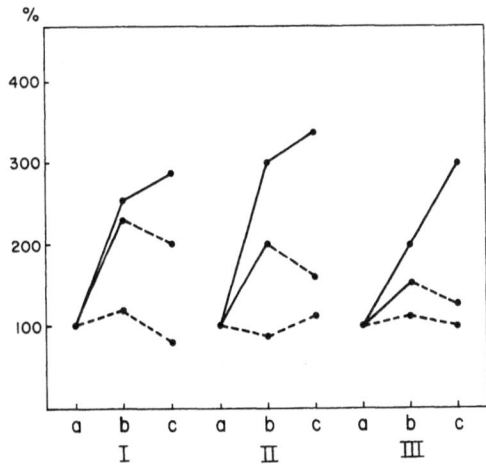

Fig. 3.5. Alteration of fibrinolytic activity and the activating plasma effect (in %) following a single and
repeated injection of tricholysin. I) Fibrinolytic activity of plasma; II) activating effect
of plasma; III) fibrinolytic activity of the euglobulin fraction of plasma. a) Before tricholysin injection; b) after a single injection of tricholysin or physiological solution; c) after repeated injection of tricholysin or physiological
solution. Solid line: after tricholysin administration; broken line: following injection of
physiological solution. See [39].

Administration of tricholysin to animals resulted in a
substantial increase in the antiplasmin level, exceeding the
initial level by 70-fold [40]. Tricholysin failed to activate the blood coagulating system: Thrombin time remained
practically unchanged, recalcification time lengthened, the
coagulation index decreased, and fibrinogen concentration was
reduced by 100 mg% [39].

Another enzyme which can both lyse fibrin and activate
human and animal plasminogen has been derived from Arthrobotrys longa, a species of fungi similar to Trichothecium. A
preparation of enzymes with specific fibrinolytic and plasminogen-activating action has been obtained by submerged cultivation of the fungus for 144 h on a synthetic medium. During incubation with the plasma, the enzyme preparation, which
was called longolytin, failed to cause a marked decrease in

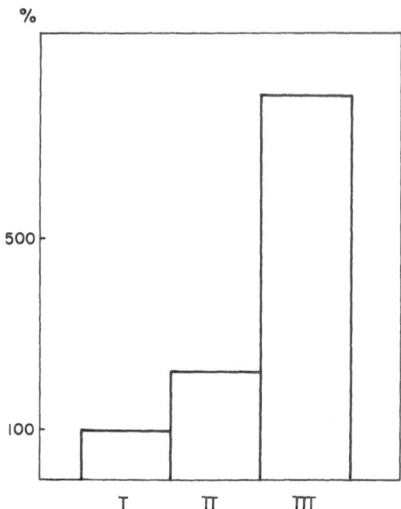

Fig. 3.6. Fibrinolytic activity (in %) of the preparation
           derived from the cultured fluid of <u>Arthrobotrys</u>
           <u>longa</u> when applied to fibrin plates. I, II) Heat-
           ed and unheated standard plates, respectively;
           III) standard plates with addition of plasminogen
           (4 mg/ml).

fibrinogen concentration despite a high fibrinolytic and activator activity induced in the plasma (Fig. 3.6). The in vitro
fibrinolytic activity of plasma increased by two- to threefold; the level of plasminogen activator increased by four-
to fivefold. Intravenous administration of longolytin has a
similar effect on the fibrinolytic system of animals [41, 42].

FIBRINOLYSIS INHIBITORS

Most of the inhibitors of proteases have the ability to
neutralize plasmin activity. At least six compounds have an
antiplasmin effect: α_1-antitrypsin, α_2-macroglobulin, antithrombin III, C1-inactivator, inter-α-trypsin inhibitor, and
α_2-antiplasmin.

α_2-Antiplasmin is the most important physiologic inhibitor of plasmin. Its concentration in normal plasma amounts
to 70 mg/liter or 1 μM [43]. Antiplasmin consists of one polypeptide chain with a MW of 65,000-70,000. In purified systems,

antiplasmin forms with plasmin a 1:1 stoichiometric complex
devoid of proteolytic and esterase activity. The complex is
formed as a result of a strong interaction between the catalytic center of the light plasmin chain and the inhibitor;
α_2-antiplasmin can also inhibit plasminogen activation. Lysine-binding sites of the proenzyme molecule bind to antiplasmin and, thus, plasminogen binding with fibrin is impaired
[43].

Its binding with plasminogen via lysine-binding sites
occurs simultaneously with fibrin formation, and adsorption
of plasminogen activators, present in the blood or released
from cells, takes place. The synthesized plasmin remains
bound to fibrin via lysine-binding sites and, therefore, is
very slowly inactivated by α_2-antiplasmin. At the same time,
plasmin, which is being released from the degraded fibrin, is
very rapidly and irreversibly inhibited by antiplasmin. In
the course of plasminogen activation, fibrin plays the role
of a regulating and catalytic surface for the reaction between plasminogen and activator, fibrin and plasmin [44].

After activation of plasmic plasminogen, the action of
forming plasmin is first reduced mainly to the binding with
antiplasmin. Only after complete plasminogen activation,
which leads to antiplasmin saturation, is the excess plasmin
neutralized by α_2-macroglobulin [43].

Antithrombin III and antithrombin III—heparin complex
have a limited physiological role as plasmin inhibitors. Their
significance increases only at decreased concentrations of
antiplasmin and α_2-macroglobulin [43].

As yet, the biological significance of other protease
inhibitors in depressing plasmin activity is little known.

Besides antiplasmins, antiactivators and inhibitors of
plasminogen activation exist in the organism. In the plasma,
urokinase and streptokinase activity is inhibited by antiuro-
kinase and antistreptokinase, respectively. α_1-Antitrypsin,
antiplasmin, and antithrombin III very slowly depress urokinase activity in purified systems. α_2-Macroglobulin is a competitive inhibitor of urokinase [43]. An inhibitor of fibrinolysis induced by urokinase and tissue plasminogen activator
and having no effect on plasmin has been isolated from aortic
intima and media, and from coronary arteries. It is assumed
to play an important regulatory role in the metabolism of fibrinogen and fibrin in the arterial wall [43].

The presence of antiactivators inhibiting the tissue plasminogen activator is a problem open to discussion. Some authors have isolated three proteins with such an action [46], while others deny the existence of a specific antiactivator [47]. Inactivation of plasminogen activator (isolated from the cells of human melanoma) in human plasma occurs (with a half-life of 90-165 min) due to formation of the complex with antiplasmin. To a lesser degree, plasminogen activator is neutralized with α_2-macroglobulin [47]. A glycoprotein rich in histidine, which has been found in blood plasma, is of great interest. It has the capability to interact with high-affinity lysine-binding sites in the heavy chain of plasmin. Therefore, the rate of the plasmin-α_2-antiplasmin reaction is noticeably decreased and the level of plasminogen binding with fibrin is lowered. About 50% of the plasminogen circulating in blood may enter into a reversible complex formation with this protein, which results in the reduction of effective plasminogen concentration in the blood and an antifibrinolytic effect. It has been suggested that this histidine-rich protein is an important regulator of enzymatic fibrinolysis [9]. In the years 1982-1985 much new data on antiactivators were published [60-63].

EFFECT OF PLASMIN ON FIBRINOGEN AND FIBRIN

Fibrin and fibrinogen are the main biological substrates for plasmin. Splitting of 160-170 peptide links in the molecule of fibrinogen and fibrin results in the formation of the four major degradation products designated as fragments X, Y, D, and E. The fibrinogen molecule is degraded asymmetrically. First, a high-molecular-weight protein and small peptides A, B, and C are formed. Fragment X has a molecular weight of 240,000-265,000 daltons and contains a degraded Aα chain, with a molecular weight of 25,000 daltons, and intact Bβ and γ chains [5]. In other forms of X fragments, the Bβ and γ chains are partially degraded. Fragment X retains the ability to slowly coagulate during the action of thrombin. At the second stage of hydrolytic splitting, fragment X forms two fragments Y with a molecular weight of 155,000 and D with a molecular weight of 85,000-100,000. In fragment Y, all of the peptide fibrinogen chains are strongly degraded. During the third stage, the splitting of fragment Y takes place, leading to the formation of the second fragment D and fragment E [1, 48].

After complete splitting of the fibrinogen molecule, fragments D and E respectively account for about 50-60% and 15-20% of the total protein content in the mixture of degradation products. Fragment E is the central inner component of fibrinogen in which six polypeptide chains are bound by disulfide bonds into firmly linked structures. Fragment D comes from the midchain and carboxy-terminal parts of fibrinogen [49]. In fibrinogen degradation products, these two form a noncovalent complex, D—E. While affecting fibrinogen, plasmin splits off the COOH-terminal components of Aα chains. Bβ chains are split more slowly, and γ chains are the most stable with respect to the effect of plasmin. These peptide links in the fibrinogen molecule, which are localized between the carboxylic groups of lysine and arginine and an amino group of the adjoining amino acid, are mainly subjected to hydrolysis. However, the late degradation products (fragments D and E) contain more than fifty peptide links that are potentially sensitive to the action of plasmin [48]. Calcium ions exert a protective effect on carboxyl-terminal γ chains in fibrinogen and protect fragment D and D-dimer from the action of plasmin. In the absence of calcium ions, fragment E and the D—E complex protect fragment D from the effect of low plasmin concentrations. Fragment D obtained in the presence of calcium ions has a strong anticoagulating effect [49].

Study of the sequence of fibrinogen breakdown by plasmin has demonstrated that the enzyme first attacks the peptide links Aα Lys 208-Met 209, Aα Lys 221—Ser 222, and Aα Lys 232—Ala 233. These bonds may be split simultaneously; therefore, the forming fragment X is heterogeneous [48]. According to more recent data, Lys 508, Lys 583, and Lys 539 are the first to break down under the effect of plasmin; these are followed by Lys 413, Lys 418, Lys 421, Lys 427, or Arg 424 [5].

The molecular weight of the Aα chain sections in the initial fibrinogen degradation is 25,000 to 79,000 daltons. These variations indicate that not only undamaged peptides with a molecular weight of 40,000-50,000 daltons are split off the Aα chain, but also other peptide links are simultaneously broken in intact fibrinogen. Certain data have given rise to the assumption that such a progressive "nibbling" of the carboxy-terminal component of the Aα chain has physiological significance during fibrinogen catabolism in the bloodstream. It is possible that such a slow fibrinogen breakdown in the plasma occurs under the action of plasmin with a limited pro-

teolytic activity due to the complex formation with α_2-macro-
globulin. The heterogeneity of fibrinogen circulating in the
blood is cited as evidence of this phenomenon [48].

There are 2 moles of fibrinopeptide A per 1 mole of early
fragment X. Fibrinopeptide B likely disppears more rapidly.
Splitting of three peptide links in Aα, Bβ, and γ chains is
required to form fragments Y and D out of fragment X. This
probably happens in the nonspiral zone of the thrice-twisted
ringlet which passes approximately through the 110 residue
of the molecule [48]. During the splitting of Bβ chains of
fibrinogen, first the sites of Bβ Arg 42—Ala 43, and then
link Bβ Lys 53—Lys 54, are attacked. The molecule of fragment
Y is broken down in the same way, i.e., by successive splitting of three polypeptide chains. The carboxyl-terminal of
fragment E is separated from fragment D by the peptide consisting of 26 amino acid residues. Fibrinopeptide A is localized
in fragment E and released during degradation by plasmin.
Eleven amino acid residues separate carboxy-terminal Bβ Lys
122 of fragment E from the amino-terminal Bβ Asn 134 of fragment D. Both sections of the Bβ chain from fragments D and
E are likely the most sensitive to the continuing effect of
plasmin.

The γ Lys 62—Ala 63 link is the first to split between
fragments D and E. As a result of the breakdown of two short
peptides, the stable COOH-terminal of fragment E is formed.
During formation of fragment D, a pentapeptide is separated
from the carboxy-terminal region under plasmin action. Consequently, the γ-γ dimer cannot be formed from the fibrinogen
D fragment [48].

The mechanism whereby plasmin affects fibrin depends on
the type of fibrin (noncross-linked or cross-linked). Rapid
cross-linking of spontaneously polymerized molecules of fibrin polymer is carried out under the effect of calcium and
activated factor XIII. It consists of the formation of intramolecular γ-glutamyl Σ-lysine bonds and formation of γ chain
dimers (γ-γ link). During slow binding of the α chain α-poly-
mers are formed, of which the molecular weight amounts to
400,000 daltons and higher. In completely cross-linked fibrin, there are 6 moles of cross-links per 1 mole of a fibrin
subunit (2 moles in cross-linked γ chains and 4 moles in α
chains). Cross-linking of fibrin molecules occurs already
prior to the formation of visible fibrin polymers and fibrin
fibers, and continues after the formation of fibrin deposits.

The number of cross-links may vary. A form of fibrin exists
that is called a partially cross-linked fibrin [50].

Clots of noncross-linked fibrin are distinguished by the
absence of fibrinopeptides A and B. Plasmin affects their
action like fibrinogen forming fragments X, Y, D, and E. Stability of the noncross-linked fibrin clot is secured by the
dimeric structure of both polymerization zones of fragments
E and D. The fibrin degradation occurs due to the breakdown
of certain peptide links in the polymerization zone or close
to it, localized in fragment D. The polymerization zone in
fragment E is structurally stable [50]. In the cross-linked
fibrin, β and γ chains first split, and then α chains. The
cross-linked polymers of α chains restrict the process of fibrin clot breakdown [1, 50]. The main distinguishing feature
of highly cross-linked fibrin degradation products is that
they are formed of various covalently bound peptide chains.
Although fragments E of fibrinogen and cross-linked and noncross-linked fibrin are similar in physical and chemical properties (sections of cross-binding are missing its chains),
certain differences have been found in immunochemical characteristics and the time of preservation in plasma [5]. Fragment D of the cross-linked fibrin is a dimer. It is stabilized by two cross-links in the carboxyl section of the γ chain
and has a molecular weight of 63,000-81,000 daltons. Cross-
linking in the γ chain prevents the further effect of plasmin
on γ chains. D-dimer forms a complex with fragment E by firm
noncovalent bonds. The complex is synthesized both in a purified system and under physiological conditions [50].

Fragment Y, formed during plasmin action on cross-linked
fibrin, also exists in the form of a dimer since it contains
the dimer of a γ-γ chain [5].

Thrombolysis, induced by an artificial increase in blood
fibrinolytic activity, occurs due to plasminogen infiltration
into the clot with subsequent plasminogen activation by the
activator mainly bound with fibrin. Early notions assuming a
possibility of "external" lysis of a thrombus by plasmin, released during dissociation of a plasmin-antiplasmin complex,
have been contradicted by recent data on the stability of the
plasmin—antiplasmin complex. Fibrin clots adsorb plasminogen
at the ratio of 1 mole proenzyme per 1 mole fibrin. High affinity of plasminogen activators to fibrin provides for rapid
formation of plasmin protected from the action of antiplasmin
by fixation with lysine-binding sites [1, 3, 50]. Whether

Glu- or Lys-plasminogen is more effective with respect to fibrinolysis remains a question. In vitro, Lys-plasminogen is
converted much quicker to plasmin, but there is likely no difference in their action in the organism [50]. It has been
reported that cross-linked fibrin with dimers of γ-γ chains
can be dissolved as quickly as noncross-linked fibrin, but
the fibrin containing polymers of α chains is resistant to
the lytic action of plasmin. An increase in complexity of α
chains of cross-linked fibrin enhances resistance of the clot.
Glu- and Lys-plasminogens and activators, however, influence
this stable bond [51].

During immunologic and physicochemical studies of the
lysis of cross-linked fibrin, a quicker and more complete
lysis of the clot was observed after the addition of streptokinase, followed by plasminogen. Stable complexes of D-dimer—
E and soluble aggregates with high molecular weights were
practically not found in the lysis products. These compounds
were present in substantial amounts, however, if the sequence
of addition was reversed, i.e., plasminogen first, and then
streptokinase. Thus, cross-linked high-molecular-weight complexes are released from fibrin first, quickly break down and
form complex D-dimer—E, which was already transforming into
pure fragments D dimer and E. It is assumed that two forms
of complex D-dimer—E exist, varying in stability. This observation is significant for the clinical application of
streptokinase and plasminogen [52].

The effect of fibrinolytic enzymes formed by nonpathogenic microorganisms has been little studied as yet.

Brinase quickly splits human fibrin in vitro to products
with high molecular weights of 310,000-230,000. Degradation
of α chains and impairment of monomeric fibrin polymerization
occurs first. Gamma and Bβ chains of fibrin are less sensitive to the effect of brinase. Lysis of Bβ chains likely
limits the rate of transformation of high-molecular-weight
products into the terminal fragments D and E. Cross-linked
α chains of fibrin resistant to plasmin action are broken
down by brinase [53, 54]. Infusions of brinase bring about
a decrease in fibrinogen concentration and emergence of high-
molecular-weight degradation products. Degradation of Aα
chains of fibrin into two main fragments retaining the cross-
links is noted. Gamma chains preserve a dimeric structure
[54]. Infusion of small doses of brinase into the bloodstream
affects fibrin only. An increased dose (to 4 mg/kg) causes

proteolysis of the blood coagulating factors [55]. The in vitro action of brinase accelerating fibrin formation is linked to the effect on protein polymerization sites. A positive ethanol test following infusion of the enzyme points to the formation of aggregates consisting of the degradation products of fibrin and fibrinogen [56]. Purified plasminogen is partially degraded in the presence of brinase, but this process is not accompanied by the generation of proteolytic activity [55].

Ocrase exerts a hydrolytic effect on both fibrin and fibrinogen, splitting a great number of peptide links in the protein, which results in the formation of degradation products with a lower molecular weight than those resulting from the action of plasmin [57].

Tricholysin and longolytin have a plasminogen-activating effect. The forming plasmin causes lysis of fibrin. These enzymes also exert a direct fibrinolytic effect on fibrin. The effect on fibrinogen is less marked [39-41].

Enzymatic fibrinolysis is used widely in clinical practice to lyse thrombi. Along with injection of plasminogen and plasmin in the bloodstream of patients, new methods for utilization of thrombolytic agents are being developed.

A new method of applying acylated plasmin and streptokinase complex for thrombolysis has been suggested. Acyl-plasmin is catalytically inert and thus unable to degrade plasma proteins and interact with α_2-antiplasmin, but can still bind to the clot's fibrin via lysine-binding sites. During subsequent deacylation occurring in the bloodstream, active plasmin bonded with fibrin is released [58].

Streptokinase immobilization on dextran has proved to be a promising method of using the enzymatic fibrinolytic system in clinical practice. The major advantages of immobilized streptase are: less acute and chronic toxicity, large single doses, markedly prolonged action, and decreased antigenicity [59].

REFERENCES

1. G. V. Andreenko, Fibrinolysis (Biochemistry, Physiology, Pathology), Moscow University Publishers, Moscow (1979).

2. E. I. Chazov and K. M. Lakin, Anticoagulants and Fibrin-
    olytic Agents, Meditsina, Moscow (1977).
3. B. V. Petrovski, E. I. Chazov, and S. V. Andreeva (eds.),
    Topical Problems of Hemostasiology: Molecular Biologi-
    cal and Physiological Aspects, Nauka, Moscow (1981).
4. P. J. Gaffney and S. Balkov-Ulutina (eds.), Fibrinolysis:
    Modern Fundamental Clinical Aspects, Meditsina, Moscow
    (1982).
5. F. J. Castellino, "Recent advances in the chemistry of
    the fibrinolytic system," Chem. Rev., 81, 431 (1980).
6. B. Wiman, "Biochemistry of plasminogen conversion to
    plasmin," in: Fibrinolysis: Modern Fundamental Clini-
    cal Aspects, Meditsina, Moscow (1982), p. 56.
7. S. Thorsen, J. Clemmensen, L. Sottrup-Jensen, and S. Mag-
    nusson, "Adsorption to fibrin of native fragments of
    known primary structure from human plasminogen," Biochim.
    Biophys. Acta, 668, 377 (1981).
8. H. Saito, S. Hamilton, A. Tavill, L. Louis, and O. D.
    Ratnoff, "Production and release of plasminogen by iso-
    lated perfusion rat liver," Proc. Natl. Acad. Sci. USA,
    Biol. Sci., 77, 6837 (1980).
9. H. R. Lijnen, M. Hoylaerts, and D. Collen, "Isolation
    and characterization of a human plasma protein with af-
    finity for the lysine binding sites in plasminogen," J.
    Biol. Chem., 255, 10214 (1980).
10. S. A. Kudinov and E. V. Eretskaya, "Plasminogen activa-
    tion during its immobilization," Ukr. Biochem. J., 51,
    340 (1979).
11. D. Ogston, "Natural plasminogen activators," in: Fibrin-
    olysis: Modern Fundamental Clinical Aspects, Meditsina,
    Moscow (1982), p. 13.
12. B. R. Binder, G. Spragg, and F. Austen, "Purification
    and characterization of human vascular plasminogen acti-
    vator derived from blood vessel perfusates," J. Biol.
    Chem., 254, 1998 (1979).
13. B. Aasted, "Immunochemical characterization of human
    plasminogen activators," Biochim. Biophys. Acta, 668,
    339 (1981).
14. G. H. Barlow, Ch. W. Francis, and V. J. Marder, "On the
    conversion of high-molecular-weight urokinase to the low-
    molecular-weight form by plasmin," Thromb. Res., 23, 541
    (1981).
15. P. Kok, "Separation of plasminogen activators from human
    plasma and a comparison with activators from human uter-
    ine tissue and urine," Thromb. Haemostas., 41, 734 (1979).

16. D. A. Diaz Batista, H. G. Hernandez Solana, and J. P.
    Corral Almonte, "On plasminogen activator from human
    plasma," Thromb. Haemostas., 42, 1607 (1980).
17. C. Kluft, G. Wijngaards, and A. F. H. Jie, "The factor
    XII—independent plasminogen proactivator system of plas-
    ma includes urokinase-related activity," Thromb. Haemo-
    stas., 46, 343 (1981).
18. P. Kok, "Separation of plasminogen activators from human
    uterine tissue and comparison with activators from human
    urine and porcine tissue," Thromb. Haemostas., 41, 718
    (1979).
19. B. R. Binder, J. Reissert, and R. Beckmann, "Isolation
    and characterization of a plasminogen activator (PA)
    from human myocardial tissue," Thromb. Haemostas., 46,
    11 (1981).
20. M. B. Bernik, D. C. Rijken, and G. Wijngaards, "Produc-
    tion of two immunologically distinct plasminogen acti-
    vators by human tissue in culture," Thromb. Haemostas.,
    46, 414 (1981).
21. J. L. Markus, J. Madeja, J. L. Evers, G. H. Hobika, S. E.
    Caiolo, and J. L. Ambrus, "Plasminogen activators in
    human malignant melanoma," Thromb. Haemostas., 46, 85
    (1981).
22. E. R. Cole and R. M. Snopko, "Physical, chemical, and
    immunological properties of porcine and human plasmino-
    gen activators of various tissues and cells," Thromb.
    Haemostas., 46, 211 (1981).
23. G. Wijngaards and M. B. Bernik, "Activation and inactiva-
    tion of a plasminogen proactivator (preurokinase) in
    human tissue culture media," Thromb. Haemostas., 46, 10
    (1981).
24. S. S. Husain, V. Gurewich, and B. Lipinski, "Purification
    of a new high-molecular-weight single chain form of uro-
    kinase from urine," Thromb. Haemostas., 46, 11 (1981).
25. D. C. Rijken, G. Wijngaards, and J. Welbergen, "Biochem-
    ical and immunological characterization of plasminogen
    activator from human tissue," Prog. Chem. Fibrinol.
    Thromb., 4, 349 (1979).
26. B. R. Binder and J. Spragg, "The effect of fibrin on the
    activity of purified human vascular plasminogen acti-
    vator," in: Protides in Biological Fluids, Proceedings
    28th Colloquium, Brussels, Oxford (1980), p. 391.
27. K. C. Robbins, R. C. Wohl, and L. Summaria, "Activation
    of human plasminogen in both purified systems and plasma
    by streptokinase and urokinase activator species," in:
    Progress in Chemical Fibrinolysis and Thrombosis, Vol. 4,
    J. F. Davidson (ed.), London (1979), p. 330.

28. L. Summaria, I. Boreisha, C. Wohl, and K. Robbins, "Iso-
    lation of plasminogen with an active site from the dis-
    sociated plasminogen—streptokinase complex," Circula-
    tion, 62 (Suppl. III), 334 (1980).
29. S. A. Ceredholm-Williams and A. J. Sharp, "The formation
    and isolation of streptokinase plasminogen activator
    freed of streptokinase," in: Progress in Chemical Fib-
    rinolysis and Thrombosis, Vol. 4, London (1979), p. 339.
30. K. W. Jacson, N. Esmon, I. Ferlan, and I. Tang, "Struc-
    ture and plasminogen activation activity of streptoki-
    nase and staphylokinase," in: Abstracts, V International
    Conference on Synthetic Fibrinolytic Agents, No. 7,
    Malmö (1980).
31. Y. Takada and A. Takada, "Studies on SK-potentiator of
    plasminogen in human plasma," in: Abstracts, VII Inter-
    national Congress on Thrombosis and Haemostasis, London,
    1979; Thromb. Diath. Haemorrh., 42 (1979).
32. A. Takada, K. Mochizuki, and Y. Takada, "Further charac-
    terization of SK-potentiator of plasminogen," Thromb.
    Haemostas., 46, 10 (1981).
33. B. A. Kudrjashov, G. V. Andreenko, N. S. Egorov, S. M.
    Strukova, and N. S. Landau, "Fibrinolytic agents isolat-
    ed from cultures of certain saprophytic fungi," Dokl.
    Akad. Nauk SSSR, 153, 939 (1963).
34. S. Andreev, A. Kubatiev, and N. Koltsova, "The possible
    use of terrilytin, an Aspergillus terricola proteinase,
    in the treatment of pulmonary thrombosis," in: Abstracts.
    VII International Congress on Thrombosis and Haemostasis,
    London, 1979, Thromb. Diath. Haemorrh., 42 (1979).
35. H. P. Klocking and F. Markwardt, "Thrombolytische wirk-
    ung einer protease aus Aspergillus ochracaus," Folia
    Haematol., 95, 179 (1971).
36. N. S. Egorov and N. S. Landau, "Proteases with fibrino-
    lytic action: biosynthesis by certain microorganisms,"
    in: Streptokinase and Other Thrombolytic Enzymes, N. E.
    Savchenko and V. S. Votyakov (eds.), Minsk (1979), p.52.
37. V. S. Egorov, S. N. Vybornykh, V. I. Ushakova, and K. A.
    Vinogradova, "Proteolytic enzyme formation during sur-
    face cultivation of brown Actinomyces with respect to
    their fibrinolytic activity," Nauk. Dokl. Vyssch. Shkol.
    Biol., 7, 97 (1971).
38. Y. V. Andreenko, A. B. Silaev, R. A. Maximova, and T. N.
    Serebryakova, "Fibrinolytic enzyme from Trichothecium
    roseum LK, EKFR," in: Proceedings First International
    Congress International Association of Microbiological
    Societies, Vol. 5, T. Hasegawa (ed.), Tokyo (1974), p.
    190.

39. T. N. Serebryakova, "Study of fibrinolytic and thrombolytic properties of Trichothecium roseum LK complex proteases," Abstract of Ph.D. Thesis, Moscow State University, Moscow (1981).

40. Y. V. Andreenko, T. N. Serebryakova, and R. A. Maximova, "Activation of plasminogen by tricholysin — a protease complex synthesized by the saprophyte fungus Trichothecium roseum," in: Progress in Chemical Fibrinolysis and Thrombolysis, J. F. Davidson (ed.), Vol. 4, London (1979), p. 188.

41. R. A. Maximova, T. S. Sharkova, A. B. Silaev, G. V. Andreenko, T. N. Serebryakova, and T. V. Teplyakova, "Arthrobotrys longa strain: fibrinolytic activity producer with activator properties," Patent No. 2638407/13 (1980).

42. G. V. Andreenko, T. N. Serebryakova, R. A. Maximova, S. G. Tsymanovitch, T. S. Sharkova, N. S. Murashova, and M. A. Kozlova, "Properties of fibrinolytic enzyme preparation (FEP) isolated from the cultural fluid of Arthrobotrys longa," Vestn. Mosk. Gos. Univ., Ser. Biol. (1982).

43. D. Collen and B. Wiman, "Physiological inhibitors of fibrinolysis," in: Fibrinolysis: Modern Fundamental and Clinical Aspects, Meditsina, Moscow (1982), p. 28.

44. D. A. Lloyd, S. A. Cederholm-Williams, and A. A. Sharp, "Binding of plasminogen and vascular plasminogen activator and the fibrin alpha chain," Thromb. Haemostas., 46 (1981).

45. O. Takashi, "A histochemical and biochemical study of fibrinolysis inhibitor of the human arterial wall," Fukuoka Acta Med., 72, No. 2, 3 (1981).

46. J. E. Walker and D. Ogston, "The inhibition of tissue activator and urokinase by human plasma," Thromb. Haemorrh., 46, 28 (1981).

47. C. Korniger and D. Collen, "Neutralization of human extrinsic (tissue type) plasminogen activator in human plasma: no evidence for a specific inhibitor," Thromb. Haemostas., 46, 662 (1981).

48. M. Furlan, "Interaction of plasmin with fibrinogen," in: Fibrinolysis: Modern Fundamental and Clinical Aspects, Meditsina, Moscow (1982), p. 104.

49. W. Nieuwenhuizne, A. Vermond, and F. Haverkate, "Factor influencing the structure of terminal plasmin degradation products of human fibrinogen and fibrin," Biochim. Biophys. Acta, 667, 321 (1981).

50. P. J. Gaffney, "Interaction of plasmin with fibrinogen," in: Fibrinolysis: Modern Fundamental and Clinical Aspects, Meditsina, Moscow (1982), p. 115.

51. P. J. Gaffney and A. N. Whitaker, "Fibrin cross-links
     and lysis rates," Thromb. Res., 14, 85 (1979).
52. P. J. Gaffney, F. Joe, E. A. Rose, and A. N. Whitaker,
     "The influence of various combinations of plasminogen
     and streptokinase on fibrinolysis," Haemostasis., 11,
     2 (1982).
53. P. J. Gaffney, K. Lord, and R. D. Thornes, "The action
     of brinase in vitro and in vivo," Thromb. Diath. Haem-
     orrh., 34, 941 (1975).
54. D.Nyman, "The effect of brinase on fibrinogen in vivo,"
     Thromb. Diath. Haemorrh., 33, 217 (1975).
55. P. H. Vanhove, M. B. Donati, S. H. Claeys, R. Verhaeche,
     and J. Vermylen, "Action of brinase on human fibrinogen
     and plasminogen," Thromb. Haemostas., 42, 571 (1979).
56. E. P. Frisch and M. Blomback, "Blood coagulation studies
     in patients treated with brinase," in: Progress in
     Chemical Fibrinolysis and Thrombolysis, Vol. 4, J. F.
     Davidson (ed.), London (1979), p. 184.
57. H. P. Klocking and F. Markwardt, "Uber die fibrinolyt-
     ische wirkung einter aus Aspergillus ochraceus isoliert-
     en protease," Acta Biol. Med., 26, 35 (1974).
58. R. A. Smith, R. J. Dupe, P. D. English, and J. Green,
     "Fibrinolysis with acyl-enzymes: a new approach to
     thrombolytic therapy," Nature, 290, 505 (1981).
59. E.I. Chazov, A. V. Mazaev, V. P. Torchilin, and V. N.
     Smirnov, "Use of biocompatible preparations of immobil-
     ized enzymes and physiologically active compounds with
     prolonged effect for treating thromboses," in: Topical
     Problems of Hemostasis, B. V. Petrovski, E. I. Chazov,
     and S. V. Andreev (eds.), Nauka, Moscow (1981), p. 32.
60. D. Collen, "Mechanisms of inhibition of tissue-type plas-
     minogen activator in blood," Thromb. Haemostas., 50, 678
     (1983).
61. L. A. Erickson, M. H. Ginsberg, and D. J. Loskutoff, "An
     inhibitor of plasminogen activator in human platelets,"
     Haemostasis (Abstracts) (1984), p. 39.
62. D. C. Rijken, I. Juhan-Vague, and D. Collen, "Complexes
     between tissue-type plasminogen activator and proteinase
     inhibitors in human plasma, identified with an immuno-
     radiometric assay," J. Lab. Clin. Med., 101, 285 (1983).
63. G. V. Andreenko, "Antiactivators and activators of plas-
     minogen," Usp. Sovrem. Biol. (1986) (in press).

LEECHES *Hirudo medicinalis* AS A SOURCE
OF PROTEOLYTIC ENZYME INHIBITORS

I. P. Baskova

Laboratory of Physiology and Biochemistry of Blood Coagulation
M. Lomonosov Moscow State University, Moscow

ABSTRACT

The purpose of the present study was to determine the inhibitory potential of extracts obtained from various sections of the medicinal leech body. Attention was focused on hirudin, a specific inhibitor of thrombin. Highly purified hirudin preparation, isolated from the head region of leeches, with an activity of 15,600 NIH AT-U per mg, was found to contain isoleucine as an N-terminal and one arginine residue; the molecular weight of hirudin was 7100. Two isoforms differing in activity with a pH of 3.8 and 3.9 were obtained from hirudin purified by isoelectric focusing. A novel polypeptide lacking biological activity with valine as an N-terminal and a molecular weight of about 5000 was isolated from the bodies of medicinal leeches. This polypeptide was termed "pseudohirudin" because it followed hirudin at all stages of isolation and purification. A high content of pseudohirudin in hirudin preparations isolated from whole leeches caused a weaker activity than hirudin obtained from the head region. The occurrence of "pseudohirudin" in these preparations was proven by the results of N-terminal amino acid analysis. Data on antithrombin, antitryptic, and antichymotryptic activities of various sections of the leech's body are presented. Recognizing the necessity of using medicinal leeches as a source of biologically active preparations of proteolytic enzyme inhibitors, it was shown to be expedient to use extracts from

93

the body sections (the richest in inhibitors) of leeches
rather than whole leeches. Animal experiments using hirudin
to cause certain physiological reactions associated with
thrombogenesis demonstrate that the stimulating effect of
adrenocorticotropin, versus epinephrine, on nonenzymic fibrinolysis was not produced through thrombogenesis. Hirudin,
in complex with thrombin, inhibits the physiological function
of thrombin as expressed by the ability to activate the anticoagulation system.

INTRODUCTION

The therapeutic effect of the leech genus _Hirudo medicin-_
alis has been known for a long time. Hirudotherapy (from the
Latin _hirudo_ meaning leech) and bloodletting used to be considered essential in the treatment of various diseases.
Throughout history, from ancient times to the present, the
interest in hirudotherapy has continued to change. In 1884,
Hycraft [1] discovered in medicinal leech salivary glands a
substance, later termed hirudin, which inhibited blood coagulation _in_ _vitro_. This was the begining of a scientific approach
to hirudotherapy. Further numerous investigations showed,
however, that hirudin apparently does not occur in blood flow,
since the main coagulation system components were virtually
unchanged with hirudotherapy.

Most authors now are inclined to explain the therapeutic
effect of medicinal leeches by the outflow of capillary blood,
realizing their psychotherapeutic effect as well. Leeches
are no longer in the forefront of clinical practice, having
yielded to modern medical approaches to therapy.

The past decade, however, has changed this situation:
Medicinal leeches have been found to be the source of hirudin,
a highly specific inhibitor of thrombin and of the inhibitors
of other proteolytic enzymes such as trypsin, plasmin, chymotrypsin, and related proteases [2, 3]. The problem of using
these medicinal leeches on a large scale to isolate these inhibitors, since they might be of great importance in clinical
practice, faces us today.

The question also arises of whether these proteolytic
enzyme inhibitors penetrate into the human organism in hirudotherapy and are capable of producing a local or general effect. In biting, the leech is likely to also secrete certain

physiologically active compounds. It is thus extremely important to study the fine mechanisms of the salivary gland secretory effect and to study the extracts of various sections of leeches on the coagulation system. Study of the composition and properties of biologically active extracts of various sections of leeches, to which this chapter is devoted, is a necessary stage in understanding this problem.

HIRUDIN AS AN INHIBITOR OF THROMBIN

Although hirudin was discovered by Hycraft [1], study of the properties of this inhibitor is associated with Markwardt [4]. Markwardt developed a technique for isolating hirudin from the head region of medicinal leeches [5, 6], and he showed that hirudin is a specific inhibitor of thrombin. With thrombin, hirudin forms an inactive, stable, noncovalent stoichiometric complex with a dissociation constant of $K = 0.8 \cdot 10^{-10}$ M[7], $6.3 \cdot 10^{-11}$ M[8]. Hirudin differs greatly from other natural inhibitors of thrombin — antithrombin III, heparin, and α_2-macroglobulin — in its extremely high specificity for thrombin [7]. Complexing between hirudin and thrombin proceeds in no time at all, i.e., faster than the reaction between thrombin and fibrinogen [4]. Due to these properties, hirudin is a perfect inhibitor of thrombin, which cannot be said for the numerous synthetic chemical inhibitors of thrombin [8-15].

The inhibitory effect of hirudin is seen not only in the slowing down, but also in the complete blocking, of fibrinogen coagulation by thrombin. The thrombin-induced activation of coagulation factors V, VII, and XIII is arrested in the presence of hirudin. Hirudin also inhibits the binding of thrombin to platelets [16, 17], thus hindering both platelet release and aggregation [18]. Because thrombin has a higher affinity for hirudin than for highly affinitive receptors on platelets [19], hirudin can induce dissociation of the thrombin complex with specific protein receptors from platelets [20]. Hirudin analogously affects thrombin bound to protein receptors of endothelial cells and fibroblasts [19].

Intravenous injection of hirudin in animals prolongs total clotting time, and thrombin and partial thromboplastin time of plasma [21, 22]. It also prevents thrombogenesis induced by lethal doses of thrombin [21, 22], endotoxin, etc. [23, 24].

We showed that hirudin decreases the effect of endogenous thrombin, the formation of which is induced by intravenous injection of epinephrine in animals. According to Kudrjashov [25], the anticoagulation system is activated by
thrombin occurring in the blood, which leads to an increase
in the anticoagulation and fibrinolytic potential of plasma
and, in particular, to the growing nonenzymic fibrinolytic
activity specific for the activated state of the anticoagulation system. The increase in this activity results from the
release of heparin and the appearance in the blood of certain
complexes of heparin with blood proteins and other compounds
[25]. This specific effect was found to decrease sharply in
rats in response to simultaneous intravenous injection of
epinephrine and hirudin [26]. This is an example of how hirudin can be used for selective blockade of endogenous thrombin during the reactions leading to thrombogenesis in the
organism.

Another use of hirudin is to clarify the fine mechanisms of physiological reactions. Nonenzymic fibrinolytic
activity has been shown to increase in animals under immobilization stress as a result of the increase in adrenocorticotropin (ACTH) titer in the blood [27]. Experiments with
simultaneous administration of ACTH and hirudin in rats under
immobilization stress demonstrate that ACTH has a stimulating effect on nonenzymic fibrinolytic activity, despite the
inhibition of endogenous thrombin by hirudin [26]. Therefore, it can be concluded that ACTH induces blood fibrinolytic potential but not through thrombinogenesis, which con-
forms well to the ACTH-induced release of heparin [28].

Since thrombin is a polyfunctional enzyme that affects
hemocoagulation in the organism [19], it is necessary to determine to what extent the physiological functions of thrombin are inhibited by its complexing with hirudin. We therefore studied the influence of the hirudin—thrombin complex
on activation of the anticoagulation system [25]. The hirudin—thrombin complex deprived of both coagulative and amidolytic activity, unlike native thrombin, appeared to lose
its capability to activate the anticoagulation system [29].
Hirudin thus permitted us to realize the importance of the
substrate binding site of the thrombin molecule in activation of the anticoagulation system.

The mechanism of thrombin inhibition by hirudin has not
been exhaustively investigated. The same regions of secon-

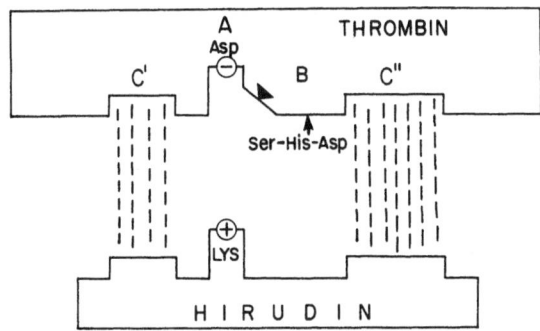

Fig. 4.1. Diagram of thrombin binding to hirudin [30]: A)
          area of primary binding to substrate; B) catalyt-
          ic region; C', C") areas of secondary binding to
          substrate. C' is the hydrophobic binding area
          and C" is the area of specific binding to proteins,
          most important in the interaction with fibrinogen
          and hirudin.

dary binding to substrate are assumed to interact in thrombin
as are those involved in the thrombin—fibrinogen interaction
[30] (Fig. 4.1).

It is not now possible to determine the part of the hirudin molecule that is responsible for the binding to thrombin.
Attempts to identify the peptide fragment corresponding to
the active site of the inhibitor [31] have failed. The synthetic analog of the peptide fragment Ac-Val-Thr-Gly-Glu-Gly-
Thr-Pro-Lys-Pro-NH$_2$ does not inhibit thrombin amidolytic
activity to the synthetic substrate Bz-Phe-Val-Arg-pNa [32].

Oxidation of the disulfide bonds leads to the loss of
antithrombin activity by hirudin [33]. Chemical modification
of free carboxyl groups in hirudin sharply decreases its affinity for thrombin [34]. This indicates the realization of
ionic interactions among molecules in hirudin—thrombin complexing.

Preliminary data on the primary structure of hirudin
have been obtained by Magnusson et al. [31, 33], and are listed below (see p. 98). In the primary structure of hirudin,
attention is attracted to a part of the amino acid sequence
-Gly-Ser-Asp-Gly-Glu- (residues 31-35) representing the in-
verse sequence of amino acids found near serine in the throm-

Val-Val-Tyr-Thr-Asp-Cys-Thr-Glu-Ser-Gly-Gln-Asn-Leu-
 1   2   3   4   5   6   7   8   9   10   11   12   13

Cys-Leu-Cys-Glu-Gly-Ser-Asn-Val-Cys-Gly-Gln-Gly-Asn-
14   15   16   17   18   19   20   21   22   23   24   25   26

Lys-Cys-Ile-Leu-Gly-Ser-Asp-Gly-Glu-Lys-Asn-Gln-Cys-
27   28   29   30   31   32   33   34   35   36   37   38   39

Val-Thr-Gly-Glu-Gly-Thr-Pro-Lys-Pro-Gln-(Ser, His,
40   41   42   43   44   45   46   47   48   49   50   51

Asx, Asx, Asx, Gly)-Phe-Glu-Glu-Ile-Pro-Glu-Glu-
52   53   54   55   56   57   58   59   60   61   62

TyrSO$_3$-Leu-Gln
 63   64   65

bin-active center: -Glu-Gly-Asp-Ser-Gly- (residues 192-196)
[35].

The primary structure of hirudin given above is inconsistent in a number of ways with the findings of other authors.
The C-terminal sequence is unlike that of -Ala-Gly-Ser-Gln-
Leu OH determined by carboxypeptides [36]. Alanine found in
hirudin [37-39] is missing. Alanine was also discovered in
a preparation of highly purified hirudin with an activity of
15,600 National Institute of Health (NIH) antithrombin units
(AT-U) per mg [40] (Table 4.1). Valine as an N-terminal of
the hirudin primary structure is also at variance with data
showing isoleucine as the N-terminal of hirudin [37, 39]. In
other hirudin preparations [40-42], isoleucine was also identified as an N-terminal amino acid. It is noteworthy that
hirudin preparations with valine as an N-terminal were isolated from the whole medicinal leech, while hirudin preparations with isoleucine as an N-terminal amino acid were obtained
from the head region of leeches.

In order to confirm these differences, hirudin preparations were obtained from whole leeches, and their head regions and bodies, and their N-terminal amino acid residues
and antithrombin activity were analyzed. Hirudin obtained
by Markwardt [8] and purified from bdellin admixtures by affinity chromatography on trypsin—Sepharose underwent isoelectric focusing at a saccharose gradient, with pH ranging
from 3 to 6. Figure 4.2 presents the behavior in isoelectric
focusing of hirudin preparations isolated from whole leeches

TABLE 4.1. Amino Acid Composition of Highly Purified Hirudin
Preparations as Determined by Several Authors

Amino acid residues	Composition by authors (molecular weight)				
	Llosa et al. [37] (13,000)	Bagdy et al. [38] (12,000)	Markwardt et al. [39] (9060)	Baskova et al. [40] (7100)	Magnusson et al. [31] (~7300)
Ala	1	3	1	2	-
Arg	-	-	-	1	-
Asp	16+1	14	10	12	9
Cys/2	6	6	6	6	6
Glu	18+1	16	13	14	13
Gly	14	13	9	9	8
His	2	2	1	2	1
Leu	7	5	4	4	4
Ile	4	3	2	4	2
Lys	5	5	3-4	4	3
Met	3-4	-	-	-	-
Phe	2	2	2	1	1
Pro	8+1	6	3	5	3
Ser	8	6	4	4	4
Thr	8	6	4	5	4
Thr	8	6	4	5	4
Trp	-	-	-	-	-
Tyr	2-3	3	2	2	2
Val	7	7	3	3	4

(B), head regions (A), and bodies (C). Fractions possessing
antithrombin activity are shaded. Figure 4.2 shows that the
behavior of the three preparations is similar. This mainly
concerns fractions I, II, and III with isoelectric points at
pH 3.8, 3.9, and 4.0, respectively. Summarizing the antithrombin activity of these fractions, Table 4.2 shows that
hirudin preparations from the head region at pH 3.9 possess
the highest activity (4250 NIH AT-U/mg).

The preparation isolated from leech bodies has virtually no antithrombin activity, but its light tracks (about 1%)
are due to hirudin, since the change in this activity during
purification is like that of hirudin both from the head region and from whole leeches. Such preparations, offering
practically no antithrombin activity, are termed "pseudohirudin" [42].

100   I. P. BASKOVA

TABLE 4.2. Antithrombin Activity (NIH AT-U/mg of dry weight)
            of Preparations from Heads, Bodies, and Whole
            Medicinal Leeches at Various Stages of Isolation
            and Purification

Purification stage	Head region	Whole leeches	Bodies
Extraction and precipitation by acetone	350	72	10
Fractionation by ethanol	1170	165	20
Affinity chromatography on trypsin—Sepharose	1570	200	21
Isoelectric focusing:			
I fraction, pH 3.8	2200	700	48
II fraction, pH 3.9	4250	1200	52
III fraction, pH 4.0	910		

The activity of hirudin preparations with pH 3.9 isolated from whole leeches is 1200 NIH AT-U/mg of dry weight,
which is 3.5 times lower than the activity of the preparation
from the head region. The activity of the fraction with pH
3.8 isolated from whole leeches is also one third as high as
the activity of the hirudin fraction isolated from the head
region. These results allow us to assume that the lower specific activity of hirudin obtained from whole leeches, compared to that obtained from the head regions, is due to
"pseudohirudin," which lacks antithrombin activity and follows hirudin both during isolation and purification. These
data are borne out by the hirudin—"pseudohirudin" quantitative ratio (1:3-4) per leech.

Our assumption is also supported by the analysis of N-
terminal amino acid residues by means of the dansyl method
using thin-layer chromatography. Isoleucine is a dominating
N-terminal amino acid for all three fractions obtained by isoelectric focusing of hirudin preparations from the head region. Valine, and its successor valyl-valine dipeptide, dom-
inate in all three fractions of "pseudohirudin" isolated from
the leech bodies. Isoleucine and valine as N-terminal amino
acids were found in preparations of whole leeches [42].

These findings thus indicate that hirudin preparations
isolated from whole leeches have "pseudohirudin" contamina-

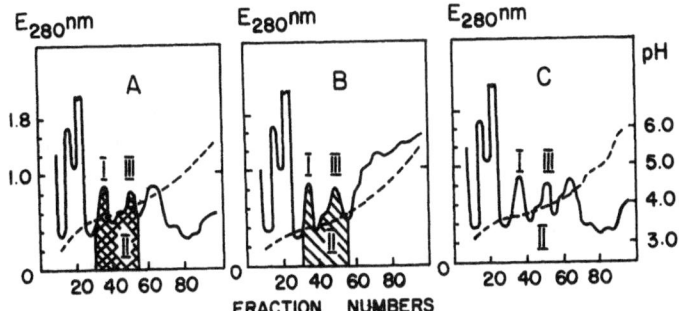

Fig. 4.2.   Isoelectric focusing of hirudin preparations isolated from the head region of leeches (A), from
whole leeches (B), and from their bodies (C). The
fractions possessing antithrombin activity are
shaded. The dotted line shows the pH gradient.

tions from the leech bodies, resulting in their lower activity compared to hirudin obtained from the head region. It
is therefore reasonable to use the head region of medicinal
leeches in order to obtain active hirudin preparations.

The data above are inconsistent with the idea that differences in amino acid N-terminals of hirudin isolated from
whole leeches or from the head region are due to the existence
of two forms of hirudin: valine and isoleucine. At the same
time, two hirudin isoforms with pH 3.8 and pH 3.9 (Figure 4.2),
with different activities (Table 4.2) and with isoleucine as
an N-terminal, were found by isoelectric focusing of hirudin
preparations.

"Pseudohirudin" was first described as a polypeptide with
a molecular weight of about 5000 [43] and with high capability for the formation of dimeric and trimeric associates with
a molecular weight of 9100 and 18,200, respectively. We also
found that the molecular weight of hirudin was 7100 and that
it had a capability for dimerization. The molecular weights
of hirudin determined by some authors [6, 37], amounting to
16,000, 15,000, and 13,000, respectively, can be expected to
refer to dimers of this inhibitor. "Pseudohirudin" differs
slightly from hirudin in its amino acid composition [43].
The content of valine is higher than that of isoleucine, and
there is a lower concentration of dicarbonic amino acids as
well as lysine and tyrosine. It is noteworthy that only two

cysteines were found in "pseudohirudin," in contrast to hirudin, indicating but one disulfide bond. On the whole, the
"pseudohirudin" molecule is 20 amino acid residues less than
that of hirudin. This finding agrees well with the difference in the molecular weight of monomers of these compounds,
which is equal to 2000 [43].

These results allow us to assume that "pseudohirudin"
can be a hirudin metabolite resulting from the cleavage of a
single arginine peptide bond in a hirudin molecule. Arginine
residue was detected by amino acid analysis of a highly purified hirudin preparation [40] (Table 4.1).

The difference in the content of cysteine residues in hirudin and "pseudohirudin," elucidated in [43], agrees with the
data in references [4 and 33]. The essential role of disulfide bridges was shown to provide stability to the hirudin
molecule conformation, which is one of the factors indicating
high inhibitory activity of hirudin, compared to "pseudohirudin." The functional role of "pseudohirudin" has not been
recognized. In addition, "pseudohirudin" possesses neither
antitryptic nor antichymotryptic activities [44].

BDELLINS AS TRYPSIN AND PLASMIN INHIBITORS

Fritz et al. [2] were the first to find, in 1969, that
commercial hirudin preparations obtained from whole leeches
were capable of inhibiting the amidolytic activity of both
trypsin and plasmin. A series of inhibitors was later isolated from hirudin by ion exchange and affinity chromatography on immobilized trypsin. These inhibitors, termed bdellins (from the greek term, meaning leech), were polypeptides
with a molecular weight on the order of 7000 (group A) and
5600 (group B) [2].

Bdellins A and B differ in the degree of their binding
to DEAE-cellulose in fractionation of the whole preparation.
Each of the groups is a mixture of inhibitors with similar
molecular weights and different amino acid compositions. So,
despite a rather high content of cysteine residues (10 for
bdellins A and 6 for bdellins B), none of the amino acids
proline, methionine, isoleucine, phenylalanine, or tryptophan
were found in bdellins B [2]. The structure of bdellin B-3
has been published in [45].

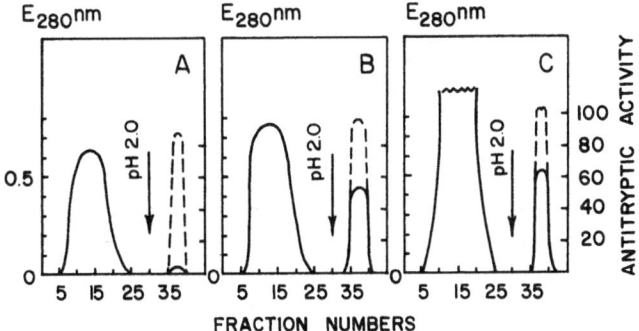

Fig. 4.3. Chromatography of hirudin preparations isolated
from the head region of leeches (A), whole leeches (B), and bodies (C) on trypsin–Sepharose. Antitryptic activity is given in μg of trypsin, connected with 0.2 ml eluate. Casein was the substrate for determining tryptic activity.

Bdellins inhibit the amidolytic activity of trypsin,
plasmin, and arosine, and form an inactive equimolar complex
with these enzymes. Lysine residue was considered to be essential in the active center of these inhibitors [2].

Bdellins follow hirudin in its isolation from leeches.
Thus, 1 gram of commercial hirudin preparation "Medimpex"
(Hungary) with an antithrombin activity of 270 NIH AT-U/mg
contains 230 mg bdellins, but only 27 mg hirudin. One gram
of hirudin preparation "Serva" with an activity of 3400 NIH
AT-U/mg contains 300 mg hirudin and 75 mg bdellins [2]. Hirudin preparation obtained and purified by the method of Markwardt [8] is bdellin-free.

Hirudin preparations which we isolated by extraction
from the head region of medicinal leeches, followed by fractionation with acetone and ethanol, also contained contaminations which inhibited trypsin caseinolytic activity [41]. Affinity chromatography on trypsin–Sepharose was used to remove these contaminations. It is noteworthy that trypsin inhibitors also follow hirudin preparations isolated from whole
leeches [40] like "pseudohirudin" preparations from leech
bodies [42] (Fig. 4.3).

These results agree well with the data obtained by Marx,
to which Fritz [2] refers, on bdellins found in all zones of

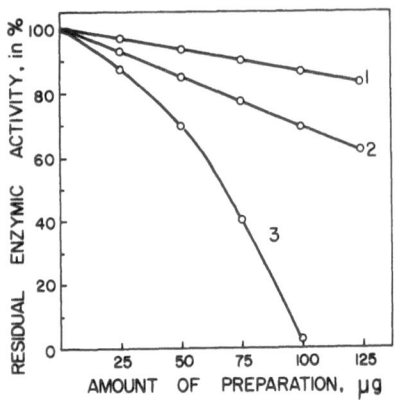

Fig. 4.4.   Antitryptic (1, 3) and antichymotryptic
(2) activities of medicinal leech salivary gland secretions and intestinal contents (1, 2).

the leech's body.   He also reported the correlation of bdellins concentrated in the region of the outer sexual organs,
with a high level of trypsin and plasmin inhibitors found in
seminal vesicles and seminal plasma of man and many animals.

High antitryptic activity was found in the salivary
gland secretions of medicinal leeches, and is expressed by
complete inhibition of tryptic caseinolytic activity (Figure
4.4).   When heated for 15 min at 80°C, the secretion loses
its antitryptic activity [44].   This activity of the leech
intestinal content (as shown in Figure 4.4) is less pronounced
than the activity of salivary gland secretions. Antitryptic activity of the intestinal content can be assumed to
depend on the antitryptic activity of salivary gland secretions, which is brought to the leech's intestine together
with sucked blood [44].

EGLINS AS CHYMOTRYPSIN INHIBITORS

Eglins are the carriers of antichymotryptic activity in
medicinal leeches.   They are a group of polypeptides with a
molecular weight of 6600-6800 which were first found, along
with bdellins, in commercial preparations of hirudin [2].
Eglins are capable of inhibiting α-chymotrypsin, subtilisin,
and neutral proteases of the human granulocytes: elastase

and cathepsin G [3]. Stable complexes with dissociation constants of $K = 2-3 \cdot 10^{-10}$ M are formed between these proteases
and eglins. Thus, their use as a therapeutic preparation is
rather promising. Highly purified eglins have been obtained
and their composition and physicochemical characteristics
have been well studied. A complete lack of disulfide bonds,
as well as methionine, isoleucine, and tryptophan, is peculiar to these polypeptides. The eglin "c" primary structure
is given below.

```
Thr-Glu-Phe-Gly-Ser-Glu-Leu-Lys-Ser-Phe-Pro-Glu-Val-
 1   2   3   4   5   6   7   8   9  10  11  12  13

Val-Gly-Lys-Thr-Val-Asp-Gln-Ala-Arg-Glu-Tyr-Phe-Thr-
14  15  16  17  18  19  20  21  22  23  24  25  26

Leu-His-Tyr-Pro-Gln-Tyr-Asn-Val-Tyr-Phe-Leu-Pro-Glu-
27  28  29  30  31  32  33  34  35  36  37  38  39

Gly-Ser-Pro-Val-Thr-Leu-Asp-Leu-Arg-Tyr-Asn-Arg-Val-
40  41  42  43  44  45  46  47  48  49  50  51  52

Arg-Val-Phe-Tyr-Asn-Pro-Gly-Thr-Asn-Val-Val-Asn-His-
53  54  55  56  57  58  59  60  61  62  63  64  65

Val-Pro-His-Val-Gly
66  67  68  69  70
```

The mechanism of chymotrypsin inhibition by eglin "c"
has been shown to involve the formation of a stable covalent
complex of chymotrypsin with the N-terminal part of the eglin
molecule, whereas the C-terminal fragment is released simultaneously. Complexing is assumed to be followed by the
cleavage of Leu (45)—Asp (46) peptide bond in the eglin "c"
molecule [46].

Eglins are isolated from the acetonic extract of whole
medicinal leeches; however, hirudin preparations obtained from
whole leeches have already been suggested [3] to possess both
bdellin and eglin contaminations.

In our experiments we showed that hirudin preparation
from whole leeches free from bdellin contaminations are capable of inhibiting caseinolytic activity of α-chymotrypsin.
The resultant hirudin fraction I and a combined fraction II +
III obtained by isoelectric focusing display different antichymotryptic activity. Thus, fraction II + III inhibits
caseinolytic activity of chymotrypsin by 88% when the amount
of the preparation is 100 μg. An additional increase in the

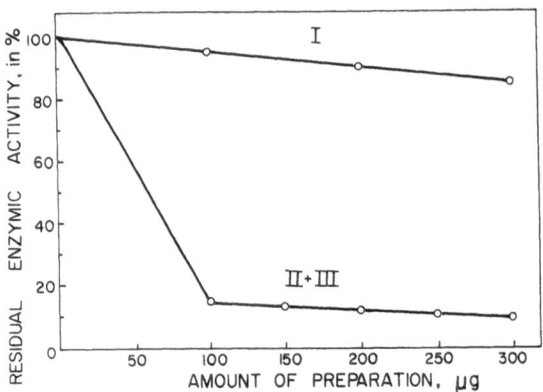

Fig. 4.5.   Antichymotryptic activity of preparations of hir-
            udin fractions I and II + III, obtained by iso-
            electric focusing.

amount of the preparation does not increase the degree of inhibition (Fig. 4.5) [44]. At the same time, 100 μg of the
fraction I preparation has practically no inhibitory activity, but an increase in the amount of preparation increases
slightly the degree of inhibition (Figure 4.5). Thus, hirud-
din fractions II + III, with a higher antithrombin activity
(Table 4.2), also have a more pronounced capability for inhibiting the caseinolytic activity of chymotrypsin as compared to fraction I (pH 3.9). This indicates a certain affinity of eglins for the most active form of hirudin.

The secretions of medicinal leech salivary glands do not
possess antichymotryptic activity at all. Yet, the intestinal content offers a pronounced capability for inhibiting
chymotrypsin caseinolytic activity (Fig. 4.4).

CONCLUSIONS

The data show that low molecular inhibitors of proteolytic enzymes of the leeches <u>Hirudo medicinalis</u> represent a
group of unique inhibitors affecting the regulation of coagulation and fibrinolysis. Eglins, which are apparently evolved in regulating the digestion of proteins in sucked blood,
form a separate group. This accounts for their high concentration in the intestinal content of medicinal leeches. Eglins are also capable of inhibiting the activity of proteinases produced by leukocytes during inflammation.

TABLE 4.3. Localization of Proteinase Inhibitors in Various
           Sections of the Leech

Leech sections	Inhibitors		
	Hirudin	Bdellins	Eglins
Salivary gland secretion	+	+	-
Intestinal content	(+)	(+)	+
Body	-	+	+?

Note: +, Many; (+), little; –, none.

The high affinity for specific enzymes or a group of enzymes is characteristic of low-molecular-weight inhibitors from
medicinal leeches. These inhibitors also have a rather high affinity for each other, resulting in a significant content of
bdellins and eglins as well as biologically inactive "pseudohirudin" in low purified hirudin preparations. Thus, in
order to separate the inhibitors isolated from extracts of
whole leeches, various multistage techniques need to be applied.

The regions of priority localization of each group of
protease inhibitors in leeches have been determined. The
secretion of salivary glands is a source of hirudin and bdellins, indicating its high antithrombin and antitryptic activity. There are no eglins in the secretion and, thus, it is
deprived of antichymotryptic activity. Alternatively, intestinal content is a source of eglins as characterized by
fairly pronounced antichymotryptic activity. At the same
time, its antithrombin and antitryptic activities are very
weak, probably due to the presence of small amounts of salivary gland secretion in the intestine, along with sucked
blood. The leech's body, beyond doubt, is a source of bdellins, and the extracts obtained from decapitated leeches show
high antitryptic activity, while the intestinal content of
these extracts scarcely inhibits trypsin activity. Since
antichymotryptic activity is found in both the extracts obtained from whole leeches and in the intestinal content, eglins need not be regarded as occurring in the leech's bodies
(Table 4.3).

It has become essential at the present time to use medicinal leeches as a source of biologically active preparations of various proteolytic enzyme inhibitors. For this use,
it is reasonable to use sections which are the richest in
either inhibitors rather than whole leeches Hirudo medicinalis.

REFERENCES

1. J. B. Hycraft, "Uber die einwirkung eines sekrets des
    offizinellen blutegels auf die grinnbarkeit des blutes,"
    Arch. Exp. Pathol. Pharmakol., 18, 209 (1884).
2. H. Fritz, M; Gebhardt, R. Meister, and E. Fink, "Trypsin-
    plasmin inhibitors from leeches. Isolation, amino acid
    composition, inhibitory characteristics," in: Proceed-
    ings International Research Conference on Proteinase In-
    hibitors, Munich, 1970, H. Fritz and H. Tschesch (eds.),
    De Gruyter, Berlin (1971), pp. 271-280.
3. U. Seemuller, M. Meier, K. Ohlsson, H.-P. Muller, and
    H. Fritz, "Isolation and characterization of a low-mol-
    ecular-weight inhibitor (of chymotrypsin and human gran-
    ulocytic elastase and cathepsin G) from leeches," Hoppe-
    Seyler's Z. Physiol. Chem., 358, 1105-1117 (1977).
4. F. Markwardt, Blutgerinnungshemmende Wirkstoffe aus
    Blutsaugenden Tieren, Fisch, Jena (1963).
5. F. Markwardt, "Untersuchungen uber hirudin," Naturwissen-
    schaften, 42, 537-540 (1955).
6. F. Markwardt, "Die isolierung und chemische charakter-
    isierung des hirudins," Hoppe-Seyler's Z. Physiol. Chem.,
    308, 147-156 (1957).
7. M. R. Dowing, J. M. Bloom, and K. G. Mann, "Comparison
    of the inhibition of thrombin by three plasma protease
    inhibitors," Biochemistry, 17, 2649-2653 (1978).
8. F. Markwardt, "Hirudin as an inhibitor of thrombin," in:
    Methods in Enzymology, Vol. 19, G.E. Perlmann and L. Lor-
    and (eds.), Academic Press, New York (1970), pp. 924-932.
9. S. Okamoto, A. Hijikata, K. Kijo, R. Kikimoto, K. Ohkubo,
    S. Tonomura, and Y. Tamao, "A novel series of synthetic
    thrombin inhibitors having extremely potent and highly
    selective action," Kobe J. Med. Sci., 21, 43-51 (1975).
10. M. E. Nesheim, F. G. Prendergast, and K. G. Mann, "In-
    teraction of a fluorescent active-site-directed inhibi-
    tor of thrombin: N-(2-ethyl-1,5-pentanediyl)amide,"
    Biochemistry, 18, 996-1003 (1979).
11. A. Hijikata, S. Okamoto, R. Kikumoto, and Y. Tamao,
    "Kinetic studies on the selectivity of a synthetic throm-

bin-inhibitor using synthetic peptide substrates,"
Thromb. Haemostas., 42, 1039-1045 (1979).

12. J. Hauptmann, B. Kaiser, F. Markwardt, and G. Novak,
    "Anticoagulant and antithrombotic action of novel spe-
    cific inhibitors of thrombin," Thromb. Haemostas., 43,
    118-123 (1980).

13. P. Walsmann, "Benzamidine derivatives — relationship be-
    tween antithrombin activity and structure," Folia Haem-
    atol., 109, 75-82 (1982).

14. A.A. Sereiskaya, V. K. Kibirev, D. M. Fedoryak, S. A.
    Poyarkova, and S. B. Serebryanyi, "Anticoagulant activ-
    ity of N-(α)-arylsulfonyl-L-arginine esters," Dopov.
    Akad. Nauk Ukr. SSR, 4, 444-446 (1977).

15. J. Hauptmann and F. Markwardt, "Studies on the anticoag-
    ulant and antithrombotic action of an irreversible throm-
    bin inhibitor," Thromb. Res., 20, 347-351 (1980).

16. P. Ganguly and W. J. Sonnichsen, "Binding of thrombin
    to human platelets and its possible significance," Br.
    J. Haematol., 34, 291-301 (1976).

17. S. W. Tam and T. C. Detwiler, "Binding of thrombin to
    human platelet plasma membranes," Biochim. Biophys. Acta,
    543, 194-201 (1978).

18. T. C. Detwiler and R. F. Feinman, "Kinetics of the
    thrombin-induced release of calcium by platelets," Bio-
    chemistry, 12, 282-289 (1973).

19. J. W. Fenton, II, B. H. Landis, D. A. Walz, D. H. Bing,
    R. D. Feinman, M. P. Zabinski, S. A. Sonder, L. J. Ber-
    liner, and J. S. Finlayson, "Human thrombin: prepara-
    tive evaluation, structural properties, and enzymic spe-
    cificity," in: The Chemistry and Physiology of Human
    Plasma, D. H. Bing (ed.), Pergamon Press, New York
    (1979), pp. 151-182.

20. S. W. Tam, J. W. Fenton, II, and T. C. Detwiler, "Dis-
    sociation of thrombin from platelets by hirudin," J.
    Biol. Chem., 254, 8723-8725 (1979).

21. F. Markwardt, "Die antagonistische wirkung des hirudins
    gegen thrombin in vivo," Naturwissenschaften, 43, III-
    II (1956).

22. F. Markwardt, "Versuche zur pharmakologischen charakter-
    isierung des hirudins," Naunyn-Schmiedeberg's Arch. Exp.
    Pathol. Pharmakol., 234, 516-529 (1958).

23. F. Markwardt, G. Novak, and J. Hoffmann, "The influence
    of drugs on disseminated intravascular coagulation (DIC).
    II. Effects of naturally occurring and synthetic throm-
    bin inhibitors," Thromb. Res., 11, 275-283 (1977).

24. A. Ishikawa, R. Hafter, U. Seemuller, J. M. Gokel, and
     H. Graeff, "Effect of hirudin on endotoxin-induced dis-
     seminated intravascular coagulation (DIC)," Thromb. Res.,
     19, 351-358 (1980).
25. B. A. Kudrjashov, Biological Problems of the Regulation
     of Blood and Its Coagulation, Meditsina, Moscow (1975).
26. F. B. Shapiro, I. P. Baskova, D. Y. Cherkasova, L. A.
     Lyapina, and M. D. Gol'dovskaya, "The effect of hirudin
     on the hormone-induced activation of nonenzymatic fibrin-
     olysis during immobilization stress," Fiziol. Zh. SSSR,
     64, 1567-1573 (1978).
27. B. A. Kudrjashov, F. B. Shapiro, E. G. Lomovskaya, and
     L. A. Lyapina, "The role of ACTH in the formation of hep-
     arin complexes in blood under immobilization stress,"
     Probl. Endokrinol., 21, 54-59 (1975).
28. B. A. Kudrjashov, F. B. Shapiro, E. G. Lomovskaya, and
     L. A. Lyapina, "Role of ACTH and glucocorticoids in non-
     enzymatic fibrinolysis under immobilization stress,"
     Fiziol. Zh. SSSR, 63, 735-741 (1977).
29. B. A. Kudrjashov, I. P. Baskova, A. S. Orlova, and F. B.
     Shapiro, "Effect of hirudin-thrombin and phenylmethyl-
     sulfonyl-thrombin preparations on some blood coagulation
     characteristics," Byull. Eksp. Biol. Med., 41, 307-309
     (1981).
30. P. Walsmann and F. Markwardt, "Biochemische und pharma-
     kologische aspekte des thrombininhibitors hirudin,"
     Pharmazie, 10, 653-660 (1981).
31. T. E. Petersen, H. R. Roberts, L. Sottrup-Jensen, S. Mag-
     nusson, and D. Bagdy, "Primary structure of hirudin, a
     thrombin-specific inhibitor," in: Protides in the Bio-
     logical Fluids, H. Peeters (ed.), Pergamon Press, London,
     Vol. 23 (1976), pp. 145-149.
32. S. Magnusson, L. Sottrup-Jensen, T. Petersen, G. Dudek-
     Wojciechovska, and H. Claeys, "Homologous 'kringle'
     structures common to plasminogen and prothrombin. Sub-
     strate specificity of enzymes activating prothombin and
     plasminogen," in: Proteolytic and Physiological Regula-
     tion, Academic Press, New York (1976), pp. 203-232.
33. D. Bagdy, E. Barabas, L. Graf, T. E. Petersen, and S. Mag-
     nusson, "New advances in hirudin," in: Methods in En-
     zymology, Vol. 45, L. Lorand (ed.), Academic Press, New
     York (1976), pp. 669-678.
34. F. Markwardt and P. Walsmann, "Die reaktion zwischen hir-
     udin und thrombin," Hoppe-Seyler's Z. Physiol. Chem.,
     312, 85 (1958).

35. S. Magnusson, "On the primary structure of bovine thrombin," Folia Haematol., 98, 385-390 (1972).
36. P. de la Llosa, C. Tertrin, and M. Jutisz, "L'enchainement C-terminal de l'hirudine," Biochim. Biophys. Acta, 93, 40-43 (1964).
37. P. de la Llosa, C. Tertrin, and M. Jutisz, "Composition en acides aminés de l'hirudine. Identification du residu N-terminal," Bull. Soc. Chim. Biol., 45, 69-74 (1963).
38. D. Bagdy, E. Barabas, and L. Graf, "Large-scale preparation of hirudin," Thromb. Res., 2, 229-238 (1973).
39. F. Markwardt and P. Walsmann, "Reindarstellung und analyse des thrombininhibitors hirudin," Hoppe-Seyler's Z. Physiol. Chem., 348, 1381-1386 (1967).
40. I. P. Baskova and D. U. Cherkesova, "Comparative properties of hirudin from whole leeches and from leech heads and bodies," Biokhimiya, 45, 226-272 (1979).
41. I. P. Baskova, D. U. Cherkesova, and V. V. Mosolov, "Hirudin preparation by isoelectric focusing," Biokhimiya, 41, 939-941 (1976).
42. D. U. Cherkesova, I. P. Baskova, V. V. Mosolov, N. A. Aldanova, and N. A. Potapenko, "Pseudohirudin from the bodies of medicinal leeches," Dokl Akad. Nauk SSSR, 241, 720-722 (1978).
43. I. P. Baskova, D. U. Cherekesova, V. V. Mosolov, E. L. Malova, and L. A. Belyanova, "Comparative study of hirudin and pseudohirudin," Biokhimiya, 45, 463-467 (1980).
44. I. P. Baskova, G. I. Nikonov, and D. U. Cherkesova, "Antithrombin, antitrypsin, and antichymotrypsin activities of salivary gland secretion and intestinal chyme of medicinal leeches. Antichymotrypsin activity of partially purified preparations of hirudin and pseudohirudin," Folia Haematol. (Leipzig), 111, 831-837 (1984).
45. H. Fritz and K. Krejci, "Trypsin-plasmin inhibitors (bdellins) from leeches," in: Methods in Enzymology, Vol. 45, L. Lorand (ed.), Academic Press, New York (1976), pp. 797-806.
46. U. Seemuller, M. Eulitz, H. Fritz, and A. Strobe, "Structure of elastase-cathepsin G inhibitor of the leech hirudo medicinalis," Hoppe-Seyler's Z. Physiol. Chem., 361, 1841-1846 (1980).

DEVELOPMENT OF BIOCOMPATIBLE
PREPARATIONS OF IMMOBILIZED
ENZYMES AND CLINICAL RESULTS
WITH IMMOBILIZED STREPTOKINASE
(STREPTODECASE)

V. N. Smirnov, Yu. I. Voronkov, V. P. Torchilin,
and A. V. Mazaev

Cardiology Research Center
Academy of Medical Sciences of the USSR, Moscow

INTRODUCTION

The role of enzymes in the practice of medicine is increasing more and more. Enzymes such as fibrinolysin, urokinase, and streptokinase are already used, although in a limited way, in thrombolytic therapy. Asparaginase has been effective in treating some malignant tumors, and digestive enzymes are widely used today [1].

It has recently been established that a number of diseases are associated with disturbances in lysosomal enzyme activity, and a curative effect is only provided by the injection of a deficient enzyme into the organism.

Unfortunately, the clinical use of enzymes is limited by a number of factors, including the high price and relatively low availability of pure forms; rapid inactivation under physiological conditions and release from the organism, thus significantly increasing the cost of therapy; enzyme antigenicity as heterologous proteins; frequent nonspecific activity (toxicity); accessibility to destruction by endogenic proteases; and the impossibility of creating high local concen-

trations sufficient for a therapeutic effect (at the site of
the specific effect on the substrate) without increasing the
systematically injected doses [2].

It is therefore necessary to develop enzymes which, within the organism, possess high stability and low negative action and which could be used more simply. The solution to
these problems can be significantly facilitated by techniques
developed in a new field of biochemistry — enzyme engineering
— which deals with enzyme stabilization in relation to different inactivating effects and the practical use of stabilized
(immobilized) enzymes in various industries. The general
principles that have been developed regarding enzyme stabilization against inactivation, as determined by the unfolding
of the protein-active globule, include protein multipoint attachment to the matrix of soluble or insoluble polymer carriers; in-polymerization of protein modified by a monomer
analog in gel; and protein mechanical entrapment in a three-
dimensional polymer gel [3]. Each of these methods has its
advantages and drawbacks.

Two possible directions can be taken in the creation of
immobilized therapeutic enzymes and other proteins. If a
drug is intended for prolonged circulation in the blood, or
if its presence is necessary in different organs, it is expedient to obtain water-soluble, stabilized forms with enhanced
stability to different inactivating effects and with increased
circulation time in the organism. These preparations include
enzyme-filled artificial cells, in particular microcapsules,
liposomes, and erythrocyte ghosts. If a drug is to be used
for treatment of local lesions and its presence in other
organs is not necessary, then it is reasonable to create enzyme polymer derivatives that can be localized by standard
means in a particular place and retained there for the time
required, continuously releasing an active enzyme into the
surrounding medium [4].

Given that it is sometimes necessary in thrombolytic
therapy to use drugs of both types either separately or simultaneously, we studied in detail the possibilities of creating
such drugs and we evaluated their use in experimental animals
and in the clinic. This study of fundamentally new approaches
to the treatment of cardiovascular diseases caused or complicated by thromboses or thromboembolism was carried out in two
experimentally independent directions.

THE FIRST DIRECTION

The first experimental direction was conditioned mainly by the use of angiography and catheterization, which permit localization of the drug effect as close to the pathological site as possible. For example, the high effectiveness of fibrinolysin intra-arterial injection was confirmed by experimental and clinical observations [5]. These studies demonstrated that the local effect on the site of thrombosis was promising. Lysis of a coronary artery thrombus in acute myocardial infarction was first performed in the U.S.S.R. in 1976 [6]. This direction, later termed endovascular (catheter) therapy, includes embolization of the vascular bed for limited perfusion of a certain organ (pulmonary hemorrhage, hemangiomas); injection of thrombolytics for pulmonary artery embolism; prolonged catheterization of the internal carotid artery in the treatment of brain tumors; and regional intra-arterial treatment of acute pancreatitis.

Such therapy does not permit a maximal drug effect although injection of the preparation is maximally close to the pathological site. Increase of the systemic concentration of the therapeutic agent is limited and, because of its protein nature, is associated with the organism's immune response and the rapid inactivation of the circulating preparation under physiological conditions.

Biosoluble, microspheric preparations of polysaccharides containing immobilized enzymes (chymotrypsin, streptokinase, urokinase) have been created to provide for regulated deposition of the therapeutic agent in organs or tissues without increasing the systemic concentration. These preparations can possess sizes adequate for the vascular bed (15-120 μm) and for the regulated time of a biosolution (from several hours to days and more) under physiological conditions.

Experimental investigations were based on the task established, i.e., the study and treatment of cardiovascular diseases. Several stages were involved in the development of a fundamentally new clinical approach based on the use of drug microsphere carriers [7]:

• Creation and in vitro study of fibrinolytic agents (fibrinolysin, streptokinase)

• Study of the behavior of microspheric polysaccharide carrier covalently bound to protein in vivo
  and experimentally.

• Demonstration of a curative effect of low doses of
  enzyme immobilized on modified Sephadex and injected in direct proximity to the thrombus.

• Evaluation of maximal sizes, amount, and technique
  for the preparation of microsphere suspensions for
  intracoronary injection.

• Objective assessment of changes in the myocardial
  microcirculation as a result of the deposition of
  Sephadex microspheres in an amount possessing an
  adequate curative effect for the ischemic area.

Creation and Study
of Fibrinolytic Agents

During the first stage of study, different types of biocompatible carriers were created based on cross-linked polysaccharides for enzyme immobilization. The rate of enzyme
release from such biodestructible carriers in the surrounding
medium was determined by the rate of destruction of the carrier itself under physiological conditions and by the enzyme as it
starts to act covalently bound to the carrier fragment. This
process should not only stabilize the carrier, but also reduce any undesirable biological reactions caused by it in the
organism. As a carrier we used Sephadex granules — a cross-
linked derivative of the biocompatible natural polysaccharide
dextran. After addition of carbonyl groups that are reactive to the protein into the carrier following perioxidant
oxidation, Sephadex granules acquire the ability to slowly
dissolve in water media, the rate of which depends on the degree of carrier modification; this can range from several
hours to several days, a result of the partial oxidative destruction of cross-linkages.

A number of therapeutic enzymes (fibrinolysin, streptokinase, urokinase, etc.) have been immobilized on such carriers. It is possible to bind from 10 to 80 mg of enzyme per
1 g of carrier. In the subsequent complete transformation from
states bound to a carrier into solution, practically all
of the enzymes manifested invariable biological activity to
both low-molecular-weight and high-molecular-weight substrates
and inhibitors; this was demonstrated by the invariability of the
corresponding constants, the Michaelis constant and the inhibition

catalytic constant. At the same time, for all of the enzymes immobilized on such carriers, we observed a sharp increase in stability in relation to autolysis and thermo- and pH-denaturation.

Immobilization principles and schemes developed on model enzymes appeared to be successfully applicable to enzymes for therapeutic use which could be subsequently obtained as microspheric preparations suitable for local deposition.

The main advantage of drug deposition in the vascular bed is the possibility of obtaining significantly higher concentrations of active preparations at the site of the pathological process, rather than systemic injection of substances that are not immobilized on a carrier.

Behavior of Microspheric Polysaccharide Carriers

Study of the behavior of microspheric polysaccharide carriers in a chronic experiment was the next stage. To control the site of deposition, we used preparations containing radioactive label. Human serum albumin labeled by ^{131}I was covalently bound to Sephadex. It has been shown previously that the dissolution rate of Sephadex containing bound protein does not depend either on the protein type or the presence of enzymatic activity in this protein.

Microspheres in the amount of 2-6 mg (2-6 million microgranules) and 20-40 μm in diameter were injected into the left coronary artery of a dog. The change in labeled Sephadex content was evaluated above the abdominal cavity and the heart by a radioactivity index. The ratio obtained compensates for the error brought about by circulation of the radioactive indicator in the blood bed.

Figure 5.1 presents data on the rates of ^{131}I-albumin concentration leveling in the heart and circulatory system by the injection of a preparation that is both native and covalently bound to Sephadex with different degrees of oxidation. The results show that regulated deposition of drugs is realistic. The time of protein retention in the injection site is determined by the rate of carrier dissolution, i.e., less oxidized Sephadex is deposited significantly longer in the myocardium than more oxidized and rapidly soluble Seph-

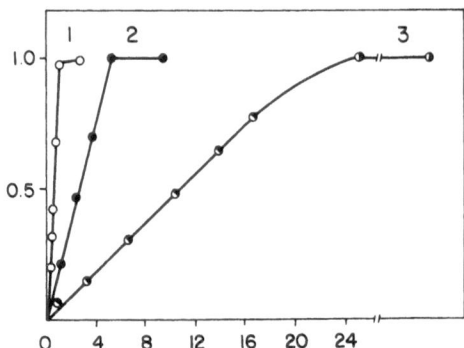

Fig. 5.1. Dependence of radioactivity index on time after
injection of the preparation. 1) Human native
serum albumin; 2, 3) serum albumin covalently
bound to aldehyde Sephadex (2 — high degree of
oxidation, 3 — low degree of oxidation). Abscissa is time (h), and ordinate is radioactivity index (RI), the ratio of radioactivity above the
heart region and above the abdominal cavity. This
ratio becomes constant (in our terms equal to 1)
after an even distribution of the labeled protein
in dog.

adex. According to our data, the dissolution rate of polysaccharide derivatives is somewhat higher in the organism
than in a chemical experiment; this difference is associated
with the action of dextrase in human and animal tissues.

Using the method of Torchilin et al., we obtained a
sample of modified aldehyde-containing Sephadex G-25 with a
granule size of 40-60 μm in the swollen state. The degree of
carrier modification provided for complete dissolution at 37°C
in phosphate buffer at pH 7.4 within 2.5 h. Fibrinolysin
(Koch-Light, England), 30 mg at 4°C, was mixed with a suspension of 100 mg of aldehyde Sephadex in 0.05 M phosphate buffer, and the reaction was continued for several hours. Non-
bound enzyme was successively washed off by ice phosphate
buffer, 0.001 N hydrochloric acid solution, 1 M sodium chloride solution, water and mixture of water and acetone, and the
preparation was then dried by acetone. According to spectrophotometric data on protein uptake at 280 nm, we managed to
bind up to 200 mg of enzyme per 1 g of carrier. Preservation
of enzyme activity was more than 85%, measured by TTT-1c pH-

state (Radiometer, Denmark) using a low-molecular-weight
substrate, ethyl ether—benzoylarginine.

Experiments were carried out on mongrel dogs weighing
10-15 kg. Using morphine-Barbamyl* anesthesia, the left and
right femoral arteries were isolated and shunted by a system
with a ball counter for measuring blood flow on a Mingograph-
34 apparatus. All of the branches adjacent to the artery
isolation were ligated, and blood flow was measured. For
animal heparinization, 1000 units of heparin per 1 kg of body
weight were used. The amount of thrombotic masses injected
into each artery (1-2 g) depended on the size of the artery.
Red thrombus was prepared from dog venous blood taken 1 h before the experiment. After embolization of the arteries by
an equal amount of thrombotic masses, contrasting was performed by 60% GIPAK† and the artery was ligated distally to
the expected thrombosis. Ten mg of preparation containing no
less than 1000 units of fibrinolysin were injected 30 min
after administration of the thrombus in the right femoral artery. After restoration of blood flow in this artery, a final
arteriography of both arteries and an x ray were performed.
Six experiments were performed.

Three experiments were conducted to evaluate the degree
of "damage" due to microembolization of the arterioles by undissolved carrier. Mongrel dogs weighing 10-15 kg and anesthetized by morphine-Barbamyl were used. The right femoral
artery was isolated and shunted by a system with a ball counter. The animals were heparinized and initial blood flow in
the femoral artery was determined. This blood flow was
achieved using a perfusion pump synchronized with the cardiac activity; the initial perfusion pressure was recorded.
Microspheres at consecutive doses of 5 to 10 mg per 1 kg of
posterior extremity weight were then injected into the artery
and perfusion pressure at the set blood flow was again recorded for 5 min. Perfusion was subsequently stopped and
blood flow was fixed by a ball counter.

In all of the experiments we observed complete lysis of
the thrombus with restoration of blood flow and normalization
of the angiographic picture in the treated artery. Figure

---

*Soviet trade name for a neuroleptic derivative of barbituric
 acid.
†UK trade name for contrasting agent used in angiography.

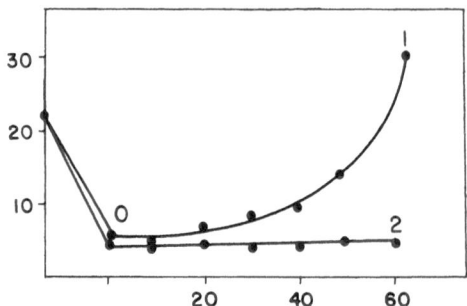

Fig. 5.2.   Curve of blood flow recording: 0) at the moment
of immobilized fibrinolysis injection; 1) in the
treated artery (right); 2) in the nontreated ar-
tery(left). Mean data of three measurements are
presented. Abscissa: time (min). Ordinate:
blood flow (ml/min).

5.2 presents the curve of blood flow during the experiment.
Blood flow was completely restored 1 h after injection of the
immobilized fibrinolysin, and was confirmed by angiographic
data. Artery permeability was accompanied by blood flow restoration and, thus, microspheres possessing lysing activity
were unable to significantly block the distal vascular bed.
It was also shown that these doses did not change the perfusion pressure or blood flow (Table 5.1).

The stability of these indices, together with blood flow
restoration, demonstrate the rapid compensation of microembol-
ized arterioles and the absence of a damaging effect. These
results suggest the following conclusions: (a) practically
all of the enzyme is concentrated at the site of thrombosis
and its circulation in the vascular bed is absent, (b) the
enzyme directly acts on substrate (thrombus), and inactivation
by natural inhibitors is sharply limited, (c) a "curative"
dose is 100-fold lower than the dose required for lysis of
the thrombus under similar conditions using systemic injection of native fibrinolysin, and, (d) the injection is single.

Creation of biosoluble microsphere preparations containing immobilized drugs (in particular, fibrinolysin) permits
the realization of an absolutely new approach in the treatment of thromboses, thromboembolism, and, possibly, ischemic
heart disease. Selective angiography provides for accurate
diagnosis of the thrombosis site and enables delivery of an
immobilized preparation to this site.

TABLE 5.1. Change in Blood Flow (ml/min) and Perfusion Pressure (mm Hg) in Intra-arterial Injection of Native Microspheres

Experiment No.	Perfusion pressure		Blood flow in femoral artery	
	initial	after injection of microspheres	initial	after injection of microspheres
1	150	120(78)	11	13(118)
2	170	130(76)	14	22(143)
3	100	76(75)	18	20(110)

Note: Parentheses indicate % of initial values.

Specific Effect of Immobilized Enzyme

The aim of the next stage was to evaluate the specific effect of deposited fibrinolysin immobilized on microsphere carriers on the blood fibrinolysis system in experiments [8].

According to Torchilin et al. [3], we obtained fibrinolysin immobilized on modified Sephadex with a carrier granule size of 20-40 μm and a time for complete biological dissolution under physiological conditions of no more than 3 h. Using this technique, we obtained preparations of immobilized enzyme with 10-80 mg of active protein per 1 g of carrier.

Nine mongrel dogs weighing 15-20 kg were anesthetized by morphine-nembutal. The femoral artery and vein were isolated, and distal and proximal vein sites and the abdominal aorta were catheterized. Ten mg of preparation containing no less than 150 units of fibrinolysin diluted in 3 ml of rheopolyglucin* were injected in the distal part of the femoral artery. Blood samples were taken from the aorta, superior vena cava (systemic blood flow), and femoral vein (blood flowing off the site where the preparation was deposited).

---

*Soviet trade name for dextran-40 in physiological saline solution.

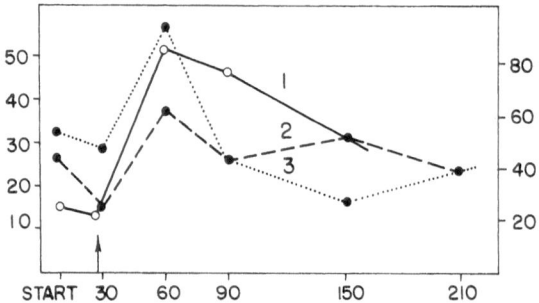

Fig. 5.3. Fibrinolytic activity of whole blood flowing from
the site of fibrinolysin deposition (1), in systemic blood flow (2), and plasmin activity in arterial blood (3). The arrow shows the time of injection of the preparation. Mean data of three
measurements are presented; scattering is no more
than 10%. Abscissa: time (min). Ordinate:
(left) plasmin activity (mm^2), (right) fibrinolytic activity (%). (Reproduced from [8].)

    The next step was to study local and systemic fibrinoly-
sis under conditions of immobilized fibrinolysin deposition
on an artificially formed thrombus of the femoral artery.
Thrombosis of the femoral artery was induced as follows: Part
of the artery 1.5-2 cm in length was ligated distally and
proximally, having previously implanted a fixator (a slow
spring of a corresponding cross section with 6-10 coils).
Then, 0.1-0.2 ml of a freshly prepared solution of thrombin
was injected into this vessel part; 10 min later the proximal
ligature was removed while the distal ligature was retained
for 1 h. Complete thrombotic occlusion was controlled by an
ultrasonic counter (NASA, USA). Ten mg of immobilized fibrinolysin was then injected directly into the thrombus site.

    The final part of the experiment consisted of the follow-
ing. The animals were anesthetized, the femoral artery and
vein were isolated, and the left coronary artery and venous
coronary sinus were catheterized selectively under the control of an electron-optic transformer and contrasting (Vero-
graphin* 76%). Intracoronary injection of 2-4 mg of biosoluble microspheres with fibrinolysin in 2-3 ml of Rheopoly-

*Yugoslavian trade name for x-ray contrasting agent.

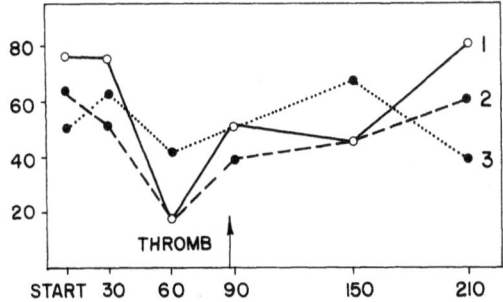

Fig. 5.4. Injection of immobilized fibrinolysin in experi-
           ments with artificial thrombosis of the femoral
           artery. Changes are recorded in the fibrinolytic
           activity of blood flowing from (1) and flowing in-
           to (2) the site of thrombosis (thromb) and in the
           systemic blood flow (3). Abscissa: time (min).
           Ordinate: fibrinolytic activity (%).

glucin was performed (before injection, the suspension was
treated with ultrasound to exclude microparticle aggregation).
Blood samples were taken from the venous sinus (blood flowing
off of the fibrinolysin deposition site), artery, and superior
vena cava (systemic blood flow) at the following intervals:
before and 15, 30, 60, 120, and 180 min after injection of immobilized fibrinolysin.

Fibrinolytic activity, plasmin and plasminogen activator
activity, and antiplasmin inhibitory activity were determined
in studying the fibrinolysis system in the blood.

Injection of immobilized fibrinolysin in the femoral artery of a dog causes local enhancement of plasmin activity in
the artery and an increase in fibrinolytic activity in the
outflowing blood and systemic blood flow. These indices later
decrease due to the compensatory reaction in a healthy dog
(Fig. 5.3).

In experiments with artificial thrombosis of the femoral
artery, the fibrinolytic activity of whole blood sharply decreases locally against the background of thrombosis and increases after injection of immobilized fibrinolysin; fibrinolytic changes are less marked in systemic blood flow (Fig.
5.4). The data obtained show that this method of local deposition of immobilized fibrinolysin provides a higher degree

of required alteration of fibrinolytic activity at the thrombosis site. This finding supports the clinical significance
of this method for injecting thrombolytic preparations.

After intracoronary deposition of immobilized fibrinolysin we observed an enhancement in fibrinolytic activity of the
whole blood (by 16% locally, and by 10.5% in systemic blood
flow), with a subsequent decrease both locally and in systemic
blood flow. We also found a reduction in the arteriovenous
difference of fibrinolytic activity of whole blood from −40%
to −8.4% after injection of the preparation, and a more marked
decrease in fibrinogen concentration in the artery than in
the vein (by 212% and 160 mg %, respectively).

Measurement of antiplasmin inhibitory activity demonstrated that the initial high content of antiplasmins can inhibit enzyme transferred from the carrier to the blood flow;
in this case, enhancement of fibrinolytic activity probably
occurs with an adequate dose of injected fibrinolysin and the
release of a considerable amount of tissue activator in the
blood bed. Reciprocal relations of plasminogen and plasmin
activator are revealed against the background of low antiplasmin activator in the blood bed (Fig. 5.5). "Differences" recorded at the 180th min of the experiment indicate that
changes caused by the injection of immobilized fibrinolysin
are directed toward an increase in blood fibrinolytic potential. Even under conditions of plasmin activity inhibition
associated with a compensatory reaction of the organism, the
activity of plasminogen activator is elevated sharply.

Injection of immobilized fibrinolysin causes a more
marked increase of inactivator activity than in plasmin activity. A single and double increase in plasmin activity is observed in the systemic blood flow. Similar results were obtained with intracoronary perfusion of fibrinolysin with heparin in patients with acute myocardial infarction. This can
be accounted for by the binding of fibrinolysis inhibitors by
injected fibrinolysin which results in the manifestation of
activity of the plasminogen endogenic activator and/or the
contamination of the fibrinolysin preparation by a trypsin
admixture added for plasminogen activation in the process of
fibrinolysin production.

The dependence of activator activity on its initial level
should be mentioned: With zero activity of the plasminogen
activator, the injection of immobilized fibrinolysin causes a

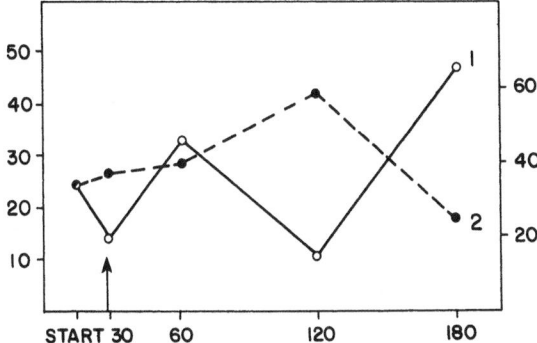

Fig. 5.5.   Reciprocal relations of the plasminogen activator
(1) and plasmin (2) activity after intracoronary
deposition of immobilized fibrinolysin. Abscissa:
time (min). Ordinate: (right) plasminogen activator (mm^2), (left) plasmin activity (mm^2). (Reproduced from [8].)

sharp increase in activity; with some initial activity, injection first leads to a reduction and then to an increase in
activator activity.

     In summary, the injection of fibrinolysin immobilized on
modified Sephadex provides for a marked increase in fibrinolysis in whole blood at the site where the preparation was
deposited due to the injected enzyme and the enhancement of
plasminogen activator activity. The results obtained suggest
that this method of applying thrombolytic preparations has
therapeutic significance.

Maximal Sizes, Amount, and Technique
for Preparation of Microspheres

     From the data presented above it is clear that the pro-
posed principle of drug deposition is associated, in some
cases, with microembolization of the vascular bed. Given a
possible damaging effect, it is very important to assess the
maximal sizes and the amount of microparticles that can be
used for injection into any vascular bed and, in particular,
the intracoronary bed. The potential for additional damage
to the pathological process (ischemic heart disease, acute
myocardial infarction) is a significant contraindication to
using this method.

Myocardial damage was appraised on the basis of ECG data, cardiospecific enzyme activities (including total creatine phosphokinase, MB-isoenzyme, and hypoxanthine content in plasma), and results of histological and histochemical studies. Dogs were given an intracoronary injection of 0.02-0.1 mg of microspheres 40-50 μm in size with a complete biosolution time of 2.5-3 h. The preparation was diluted in 3-5 ml of rheopolyglucin sterile solution. The ECG was recorded in the standard leads before injection of the microspheres, at the moment of injection, every hour for 10 h, after 24 and 48 h, and, in four cases, 2 weeks after the beginning of the study. The criteria of damage accepted for the dogs were used for analysis of the ECG. Histological study was performed on myocardial samples cut from different parts of the heart: the anterior and posterior walls of the left and right ventricles at two levels, and the left and right atria. The activity of enzymes that were most sensitive to early ischemic lesions — malate dehydrogenase and diphosphatase — was detected in histochemical study. The animals were subjected to euthanasia at 1 h (2 dogs), 24 h (3 dogs), 48 h (10 dogs), and 2 weeks (4 dogs) after intracoronary injection of the microspheres; two dogs served as controls. Blood was taken from a vein every hour for 8-10 h to determine creatine phosphokinase activity. Quantitative determination of hypoxanthine in the plasma was performed by high-pressure liquid chromatography using the Varian apparatus (USA).

Following the above methods for verifying myocardial necrosis, we found that the deposition of 0.02 mg of Sephadex (20,000 particles per 1 g of myocardium) did not induce damage of the intact heart. Morphological peculiarities of the myocardial microvascular system most likely provide adequate redistribution of blood flow, thereby compensating for the embolized arterioles.

Assessment of Changes
in Myocardial Microcirculation

The previous stage was logically continued by the experimental assessment of the diffusive ability of the cardiac capillary bed in ischemia and its alteration as a result of the deposition of biosoluble microspheres in the coronary bed [9]. The exchange transport function is realized at the capillary level. It was necessary to define the amount of microparticles by which embolization of the myocardial arteriolar bed

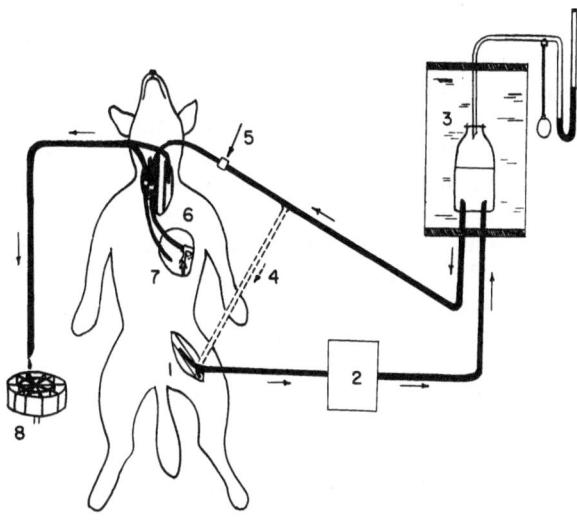

Fig. 5.6.   Diagram of the experiment on controlled coronary
            autoperfusion (arrows show direction of blood
            flow): 1) femoral artery; 2) perfusion pump; 3)
            thermostatic vessel with manometer; 4) shunting
            magistral for switching blood flow in perfusion of
            the left coronary artery; 5) counter; 6) left cor-
            onary artery; 7) venous coronary sinus; 8) collec-
            tor for samples of outflowing blood. (Reproduced
            from [8].)

did not cause additional exclusion of some part of the myocardial capillary surface from active blood flow.

    An objective assessment of the state of tissue or organ
microcirculation can be accomplished using the PS index (permeability multiplied by capillary surface area) determined by
the double-indicator method. Indicator extraction by the myocardial capillary bed was determined by the difference in concentration of the indicator penetrating and nonpenetrating
through the capillary wall in blood flowing from the heart.
Using corresponding formulas, the PS index was calculated from
the blood flow value, reflecting in stable permeability the
amount of perfused capillaries and distribution of blood micro-
flow in the myocardium.

    A diagram of the experiment is presented in Figure 5.6.
Adequate isolated perfusion of the left coronary artery was

achieved in closed-chest dogs. Outflow from the venous coronary sinus at the rate of one sample per 1 sec was collected
in a special collector. Myocardial ischemia was induced by
intravenous injection of Izuprel*.at the rate of 4.4-11.3
µg/min for 9-10 min under conditions of an unchanged blood
flow. Microspheres (2-6.6 mg per 100 g of myocardium) were
infused against the background of ischemia. Electrocardiographic signs of ischemia were manifested by ST-segment elevation by 4-7 mm and a decrease in the R-wave amplitude in
the second standard lead. As known, Izuprel promotes an increase in the myocardial demand for oxygen and exerts a
direct cardiotoxic effect. However, the dose which is directly damaging to myocardial cells is about twofold higher than
the dose we used and, therefore, ischemic changes were due
to tachycardia and the peripheral coronary vasodilatory action of the preparation. In all of the experiments the PS
index decreased on an average by 23.9% (p < 0.001). In experiments with intracoronary deposition, the decrease in the
PS index averaged 19% of the initial level (p < 0.05). There
was no significance in the difference between the PS index
in ischemia with and without deposition of Sephadex microspheres.

The decrease in the PS index at first seems unexpected
since Izuprel promotes vasodilation of the large and terminal
arterioles and, therefore, the diffusive ability of the capillary bed should increase. There are no data on the PS index
in myocardial ischemia in the literature. However, it is
known that active vasomotion provides for adequate distribution of blood flow in the organ according to local demands.
The effectiveness of regulation depends on the functional
state of the microvascular system. Ischemia and, therefore,
distal vasoplegia, exclude adequate vasomotion and optimal
distribution of blood flow. Arterioles and capillaries of
ischemic tissue are transformed into passive pathways of
blood flow, which is distributed according to the perfusion
pressure gradient. Blood flow naturally chooses the shortest
way anatomically and its larger part passes through a smaller
amount of capillaries. This phenomenon can be considered as
functional shunting of the ischemic area. By comparing the
central hemodynamic indices, contractility, degree of ST-segment elevation, and PS index, one can conclude that intracoronary injection of an adequate amount of carrier particles

---

*UK trade name for isopropylnoradrenaline.

of corresponding diameter does not aggravate changes in these indices against the background of ischemia. This finding is associated with the possible rapid switching of blood flow from embolized magistrals to others, since the amounts of the preparation used did not exhaust the reserves of functional shunting. The possibility that some of the microspheres penetrated the area without active blood flow should not be excluded completely.

Thus, in myocardial ischemia, one can observe irregular microcirculation in the corresponding area, which predetermines the possibility of carrier deposition with an active drug. Based on the analysis of the results of experimental studies in the first direction, use of the new principle of regulated deposition of drugs within an organism can be considered clinically realistic.

THE SECOND DIRECTION

The second experimental direction dealt with the development of water-soluble preparations of immobilized enzymes intended for prolonged retention in the circulatory system. In principle, water-soluble, stabilized enzymatic preparations can be obtained without using a carrier or binding them to water-soluble carriers. Enzyme stabilization against denaturing effects without using a carrier is realized by both the chemical modification and creation of intramolecular cross-linkages. This effect can also be achieved by enzyme auto-stabilization in injections of potentially reversible cross-linkages of different length into the enzyme molecule [10].

The method for creating water-soluble, stabilized protein preparations by their modification with soluble polymers has been developed in more detail. Among the great number of such polymers, the most preferable as carriers are the polysaccharides, and particularly dextrans because of their high degree of biocompatibility and utilization in the organism. Some nontoxic and nonimmune synthetic polymers are also suitable for immobilization of therapeutic enzymes (for example, polyvinylpyridine and polyvinylpyrrolidine derivatives, polyethyleneglycol, polyvinyl alcohol, polymers with sulfoxyl groups, polymer hydrophilic esters and substituted methacrylamides or acrylamides, and some polypeptides).

Undoubtedly, it is even more preferable to use carrier
substances peculiar to the organism itself (but not possessing antigenic properties) or natural compounds that are able
to enhance the therapeutic action of the enzyme bound to them
or possessing their own useful biological activity. We suggest the use of heparin for immobilization of proteolytic
and, especially, thrombolytic enzymes for therapy, usually
conducted against the background of general heparinization.

Protein binding to soluble carriers is mainly realized
by conventional methods of immobilization.

Modification of physiologically active substances of a
protein nature by water-soluble polymers leads to an increase
in their conformational stability and their stability to protease action. This is achieved by multipoint binding to the
carrier and modification by volumetric substitutes, creating
steric obstacles for the hydrolysis of lysine residues that
are sensitive to digestion by trypsinlike proteases.

An increase in circulating time of the preparation in
the blood as a result of a decrease in its filtration rate
through the kidneys is another important consequence of modification. This effect is conditioned by an increase in the
molecular mass of the preparation and prevention of its interaction with surface receptors of cells participating in the
removal of protein from the circulation. This also leads to
a reduction in the immunologic and allergic reaction of the
organism to the injection of enzyme forms modified by polymers. This effect is associated with a decreased ability of
modified enzymes, as compared to native forms, to stimulate
antibody formation and to react with them due to serious steric obstacles for specific antigen—antibody interaction.

Using the carriers developed and binding methods, we obtained a whole set of enzymes immobilized on different carriers, protein inhibitors, and hormones, and a number of other
compounds that can be transformed into effective drugs after
biomedical and pharmacological tests.

Streptokinase was chosen as the first preparation suitable for medical application. This choice was conditioned by
several factors: (a) it is expedient to create thrombolytic
preparations given the prevalence and danger of thrombotic
diseases and complications, and the presence of natural processes and thrombolysis enzymes; (b) among thrombolytic en-

zymes such as fibrinolysin, urokinase, and streptokinase, only the latter is widely used because of its relatively low price; and (c) drawbacks in enzyme therapy have been seen most clearly with streptokinase (these include various side effects, difficulties in application, and, thus, the limited possibilities for using this type of therapy, and marked immune reactions when streptokinase was used as a microbic enzyme).

Streptokinase immobilization of activated, partially oxidized dextran enabled us to remove these drawbacks to a large degree. The enzyme became more stable and was maintained longer in the blood; it caused less toxic and allergic reactions, and its antigenicity was reduced by about 30-fold. Based on these data, a new technology has been developed for factory production to obtain a new preparation, including stages of carrier activation, enzyme immobilization, consolidation of enzyme-carrier link and removal of excess carrier reaction groups, product purification, sterile filtration, bottling in vials, and lyophilization. The new preparation, named "streptodecase," has been successfully tested clinically, and the U.S.S.R. Pharmacological Committee has given permission for its clinical use in the treatment of thrombotic diseases.

CLINICAL STUDY OF STREPTODECASE

What has been shown by the clinical studies to elucidate the therapeutic and biochemical action of streptodecase?

Effective, simple, and safe thrombolytic therapy is ideally necessary for clinicians. Despite initial enthusiasm and optimism, the introduction of fibrinolytics of enzymatic origin (streptase, urokinase, fibrinolysin) in practical medicine has not adequately solved the problem. Why? Clinical results indicate that a positive specific effect (lysis of a thrombus) is very often accompanied by complications (up to 80% hemorrhages, 30-60% recurrent thromboses and embolism due to thrombi fragmentation, and 30% allergic reactions) [11]. All of these complications are caused by a sharp dissociation of the hemostasis system under the effect of these preparations, leading to unsatisfactory final results.

Yet, the incidence of thromboses and embolism continues to increase throughout the world, thus stimulating the search for new approaches to treatment and for thrombolytic drugs that are optimal for clinical use. Newly discovered properties of immobilized streptokinase allow us to hope for optimized thrombolytic effects under clinical conditions.

We have analyzed the results of treating 150 patients according to the following: methods of application, clinical effectiveness, effect on hemostasis, and complications and tolerance to streptodecase. Streptodecase was compared with the known properties of a native enzyme — a "streptase" preparation. Streptodecase therapy was conducted in patients with acute massive thromboembolism of the pulmonary artery, thromboses of peripheral vessels, acute myocardial infarction, intraocular hemorrhages of different etiology, and thrombosis of the central vein of the retina and its branches [12-16].

Method

One vial of the preparation containing 1 or 1.5 million units of streptodecase was diluted in 10 ml of physiological solution, and 300,000 units (2-3 ml) were injected intravenously in a stream. One hour later, after biological testing, the remaining 2.7 million units, diluted in 18-28 ml of physiological solution (sometimes in 50 ml), were injected intravenously over 3-5 min. The total therapeutic dose, except in patients with hemophthalmia, was 3 million units (and sometimes 6 million units). Heparin and antiaggregating agents were obligatory 6-24 h after injection of the main dose (several patients did not receive heparin and antiaggregating agents, however). For treatment of hemophthalmia, streptodecase (30,000-45,000 units) was injected retrobulbarly, subconjunctivally, or intravitreously (0.2-0.3 ml preparation, one to four injections). We did not study the state of the hemocoagulation system in these patients, since local injection of such "small" doses could not affect systemic hemostasis.

Results

Method of Application. A single intravenous injection of the entire therapeutic dose significantly simplifies therapy and enables its use, even in emergency service, which will improve the final results of treatment.

Tolerance. It is known that streptase causes allergic, pyrogenic, and toxic reactions in about 20-30% of the cases, thus limiting the possibilities for its use. Streptodecase possesses a lower immunogenicity by severalfold, due to the alteration of the protein molecule conformation by the polysaccharide matrix. Only in 6% of the cases did injection of streptodecase significantly cause hyperthermia, which was eliminated by a single dose of analgesics. It should be noted, however, that these patients were primarily those with acute pathological and initial resorptive fever, which complicated verification of the hyperthermia; one patient had a pain syndrome in the loin (lumbago) which was eliminated by narcotics. Changes in rhythm and blood pressure were of short duration and were clinically insignificant.

Complications of Therapy. The actual complications with streptodecase (several cases of small subcutaneous hematomas at the site of the puncture of the large veins and arteries, one case of thromboemboli fragmentation) differ considerably from the characteristic complications observed in treatment with streptase (70-80% have hemorrhages and 8-12% have rethromboses).

These findings testify to the safety of the preparation and allow us to expect wide application of streptodecase in thrombolytic therapy.

Effect on Hemocoagulation System. Figure 5.7 presents the dynamics of activator and plasmin activity (fibrinolysis) and partial thromboplastin time (PTT) of coagulation against a background of continuous intravenous drop injection of streptase to patients with acute myocardial infarction. This figure shows that sharp suppression of coagulation (PTT was practically undetermined in the first hours) with high fibrinolytic activity of the blood causes a hemorrhagic syndrome. Cases of hemorrhage often induced the withdrawal of streptase injections. By the end of the second day, fibrinolytic activity dropped and PTT decreased sharply (hypercoagulation). These changes, together with other characteristic shifts in hemostasis (high content of fibrinogen degradation products and soluble fibrin, acid-labile inhibitors of fibrinolysis and antiplasmin), caused a predisposition to rethrombosis. A sharp drop in the coagulating link on the first and second day of streptase treatment practically excluded heparin therapy, which could also promote the development of hypercoagulation.

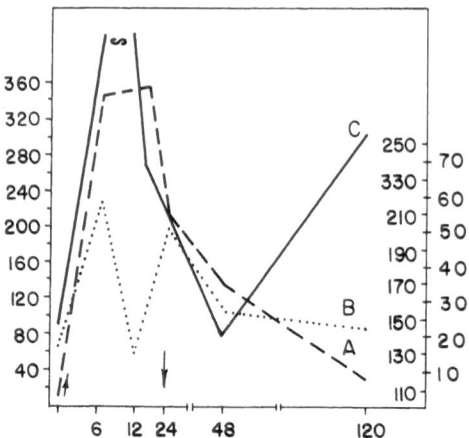

Fig. 5.7.   Dynamics of plasminogen activator (A), plasmin (B),
and PTT (C) activity in treatment with streptase
of patients with acute myocardial infarction. Abscissa: time (h). Ordinate: A, B, mm^2; C, sec.
(Reproduced from [12].)

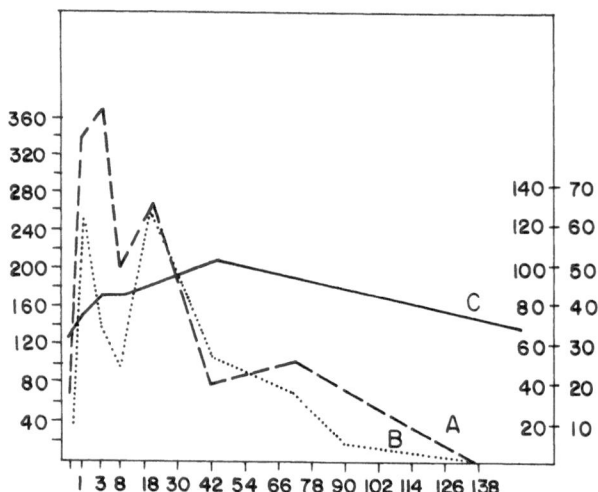

Fig. 5.8.   Plasminogen activator content (A), plasmin activity (B), and PTT dynamics (C) in treatment with
streptodecase of patients with thromboses. Abscissa: time (h). Ordinate: A, B, mm^2; C, sec.
(Reproduced from [12].)

Hemocoagulational shifts in patients after a single injection of the entire therapeutic dose of streptodecase differ quantitatively (activation and duration of fibrinolysis);
these may be associated with individual variations of the
major disease or the initial state of hemostasis. Nevertheless, the following findings were distinguished for all patients: a rather significant and prolonged (3-10 days) increase in fibrinolytic activity and minimal influence on the
coagulating link of the hemocoagulation system. The increase
in fibrinolytic activity in the blood was not accompanied by
a "streptase-like" suppression of coagulability (Fig. 5.8),
thus substantiating the therapeutic safety of thrombolytic
therapy using the new preparation. After injection of streptodecase we also found a decrease in the acid-labile inhibitors of fibrinolysis (in some cases we observed an inversion
of inhibition and fibrinolysis activation), a decrease or insignificant increase in antiplasmin activity compared to
streptase, and a less prolonged and significant increase in
fibrin degradation products.

In enhanced fibrinolysis, especially on the first day
after injection of streptodecase, one observes degradation
of fibrin, fibrinogen, and fibrin complex compounds, which is
accompanied by a reduction in residual fibrinogen by 50% and
soluble fibrin by 40%, and by an increase in fibrin degradation products. Residual fibrinogen is a protein found in the
plasma sample after incubation for determination of fibrinolytic activity. Changes in soluble fibrin content, detected
by protamine—sulfate precipitation, with subsequent quantitative determination of fibrin were unidirectional with the
dynamics of residual fibrinogen concentration. Figure 5.9
illustrates the dynamics of the indices in patients not receiving heparin. After the significant decrease on the first
day, soluble fibrin concentration was still above the initial level. Soluble fibrin represents fibrin-monomer complexes with fibrinogen and early products of their degradation, which are not released by the kidneys and can turn into
insoluble fibrin, complicating the development of thrombosis.
These data, and the need for antiplasmin inactivation, condition the use of streptodecase in combination with heparin;
the nature of the changes in the coagulating link of hemostasis permits such application.

Against the background of streptodecase action, we observed a rapid decrease in the fibrinogen level from the initially high level (1.5- to 2.0-fold higher than normal) to an

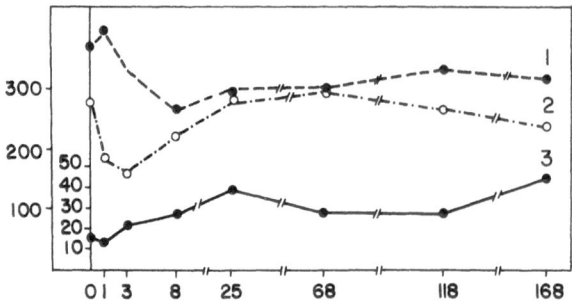

Fig. 5.9.   Concentration of total and residual fibrinogen and
            soluble fibrin against the background of a strepto-
            decase effect in patients with thromboses of the
            pulmonary and peripheral arteries: 1) total fib-
            rinogen; 2) residual fibrinogen; 3) soluble fib-
            rin. Abscissa: time (h). Ordinate: fibrinogen
            (mg %); soluble fibrin (mg %).

almost normal level (not lower than 16-30%), with a gradual
increase to the initial concentration on the third to tenth
day. A maximal relative "crisis" (not higher than 10% of the
value before treatment) was recorded on the fifth day after
injection of streptodecase only in patients with acute myocardial infarction. Enhancement of fibrinolysis was accompanied by a decrease in plasminogen level (in some cases to
zero) which soon (12 h later) increased to the initial value.

     In the first hours we observed some decrease in the
amount of platelets (no more than by 35% of the initial value)
but not lower than the normal indices. The aggregation rate
dropped and then increased by the end of the first day, but
the disaggregation process was sufficient and its subsequent
regulation by antiaggregating agents was adequate.

     As for the prolonged action of the new thrombolytic agent,
special attention should be given to its effect on hemostasis
in patients with chronic thromboses. In these patients the
specificity of the preparation can be detected more reliably
since the activating factors of acute thrombosis are mostly
excluded. The dynamics of fibrinogen concentration are two-
phase (on day 1 and on days 4-6, without a significant drop
and recurrent "crisis"), and the second phase is more prolonged (Fig. 5.10). Fibrin degradation products were deter-

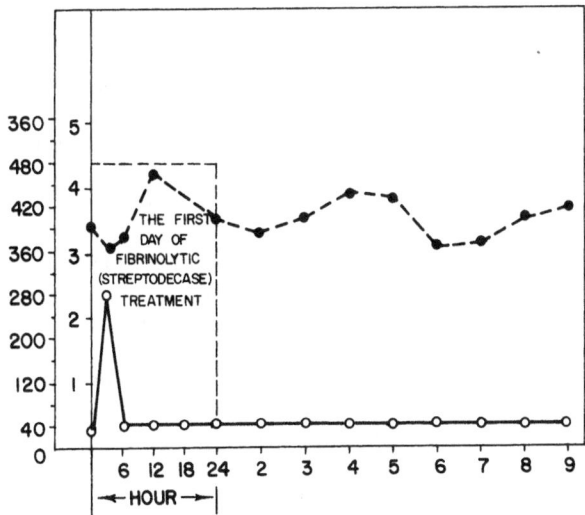

Fig. 5.10.   Dynamics of concentration of fibrinogen and fib-
            rin degradation products in patients with chronic
            thromboses of the deep veins of the lower extrem-
            ities, thrombotic arteriopathy, and ischemic
            cardiopathy after a single injection of 3 million
            units of streptodecase. Abscissa: time, 6-24 h;
            days 2-9. Ordinate: PDF (mg %); fibrinogen (g/
            liter).

mined only in the first 6 h after injection of streptodecase.
Such two-phase dynamics coinciding in time are observed in
indices of the leukocytic formula and protein fractions, which
are also characteristic of native enzyme (i.e., only the first
phase).

     The demonstrated hemocoagulation effects suggest the fol-
lowing mechanism of action of streptodecase. On the first
day, the effect is due to the molecule itself (a part of the
free, i.e., not immobilized by dextran, streptase terminants
are specific activators of fibrinolysis) and to a small amount
of free streptase (1% by technical conditions). Evolution of
fibrinogen, fibrin degradation products, leukocytes, antiplasmins, plasmin activator, and plasmin in the first hours
is conditioned by free streptase, possibly to a larger degree
than by the streptodecase molecule. Penetration of streptase
inside the thrombus (endothrombolysis) is not excluded during
this time. Between days 1 and 4 (of course, there can be in-

dividual variants), an equimolar amount of active "streptase"
terminants on the molecule surface leads to the formation of
a plasminogen proactivator complex and catalyzes its direct
transformation to plasmin. Quite likely, plasmin circulates
in a bound state in this case, and to a larger degree not
with physiological inhibitors of fibrinolysis but with a proactivator complex or with active, but streptase-free, centers
of modified dextran. This can account for the following findings: absence of a sharp fibrinogenolytic effect, absence
or insignificant increase in degradation products of the fibrinogen—fibrin complex, minimal effect of plasmin as an effective free substance suppressing fibrin polymerization, transformation of factors I, V, and VIII, and an antithrombin effect. This is, of course, a "pure" variant of the action of
streptodecase, which is not possible in acute thrombotic pathology since there is endogenic stimulation of fibrinolysis.

In direct proximity to the thrombosis the streptodecase
complex dissociates from plasmin (we observed an increase in
macroglobulins over several days) and the separated plasmin
dissolves fibrin located on the surface. The "behavior" of
inhibitors, especially the inversion of acid-labile inhibitors
of plasmin and its activators, which we have observed, is almost completely contrary to that observed with native enzyme.
It seems likely that the immobilization of native enzyme promotes plasmin binding by nonreacted dextran active groups or
a conformationally changed protein molecule is recognized by
antibodies to a smaller extent.

These lysis processes occur mainly on the thrombus surface although the streptase "classic" effect is also superim-
posed during this period due to gradual biodegradation of the
dextran molecule under the effect of endogenic dextranases
and oxidases. As a result, endothrombolysis occurs, with the
inactivation by antiplasmins being minimal due to the described
mechanism. This is favored by prolonged activity, fibrinolytic activity, and probably small amounts of degradation
products of the fibrinogen/fibrin complex.

Between days 4 and 6 (variants are possible) biodegradation acquires a relatively avalanchelike nature (many polysaccharide bonds are destroyed and it crumbles). The molecule fragments formed have a molecular weight sufficient to
penetrate inside the thrombus, which can continue the thrombolytic process against the background of blood that is "ex-

hausted" of its inhibitory properties. It is also possible
that small pieces of dextran biodegradate are in the thrombus, releasing streptase "undamaged" active centers. Plasminogen molecules transform into plasmin on the spot due to the
formation of activators in the thrombus which are adsorbed
in fibrin fibers. Thus, the streptodecase molecule functions
to prepare the way for intensive and rapid lysis, especially
of fresh thrombi since they have little or no antithrombin.

The repeated changes in the dynamics of all of the hemocoagulation indices and the leukocytic formula, which are
characteristic of native enzyme during intravenous drop injection, are likely a result of the avalanche of streptodecase molecule biodegradation.

All of these findings bring streptodecase closer to an
"ideal" variant of a selective thrombolytic agent that can
provide for lysis without the side effects on hemostasis. It
should be emphasized that the above description is only probable and should be further verified in biochemical studies,
both experimental and clinical.

EFFECT OF STREPTODECASE ON THE CLINICAL COURSE OF VARIOUS DISEASES AND VERIFICATION OF THE THROMBOLYTIC EFFECT

Massive Pulmonary Embolism

In the past decade, fibrinolytic therapy has been firmly
placed in the treatment of pulmonary artery thromboembolism
(PATE). Activating types of preparations are most widely used
among the different drugs available for therapeutic thrombolysis. Standard high doses of lytic agents are usually used.
These do not always provide for sufficiently rapid elimination of the embolic occlusion and often cause dangerous hemorrhagic and allergic complications. In order to achieve a curative effect, it is necessary to use continuous drop infusion
of available drugs (sometimes for several days), which significantly complicates the conduct of thrombolytic therapy.

In connection with these disadvantages, the improvement
of methods for treating thromboses and embolism using fibrinolytics has taken three main directions: (a) increase of
thrombolysis effectiveness, (b) reduction of hemorrhagic and
allergic complications, and (c) simplification of treatment

technique. Until recently, however, none of the available
methods has met these requirements — high effectiveness, safety, and simplicity of application.

We present below the results of streptodecase treatment
in nine patients with massive PATE, carried out at the N. I.
Pirogov surgical clinic under the supervision of Academician
V. S. Saveliev (the physician was V. N. Iljin). Thrombolytic
therapy was conducted in four men and five women aged 21-81
years with a duration of disease from 3 h to 9 days from the
moment of embolization to the beginning of treatment. All
patients had multiple nonocclusive lesions in the pulmonary
artery system with localization of emboli in branches of
varying diameter. Three patients had embolism of the right
main pulmonary artery, three patients had embolism of the
left main pulmonary artery, two patients had emboli in both
main pulmonary arteries, and one patient had emboli in all of
the lobar branches of the pulmonary artery. In five patients
embolization was caused by flotating thrombi in the iliac
venous segments, in two patients by emboli in the inferior
vena cava, and in two patients the source of embolization was
not found.

In seven cases, an intravenous filter was implanted in
the inferior vena cava immediately before streptodecase injection in order to prevent recurrent pulmonary embolism.
Surgical intervention was not performed in two patients who
had occlusive thrombosis of the deep veins of the lower extremity and iliac venous segment.

The preparation was injected according to the above method (3 million units were injected into seven patients, and 6
million units were injected into two patients because of massive thromboembolism). The patients also received heparin,
rheopolyglucin, trental, nicotinic acid, and phenylin.

The results were evaluated on the basis of repeated scan-
ographic and angiographic studies and measurement of pulmonary artery pressure. Lung scanning was performed when the
patients were admitted to the hospital and, then, at 1, 2,
and 4 weeks, at 3-6 months, and at 1 year after the first
examination. Localization, perfusion deficit, and degree of
radioactivity reduction were determined in the scanogram
analysis.

TABLE 5.2

A. Changes in the Miller Index with Streptodecase Therapy

Group		Miller index (M ± m)			
		total		perfusion	
		before treatment	after treatment	before treatment	after treatment
I	7-48 h	23.6±1.7	9.012.0	9.8±0.9	3.0±0.8
II	6-9 days	22.5±1.7	16.7±2.7	8.0±0.8	6.7±0.7

B. Dynamics of Perfusion Deficit with Thrombolytic Therapy (M ± m)

Before treatment	After beginning of treatment				
	1 week	2 weeks	1 month	3-6 months	1 year
40.9±2.2	24.7±3.1	17.5±5.2	16.0±5.3	4.7±3.1	4.2±3.0

Angiography of the pulmonary artery was performed before the injection and at 1, 2, and 4 weeks after the injection. The degree of lesion in the lung vascular bed was determined on the angiogram by calculating the total and perfusion Miller indexes.

Blood samples for analysis of hemostasis were taken from the catheterized pulmonary artery before and at 1, 3, 5, and 10 days after the beginning of treatment. A thromboelasto-gram (TEG) and determination of fibrinogen concentration were carried out every 3 h for the 4 days after streptodecase injection.

The study results showed that the use of streptodecase in patients with PATE provided for prolonged activation (14 days) of fibrinolysis with a simultaneous absence of hyper-coagulation and hyperaggregation of thrombolytes. Table 5.2 shows the dynamics of the quantitative indices of thrombosis according to antiography and lung-scanning data.

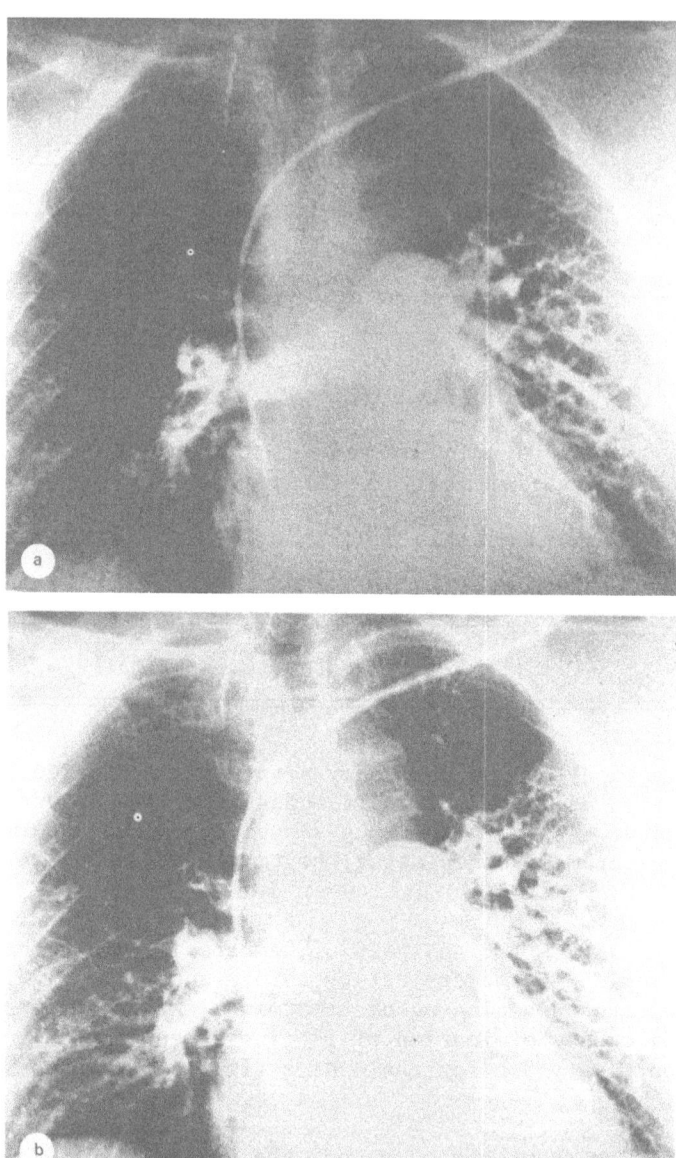

Fig. 5.11.   Angiograms of a patient with massive thromboembolism of the pulmonary artery treated with streptodecase: a) before injection; b) on the sixth day
after a single intravenous injection of 3 million
units of streptodecase (see text).

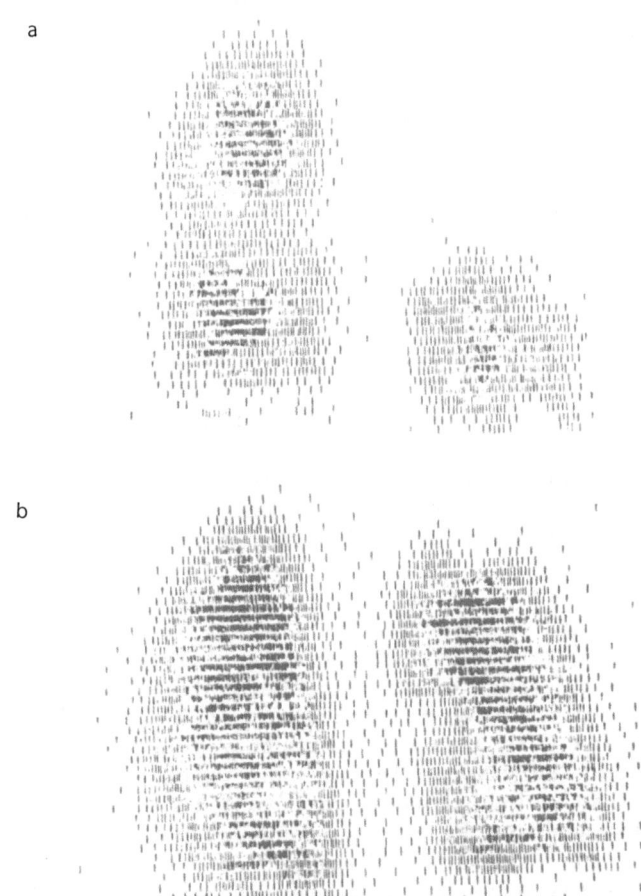

Fig. 5.12. Scanograms of a patient with massive thromboembolism of the pulmonary artery (posterior projection): a) before injection (perfusion deficit about 47%); b) two weeks after injection of
6 million units of streptodecase (see text).

Based on these data, one can conclude that streptodecase not
only possesses its own lytic action on pulmonary emboli, but
also stimulates spontaneous thrombolysis, as confirmed by
prolonged enhancement of blood fibrinolytic activity and by
progressive decreases in the Miller index and the perfusion
deficit over the long term. The effectiveness of streptodecase treatment was also shown to depend on the length of time

between the onset of disease and treatment with streptodecase
(see groups I and II in Table 5.2).

Figure 5.11a, b presents the angiograms of a patient with
nonocclusive thromboembolism of the right main pulmonary artery and left inferior-lobar pulmonary artery against the background of streptodecase treatment. The control angiogram
testifies to complete restoration of the permeability of the
left branches of the pulmonary artery and lysis of the thrombus in the trunk of the right pulmonary artery. Only the right
mid-lobar branch remained occlusive.

The next figure (Fig. 5.12a, b) shows lung scanograms
(posterior projection) of the patient with nonocclusive
thromboembolism of the left main and right medio-inferior-
lobar arteries occurring 3 h before the study. One can see
an extensive perfusion deficit of the pulmonary tissue (PD =
47%). Complete restoration of perfusion was revealed in lung
perfusion scanning conducted 2 weeks after streptodecase injection.

The success of thrombolytic therapy is affected by the
presence and duration of acute cardiac failure caused by an
embolic lesion. Four of the nine patients in our study showed
no signs of circulation insufficiency, and collapses of only
short duration were observed in another four patients. All
eight patients survived. In one case we observed prolonged
progressive arterial hypotonia resistant to medical correction.
Thrombolytic therapy in this patient was not successful, and
the patient died 4 h after the onset of disease. Autopsy
showed undissolved nonocclusive thromboemboli in both main
branches with thrombosis in the lobar and segmental arteries.
In this case, irreversible decompensation of cardiac activity
caused by massive embolism occurred before thrombolysis could
be effective. There were no recurrent pulmonary emboli in
our study. In one case we observed fragmentation of the
thrombus from its source of embolization with migration from
the iliac to inferior vena cava where it was detained by the
implanted cavofilter. Best results were obtained in cases
with nonocclusive lesion and no marked hemodynamic disorders.

The method suggested for thrombolytic therapy is simple
and is not dangerous in terms of the development of severe
hemorrhagic complications and allergic reactions. Combined
injection of anticoagulants and antiaggregating agents is a

TABLE 5.3a. Change in Plasminogen Activator and Plasmin Indices in Patients Treated with Streptodecase

Patient	before treatment	Time after injection, h							
		1	3	8	16-18	38-42	66-78	86-90	126-138
		Plasminogen activator, mm²							
R.	21	40	172	26	9	0	0	0	25
K.	65	336	364	200	260	75	84	60	0
G.	0	176	147	105	60	0	0	0	-
S.	0	189	176	84	-	25	0	16	-
		Plasmin, mm²							
R.	9	9	36	16	16	0	0	0	0
K.	16	64	36	25	64	25	16	4	0
G.	0	49	49	64	4	0	0	0	-
S.	0	100	49	16	-	0	9	0	-

Note: In donors, the level of plasminogen activator was 25 ± 5.1 mm²; plasmin level 8.0 ± 1.0 mm²; fibrin degradation products (FDP) 17.2 ± 1.1 mg %; partial thromboplastin time (PTT) 65 ± 1.3 sec.

TABLE 5.3b. Change in FDP and PTT Indices in Patients Treated with Streptodecase

Patient	before treatment	Time after injection, h							
		1	3	8	16-18	38-42	66-78	86-90	126-138
					FDP, mg %				
R.	43.7	7.2	0	8.5	29	18	25.2	22	21
K.	63	7.3	95	73	38.7	42	38.7	42	57.5
G.	95	120	105	134.5	191	138	154	170	
S.	19.2	7.2	20.7	23.7	-	16	0	8.5	
					PTT, sec				
R.	110	116	80	100	100	58	50	60	160
K.	65	76	86	86	65	75	100	140	70
G.	55	80	100	65	55	50	50	70	-
S.	85	75	135	95	-	70	130	150	-

safe measure for preventing relapses of venous thrombosis
after streptodecase injection.

Acute and Chronic Venous
and Arterial Thromboses

Streptodecase (without heparin and antiaggregating
agents in the first 10 days after injection) was used for the
treatment of four patients with thrombosis or thromboembolism
of the arteries of the lower extremity. These cases of systemic thromboses were the only ones in which streptodecase
was injected without subsequent use of heparin. There were
no side reactions or hemorrhagic complications. Analysis of
hemostasis demonstrated a "selective" effect of streptodecase:
minimal suppression of coagulability and rapid and significant activation of fibrinolysis (but shorter in time, averaging 48 h) (Table 5.3).

Change in Plasminogen Activator and Plasmin Indices.
Study of fibrinolysis inhibitors and antiplasmins showed that
their titer increased in the first hours and then returned to
normal values within 1 day. Determination of the content of
acid-labile inhibitors of activator activity revealed a significant decrease in their concentration; in three cases an
inversion of inhibition to fibrinolysis activation was revealed. An increase in fibrin degradation products (FDP) was
observed in the first 16-18 h after streptodecase injection.

Fibrinogen dynamics in these patients correlate with
those of the PATE patients: A decrease of up to 30% of the
initial level occurs on the first day, followed by a subsequent elevation (which, however, is still lower than the initial level); the changes between the third and sixth days follow a similar curve but are more prolonged.

Streptodecase treatment produced an undisputed clinical
effect in two patients (R. and K.). Signs of acute arterial
thrombolytic occlusion began to decrease in 3-6 h and almost
completely disappeared in approximately 14-16 h after injection of the preparation.

The activity of plasmin and plasminogen activator equalled
zero in two cases (G. and S.) who had high FDP content.
This may indicate the presence of blood intravascular coagulation. Injection of 3 million units of streptodecase caused a

significant increase in blood fibrinolytic activity (FA) in
1-3 h, maintaining a high level for not more than 5 days.

The content of plasminogen activator, as well as plasmin activity, increased ninefold. The content of FDP increased against the background of FA elevation and then a
tendency to normalization was observed.

Special attention should be given to the nature of FDP
and FA changes in patient G. After a short elevation in FA,
plasmin activity decreased to zero, FDP content increased,
and, within 86 h after injection of streptodecase, reached 170
mg %, with FA equal to zero. This testifies to hypercoagulation and good correlation of the biochemical indices with the
clinical picture (complete restoration of blood flow was not
observed in this case).

The effect of streptodecase on blood coagulation indices
is of special interest and significance. During the first
day partial thromboplastin time (PTT) was practically at the
initial level and, only on the fourteenth day, was coagulation
time decreased (the patients did not receive any other therapy, and so this finding reflects the state of the main disease).

The absence of a marked effect of streptodecase on hemocoagulation quite convincingly demonstrates the need for theoretically substantiated subsequent therapy with heparin, which
is supported by the possibility of a specific effect of streptodecase on hemostasis, without the fear of hemorrhagic complications characteristic of therapy with other preparations.

Effectiveness in Patients with Chronic Thromboses. This
study was conducted at the cardiological hospital in Bordeaux
(the supervisor was Professor G. Bricaud, France) in collaboration with N. A. Gratsiansky and Z. Bonnet. In this category
of patients it is possible to give a more reliable and objective assessment of the tolerance to streptodecase, its effect
on blood and internal organ function, and the specificity of
its effect on hemostasis, since autolysis is excluded in chronic thrombosis due to the endogenic activity of the fibrinolytic system ("independently" acting factors of acute thrombosis). Also, enzymatic systems and tissue and organ metabolism in these patients are rather stable and any shift is more
objectively associated with the effect of the preparation.

This study included patients with chronic thromboses of the deep veins of the lower extremities (four patients), thrombotic arteriopathy of the lower extremities (one patient), and ischemic heart disease (stable form, two patients). In one case, 6 million units of streptodecase were used, and five patients received intravenously 100 g of hydrocortisone before injection of the preparation.

Within the first 6 h heart rate increased and arterial pressure decreased (to not more than 15% of the initial level). A temperature reaction, eliminated by a single injection of hydrocortisone and analgesics, was recorded in one case. The effect on the coagulating link of hemostasis in these patients allowed us to conduct therapy with heparin and antiaggregating agents without any fear of hemorrhagic complications, and the values of biochemical changes were within accepted limits. The leukocytic formula showed changes in two phases as in therapy with streptase (on the first and fourth to sixth days).

Insignificant immune shifts were observed during treatment.

A decrease was observed in the concentration of apolipoproteins A and B, together with an increase in the content of macroglobulins in the first 3 days after injection. The changes observed in lipid metabolism appeared to be insignificant (high-density lipoproteins decreased while low-density lipoproteins increased during 12 days of observation).

Study of the indices of mineral exchange and enzymatic systems testified to the absence of a possible damaging effect of streptodecase on heart, liver, kidney, and lung function.

In the first 2-12 h, platelet content decreased (within normal levels) without a marked increase in aggregating ability or a disturbance in disaggregation. Subsequently, against the background of aspirin therapy, the aggregation indices corresponded to normal levels.

Angiographic verification demonstrated, in each case, the disappearance or reduction of thrombosis and improvement in the state of microcirculation, and in myocardial contractility in patients with ischemic heart disease. No hemorrhagic complications or thrombus fragmentation were observed.

Streptodecase therapy in this category of patients demonstrated high effectiveness, simplicity, and safety, and, moreover, permitted us to evaluate more objectively the action of the preparation on both the thrombus and the organism in general. The results obtained are very important for clinical practice and they substantiate the possibility, and even necessity, and practical safety of combined therapy with heparin.

Acute Myocardial Infarction

Thirty patients were treated with streptodecase in this study. Of these, 24 patients had macrofocal myocardial infarction and 6 had a threatening syndrome of unstable angina.

The study included 26 men and 4 women aged 47-74 years. It was the first myocardial infarction in 18 patients and a repeat infarction in 12. Localization of the lesions was as follows: Nine cases had an extensive anterior myocardial infarction, 11 had a posterior infarction, and 4 had a circular infarction. Microfocal myocardial infarction was diagnosed in three patients (2 posterior, one anterior). In 21 patients, we observed angina on effort of long duration; 4 had resting angina and 14 had hypertension requiring special treatment. In 11 cases, unstable angina was observed immediately before the development of the myocardial infarction.

In all cases, myocardial infarction was accompanied by an anginal attack with a duration of several hours to 2 days; in 20 patients, this attack could not be completely eliminated by repeated injection of narcotics. The condition of all patients was considered moderately severe or severe: Two patients were in cardiogenic shock, 2 had lung edema, 9 had x-ray signs of congestion in a small circle, and 15 had disturbances in rhythm and conduction (2 had a complete transverse heart block).

In most of the patients, streptodecase was used on the first day of disease; in eight patients the preparation was injected against the background of therapy with heparin, and in one patient with phenylin.

Concomitant therapy included narcotic analgesics, nitrates, β blockers, calcium antagonists, and antiarrhythmic drugs. Cardiac glycosides and diuretics were used in three patients.

Combined heparin therapy was also used. In eight patients, direct anticoagulant injections, which were started
at the prehospital stage, were continued according to conventional indices of coagulability. In the remaining patients,
heparin therapy was started no later than 24 h after the injection of all therapeutic doses of streptodecase throughout
the 7 days, with a gradual transition to indirect anticoagulants.

The criteria for assessing the effect of streptodecase
on the course of acute myocardial infarction were peculiarities of the clinical course, ECG study (including 11 patients
with precardial mapping), changes in the activity of enzyme
markers of myocardial damage, and central hemodynamic indices.

The clinical effectiveness of streptodecase was demonstrated by the course of myocardial infarction, tolerance,
and complications (summarized in Table 5.4). Corresponding
data on treatment with the native enzyme streptase are presented for comparison.

The dynamics of enzyme activity, serving as a witness
of myocardial damage, and the central hemodynamics correspond
to a noncomplicated course of myocardial infarction under
conditions of bed rest and activation.

This study of the effect of streptodecase on the course
of acute myocardial infarction allows us to expect broad practical prospects since the preparation is safe and simple, and
since the thrombolytic effect is prolonged and without a disturbance of the hemostasis coagulating link. Specific properties of the preparation allow one to combine therapy with
heparin. Control for this type of thrombolytic therapy is
also simple. All of these findings raise the possibility of
significantly increasing the effectiveness of treatment for
patients with myocardial infarction.

Intraocular Hemorrhages

This study was conducted at the Helmholz Research Institute of Eye Diseases, under the supervision of Professor R. A.
Gundarova (the physician was A. D. Romashchenko).

Treatment of intraocular hemorrhages is an urgent problem in modern ophthalmology. Difficulties in the treatment

V. N. SMIRNOV ET AL.

TABLE 5.4. Clinical Effectiveness of Streptodecase

Clinical criteria	Streptase (38 patients)	Streptodecase (30 patients)
Mortality (in %)	17	12
Cardiorrhexis (in %)	8.3	4
Prolonged course and repeat infarction (in %)	7.7	4
Decrease and disappearance of circulatory insufficiency (in %)	77.8 of 48.4 on admission	77 of 42
Reduction and disappearance of pain syndrome (in %)	63.9	92
Rapid ECG dynamics (in %)	42.8	60
Cardiac aneurysms (in %)	4.2	-
Complications (in %) Hemorrhages Thromboembolism Allergic reactions	75.3 11.3 25	- - -

of this pathology are associated with the pathogenesis of hemophthalmia and the insufficient effectiveness of thrombo- lytic drugs.

Traumatic Hemophthalmia. It has been found that, in traumatic hemophthalmia, heparin and fibrinolysin act most effectively on the hemocoagulating properties of the vitreous body, and fibrinolysin and streptokinase effect fibrinolytic activity. These preparations increase the anticoagulant properties of the vitreous body leading to local hypocoagula- tion and, thus, promoting a decrease in fibrin formation. At the same time, fibrinolysin and streptokinase provide the vit- reous body with fibrinolytic activity due to a decrease in inhibitors and an increase in plasminogen activators, which promote lysis of intravitreal hemorrhages.

However, all of the available thrombolytic preparations are not sufficiently effective because of the following fac- tors. Optimal local concentration sufficient for blood lysis is not achieved with native thrombolytic enzymes. Increase in a single dose can cause repeat hemorrhages and can lead to an increase in volume of the injected substance. Repeated injection of the preparation, which may be necessary for maximal effect, causes sensitivity and local allergic reaction (eyelid edema, chemosis, hyperemia, etc.), thus complicating therapy. Lastly, native enzyme is unstable and toxic.

The action of streptodecase was studied experimentally in 34 chinchilla rabbits (68 eyes) with hemorrhages in the vitreous body. Hemophthalmia was reproduced by injecting 0.5-0.6 ml of autologous blood into the vitreous body after preliminary removal of an equivalent amount. Streptodecase was injected subconjunctivally, retrobulbarly, and intravitreally at a dose of 15,000-45,000 units in 0.1-0.3 ml of physiological solution. The preparation was injected subjunctivally and retrobulbarly one to three times with an interval of 5-7 days, depending on the dynamics of resolution of traumatic hemopthalmia. Streptodecase was injected intravitreally at a single dose of 30,000 units.

Observation began several hours to 30 days from the moment of trauma reproduction. Rabbit eyes were studied on days 1, 3, 7, 14, and 30 during the course of hemophthalmia.

Biomicroscopic studies of rabbit eyes showed that maximal lysis of the blood occurred by the fifth to seventh day from the onset of disease. By the seventh to tenth day, a pink response from the fundus oculi partially appeared, and by the fourteenth day the vitreous body was practically transparent and single floating dimnesses were identified in the passing light. Blood lysis occurred more rapidly with intravitreal injection. Table 5.5 compares the data on acoustic scanning which reflect the dynamics of traumatic hemophthalmia resolution in nontreated rabbits and in rabbits treated with streptodecase. The acoustic scanning data objectively reflect the effectiveness of streptodecase.

Electrophysiological study was also performed to evaluate the functional state of the retina at different doses of streptodecase (from 30,000 units to 60,000 units per injection).

In experiments using 60,000 units of streptodecase injected intravitreally, the following results were obtained: at 5 and 30 min and at 1 h after injection of the preparation,

TABLE 5.5. Echographic Data of Rabbit Eyes Not Treated and
            Treated with Streptodecase

Nontreated	Treated with streptodecase
Day 3	
Multiple rough pellicles along the entire vitreous body. Fading force, 20 dB	Pelliculate dimnesses with lower acoustic density, occupying one third to one half of the vitreous body. Fading force, 15 dB
Day 7	
Multiple pelliculate dimnesses with tendency to consolidation, increase of acoustic density to 25 dB.	Single pelliculate dimnesses in the vitreous body.
Day 14	
Starting fibrosis of the vitreous body, acoustic density up to 40 dB. Traditional amotio retinae seen in 20% of the cases.	Single filamentous dimnesses, occupying a limited site.
Day 30	
Rough fibrosis of the vitreous body, acoustic density to 40-50 dB. Amotio retinae in 50% of the cases.	Vitreous body is acoustically transparent; single floating dimnesses.

the amplitudes of all responses (total, photopic, and scotopic) decreased 30-50% compared to the initial data. At 3 h the amplitudes of the photopic and total "b" wave of the electroretinogram (ERG) reached the initial level, and in subsequent days were within normal limits. Amplitude of the scotopic components by the thirtieth day did not reach the initial level and was only 15-30% of this value. A lower therapeutic dose of the preparation did not have a toxic effect on the retina. In contrast, an increase in the amplitude of all ERG components was observed in the immediate hours and at distant periods after the injection of streptodecase.

It is also known that the vitreous body in traumatic hemophthalmia essentially does not manifest fibrinolytic activity. This is increased by 5% with native streptokinase. A single injection of 30,000 units of streptodecase on days 2-3 of traumatic hemophthalmia increased fibrinolytic activity of the vitreous body to a small degree; by days 5-7 it increased to 33.7%, and in some cases to 76.9%, as compared to the control. Fibrinolytic activity decreased on day 14, but was still above normal.

Streptodecase was clinically tested in 46 patients (46 eyes) with intraocular hemorrhages. The preparation was injected conjunctivally, retrobulbarly, or intravitreally. The course of treatment consisted of one to three injections with an interval of 5-7 days. A single dose equalled 30,000-45,000 units (0.2-0.3 ml). All patients also received anti-inflammatory and corticosteroid therapy. Hemorrhages lasted from several hours to 20 days.

A clinical effect was usually observed by the second to third day. Maximal lysis occurred three to eight days after the injection. Hyphemia disappeared completely within 2-3 days in all patients, and blood in the vitreous body was eliminated within 7-8 days (in 60% of the patients). Visual functions were restored with light sensation to 0.2-0.7 (Fig. 5.13a, b). In 30% of the patients, blood in the anterior chamber and vitreous body dissolved completely, as confirmed by biomicroscopic and ultrasonic data, but vision was still low, which was probably associated with the nature of the trauma (central cornea scar, traumatic cataract, amotio retinae, etc.). In 10% of the cases lysis of the blood in the vitreous body was partial and occurred rather slowly. This group consisted of patients with severe penetrating wounds, the presence of an intraocular foreign body, or amotio retinae.

Use of streptodecase in patients in the intraocular hemorrhages was more effective than any other preparation used. It permitted a maximal increase in blood lysis without the specific complications observed with the injection of native fibrinolytics; it increased the functional properties of the traumatized eye and reduced the patient's length of stay in the hospital.

Retinal Central Vein and Its Branches. This study was performed at the Helmholz Research Institute of Eye Diseases,

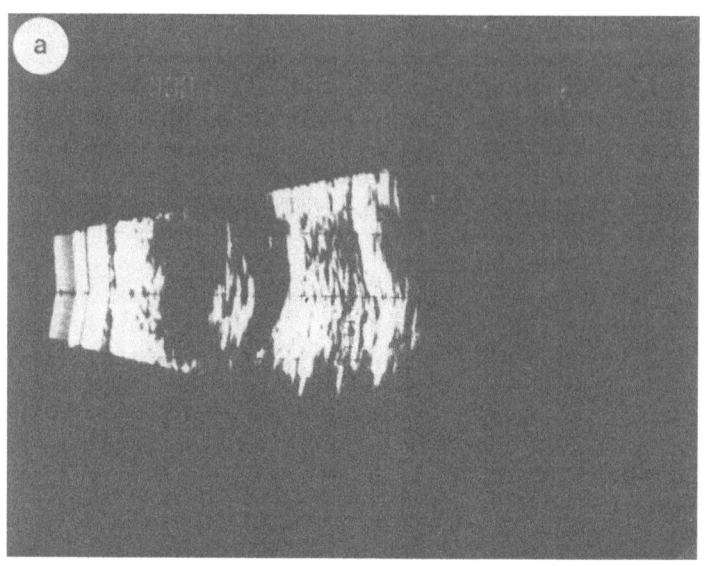

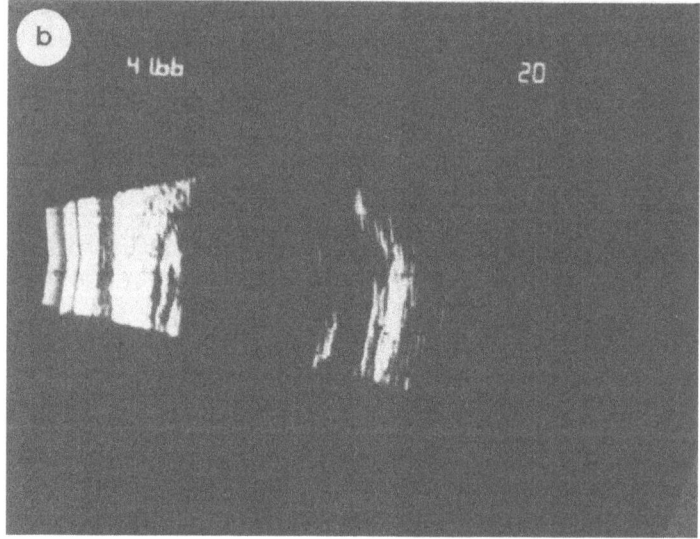

Fig. 5.13.   Eye echograms of a patient with traumatic hemo-
             phthalmia.   a) Before streptodecase treatment:
             multiple fibrin pellicles, vision function — light
             sensation.   b) On the seventh day after local (sub-
             conjunctival) injection (two injections with a 3-
             day interval) of 45,000 units of streptodecase
             (single injection): single filamentous dimnesses,
             vision function — 0.7.

under the leadership of Professor L. A. Katsnelson. The study was conducted in 10 patients (5 men and 5 women) aged 40-72 years with thromboses (rethromboses) of the retinal central vein and its branches. All of the patients had previously undergone intensive thrombolytic therapy which had not been effective.

Streptodecase was injected retrobulbarly at a dose of 30,000-45,000 units in 0.2-0.3 ml of physiological solution with an interval of 5 days. In the intervals between streptodecase injections, dexamethasone with heparin was injected retrobulbarly. Dexamethasone decreased the local allergic reaction, manifested as hyperemia and edema, which was observed in 90% of the patients after two to three injections of streptodecase. Use of dexamethasone retrobulbarly and antihistamine preparations within 1-2 days leads to complete elimination of the local allergic reaction. As shown by our results, percutaneous retrobulbar injection of streptodecase causes a lower degree of local allergic reaction than transconjunctival injection, which is accompanied by chemosis and prolonged hyperemia of the conjunctiva. The duration of treatment, 2-6 weeks (two to six injections), was determined by its effectiveness.

The following records from the patients' medical histories demonstrate the results of treatment.

1. Patient B., aged 63, right eye — thrombosis of the retinal central vein, 5 months' duration, hypertension. Visual acuity before treatment — 0.1. Treatment — two streptodecase injections. Visual acuity after treatment — 0.4.

2. Patient K., aged 60, right eye — thrombosis of retinal central vein, 2 months' duration, cyst-like edema in macular region, hypertension. Visual acuity before treatment — 0.1. Treatment — three streptodecase injections. Visual acuity after treatment — 0.8.

3. Patient B., aged 63, right eye — thrombosis of inferior-temporal branch of retinal central vein with edema in macular region, 4 months' duration, hypertension. Visual acuity before treatment — 0.1. Treatment — four streptodecase injections. Visual acuity after treatment — 0.5.

4. Patient A., aged 56, left eye — thrombosis of superior—temporal branch of retinal central vein, 4 months' duration, vegeto-vascular dystonia with hypertensive reactions.

Visual acuity before and after treatment — 1.0. Treatment —
two streptodecase injections.

5. Patient M., aged 68, right eye — rethrombosis of inferior—temporal branch of retinal central vein. Remoteness
of first thrombosis — 6 months, rethrombosis — 1 month. Hypertension. Visual acuity before treatment — 0.2. Treatment —
two streptodecase injections. Visual acuity after treatment
— 0.5.

6. Patient B., aged 52, right eye — thrombosis of superior—temporal branch of retinal central vein with edema in
macular region and marked hemorrhagic activity of 3 months'
duration. Hypertension. Visual acuity before treatment —
0.06. Treatment — three streptodecase injections. Visual
acuity after treatment — 0.2.

7. Patient K., aged 72, right eye — rethrombosis of
superior—temporal branch of retinal central vein with cyst-
like edema in macular region. Remoteness of first thrombosis
— 9 months, rethrombosis — 3 months' duration. Hypertension.
Visual acuity before treatment — 0.1. Treatment — three
streptodecase injections. Visual acuity after treatment —
0.4.

8. Patient D., aged 71, right eye — thrombosis of inferior—temporal branch of retinal central vein with cyst-
like edema in macular region of 3 months' duration. Hypertension. Visual acuity before treatment — 0.1. Treatment —
four streptodecase injections. Visual acuity after treatment
— 0.4.

9. Patient T., aged 52, right eye — rethrombosis of inferior—temporal branch of retinal central vein with dys-
trophic changes in macular region and marked hemorrhagic
activity. Hypertension. Remoteness of first thrombosis — 6
months, rethrombosis — 2 months' duration. Visual acuity before treatment — 0.2. Treatment — six streptodecase injections. Visual acuity after treatment — 0.3.

10. Patient S., aged 40, right eye — rethrombosis of retinal central vein with cyst-like edema in macular region. Remoteness of first thrombosis — 3 months, rethrombosis — 1
month's duration. Vegeto-vascular dystonia with hypertensive
reactions. Visual acuity before treatment — 0.1. Treatment
— six streptodecase injections. Visual acuity after treatment — 0.3.

In order to verify the thrombolytic effect, fluorescent angiography of the fundus oculi was performed in four of these patients before and after streptodecase treatment. Angiograms after treatment demonstrated enhanced contrast filling of the magistral veins, formation of collaterals, and restoration of perfusion in some vessels.

Thus, the results obtained on the clinical use of streptodecase in thromboses and rethromboses of the retinal central vein and its branches demonstrates its potential in the future for the treatment of these diseases and its effectiveness in cases with significant remoteness of the thrombotic process.

CONCLUSIONS

A prolonged, purposeful, and local drug effect is achieved by using biocompatible polymer carriers of a polysaccharide nature that are suitable for application. On the one hand, these carriers do not cause alterations in the activity of the preparation and, on the other hand, they do not have unfavorable side effects on the organism. The possibility of regulating the rate of carrier destruction, the amount of preparation bound to it, and the dynamics of its release opens the way for their use in clinical practice.

Study of the dissolution rates of polysaccharide carriers for prolonged drug action in chronic experiments demonstrates that the rates of destruction of Sephadex microspheres in the organism and in laboratory conditions are similar and can be changed by modifying the polysaccharide. The Sephadexes obtained are suitable for achieving firm chemical binding of significant amounts of protein compounds and can be used for maintaining a prolonged and local drug effect at the site of the pathological process.

The lysing effect of fibrinolysin immobilized on modified Sephadex has also been shown. The systemic injection of fibrinolytic agents in use at the present time has some negative properties. In our modification, practically all of the enzyme will act on substrate (thrombus) and the amount that gets into the blood bed is insignificant and not that important compared to the usual therapeutic doses. Enzyme inactivation is less probable than inactivation in its circulation in the vascular bed. Use of thrombolytic enzymes by the method suggested has essential advantages. It was also

shown that injection of fibrinolysin immobilized on microspheres of a polysaccharide nature significantly increases
fibrinolysis at the site of deposition due to both the enzyme
and an increase in blood activator activity. At present,
there are no fundamental limitations for the treatment of
peripheral thromboses and thromboembolism, including thromboses and embolism in the pulmonary artery system, with prolonged fibrinolytics.

The data obtained on the diffusive ability of the myocardial capillary bed in ischemia testify to the functional
shunting of the ischemic area and create prerequisites for
the use of drugs (thrombolytics, hyaluronidase, insulin, etc.)
in the treatment of ischemic heart disease by direct deposition in the ischemic or peri-infarct zone.

Clinicobiochemical studies of streptodecase in different
pathologies caused by thromboses demonstrate that this preparation most adequately meets the requirements of practical
medicine. High thrombolysing effectiveness is realized with
a minimal "disturbing" action on systemic hemostasis, which
predetermines the absence of clinically significant complications such as hemorrhagia, rethromboses, and thromboembolism.
The simplicity of therapy and its control also provides for
wide use of the preparation. Streptodecase also possesses
unique properties in hemophthalmia therapy.

All of the above testify to the promise of thrombolytic
enzymes in general which, together with streptodecase or in
combination with it, widen the possibilities for this type of
treatment and increase its effectiveness.

REFERENCES

1. M. Wolf and K. Raneberger, Treatment by Enzymes [Russian
    translation], Mir, Moscow (1976), p. 232.
2. V. P. Torchilin, A. V. Mazaev, E. V. Il'ina, V. S. Gold-
    macher, V. N. Smirnov, and E. I. Chazov, "Chemical as-
    pects of enzyme modification and stabilization for use
    in therapy," in: Future Directions for Enzyme Engineer-
    ing, L. B. Wingard, I. V. Berezin, and A. A. Klyosov
    (eds.), Plenum Press, New York (1979), pp. 219-240.
3. K. Mosbach (ed.), Immobilized Enzymes, Method in Enzymol-
    ogy, Vol. 44, Academic Press, New York (1976).

4. N. I. Larionova and V. P. Torchilin, "Modern status and prospects for use in medicine of immobilized physiologically active substances of a protein nature," Khim. Farm. Zh., No. 4, 21-36 (1980).

5. E. I. Chazov, Thomboses and Embolism in the Clinical Treatment of Internal Diseases, Meditsina, Moscow (1967), p. 262.

6. E. I. Chazov, L. S. Matveeva, A. V. Mazaev, M. Ya. Ruda, and G. V. Sadovskaya, "Intracoronary injection of fibrinolysin in acute myocardial infarction," Ter. Arkh., 4, 8-13 (1976).

7. E.I. Chazov, A. V. Mazaev, V. P. Torchilin, and V. N. Smirnov, "Possibilities of using biocompatible preparations of immobilized enzymes and physiologically active compounds of prolonged action for treatment of thromboses and other diseases of the cardiovascular system," in: Urgent Problems of Hemostasiology, Nauka, Moscow (1981), pp. 327-336.

8. L. A. Bessolitsina, A. V. Mazaev, A. A. Suvorov, V. P. Torchilin, V. N. Smirnov, and E. I. Chazov, "Effect of biosoluble microspheres of immobilized fibrinolysin on the fibrinolysis system," Byull. Eksp. Biol. Med., No. 1, 16-18 (1980).

9. A. V. Mazaev, E. M. Ginzburg, L. I. Serebryakova, and V. A. Bankuzov, "Microcirculation state in myocardial ischemia and its changes after deposition of Sephadex biosoluble microspheres in the coronary bed," Kardiologiya, 10, 45-48 (1979).

10. V. P. Torchilin, I. L. Rejzer, and E. V. Il'ina, "Principles of prolongation of enzymatic preparations," in: Myocardial Metabolism, E. I. Chazov and H. Morgan (eds.), Meditsina, Moscow (1979), pp. 362-372.

11. A. Sahara, "The case for fibrinolytic therapy," and J. E. Dalen, "The case against fibrinolytic therapy," J. Cardiovasc. Med., 5, 793-813 (1980).

12. E. I. Chazov, A. V. Mazaev, L. A. Suvorova, Yu. I. Voronkov, and V. P. Torchilin, "Effect of the new Soviet preparation streptodecase on the fibrinolysis system," Kardiologiya, 8, 18-21 (1981).

13. E. I. Chazov, A. V. Mazaev, Yu. I. Voronkov, and L. A. Suvorova, "Streptodecase — new Soviet thrombolytic preparation of prolonged action," Ter. Arkh., 9, 79-84 (1981).

14. E. I. Chazov and A. P. Golikov, "Use of streptodecase in acute myocardial infarction," Kardiologiya, 12, 10-14 (1981).

15. V.N. Ilyin, Yu. I. Voronkov, S. B. Rodionov, I. G. Mal-
     yutina, S. G. Leontiev, and A. V. Mazaev, "Successful
     treatment by streptodecase of massive thromboembolism
     of the pulmonary artery combined with extensive acute
     myocardial infarction," Ter. Arkh., 11, 140-143 (1982).
16. E. I. Chazov, R. A. Gundorova, A. D. Romashenko, V. I.
     Makarova, V. P. Torchilin, and Yu. I. Voronkov, "Immobil-
     ized streptokinase (streptodecase) in the treatment of
     intraocular hemorrhages," Vestn. Oftalmol., 4, 61-64
     (1982).

THROMBOLYTIC ACTIVITY OF PROTEINASES IMMOBILIZED ON WATER-SOLUBLE MATRICES

B. V. Moskvichev, T. I. Bogacheva, and O. A. Mirgorodskaya

Research Institute of Antibiotics and Enzymes
Leningrad

ABSTRACT

The effect of a chemical modification of proteolytic enzymes by water-soluble matrices (copolymers based on vinylpyrrolidine, polysaccharides, and blood plasma proteins) on their physicochemical, enzymatic, and biological characteristics was studied. It is shown that the modified enzymes have decreased affinity for inhibitors of plasma proteinases and reduced antigenic properties. Modification with antibodies to fibrin leads to an increase in proteinase fibrinolytic activity. The effect of a polymeric microenvironment of the enzymes on the mechanism by which they interact with fibrinogen and fibrin is discussed. The stability of the modified proteinases during denaturation and autolysis is examined, and it is demonstrated that an alteration of enzyme stability caused by various denaturing effects is conditioned by the nature of their interaction with the matrix and inactivation of the native and modified enzyme. The chemical modification results in a decrease of proteolytic enzyme toxicity and provides for prolonged fibrinolytic action. The probability of hemorrhagic complications is reduced when these enzymes are used for thrombolytic therapy. It is concluded that immobilization on water-soluble matrices allows for directional regulation of specific enzyme properties, and thus opens up the possibility of developing highly effective thrombolytic preparations.

Modern medicine urgently needs effective preparations with fibrinolytic action. Hope had been set on proteolytic enzymes, including proteinases produced by microorganisms. It appeared that these enzymes, which are analogous to native fibrinolytic plasmin, would be able to effect direct hydrolysis of fibrin — the basis of a thrombus. However, experimental and clinical use of brinase [1, 2], trypsin [3], and terrilytin [4] has proved that these preparations are not sufficiently effective because of their low specificity of action. In order to create highly effective medicinal fibrinolytics, the enzymatic characteristics of proteinases must be modified so that their specificity to fibrin is retained or even increased, while the undesirable side interactions with plasma proteins are reduced.

This review presents the major results of proteolytic enzyme modification by water-soluble polymeric matrices. Two proteinases (trypsin and terrilytin) were chosen as model enzymes. Both enzymes possess high fibrinolytic activity, but have different substrate specificity [5] and physicochemical and antigenic properties, and they are derived from different sources.

Trypsin and terrilytin were modified by a set of water-soluble matrices which embraces all groups of polymers appropriate for medical use. These include synthetic polymers based on vinylpyrrolidine, native polysaccharides, and blood plasma proteins. Covalent binding of the enzymes with polymers was accomplished in most cases by reductive alkylation. Aldehyde groups were either initially present in the carrier [6] or were introduced at the stage of initial activation [7]. In order to study the effect of electrochemical properties of the matrix on the enzyme, charged functional groups were introduced in certain cases into the macromolecular proteinase derivative using sodium bisulfate or hydroxylamine [8].

As a result of chemical modification, the proteolytic enzymes are completely transformed into high-molecular-weight derivatives. This is corroborated by gel-chromatography, diffusion, and sedimentation data [8]. Electrochemical properties of the proteinases are also substantially altered: The isoelectric point of the enzymes is shifted toward that of the matrix [9].

The synthesis of a whole set of polymeric derivatives of proteolytic enzymes made it possible to conduct a large-scale investigation of the effect of the polymer environment on enzymatic and physicochemical characteristics of proteinases. A comparative study of the properties of native and modified enzymes was performed in model systems in which the conditions were similar to physiological ones.

After administration of an exogenous proteinase into the vessel bed, it interacts not only with fibrinogen and fibrin, but also with a great number of other proteins. Proteinase inhibitors, which rapidly inactivate enzymes, make up the bulk of the group. Plasma proteins are also substrates of proteolytic enzymes and are subjected to proteolysis. Since proteolytic enzymes are foreign proteins, they have antigenic properties. Allergic complications often accompany the use of proteinases of microbial origin [10]. The low in vivo efficacy of proteolytic enzymes is mainly due to their distribution among numerous substrates and inhibitors.

Modification of proteinases by water-soluble polymers and biopolymers makes it possible to regulate the selectivity of enzyme interaction with blood plasma proteins. A polymeric matrix creates spatial obstacles to the approach of matrix-bound enzymes to macromolecules of physiologically active proteins. This is specifically manifested in a noticeable decrease of proteolytic enzyme affinity for plasma inhibitors. As shown in Figure 6.1, the inhibition of native trypsin in plasma is mainly due to the action of two inhibitors [11]. First, trypsin-α_2-macroglobulin complex, having esterase activity and completely inactive with respect to the high-molecular-weight substrate casein, is formed at small enzyme concentrations. As α_2-macroglobulin is exhausted, trypsin begins to interact with α_1-antitrypsin. With this complex it has neither esterase nor caseinolytic activity. Trypsin covalently bound with albumin does not interact at all with a heavy macromolecule of α_2-macroglobulin (molecular weight of 750,000), and its affinity for α_1-antitrypsin (characterized by the value of the inhibition constant) is decreased 30-fold [12].

The effect of chemical modification on the antigenic properties of terrilytin microbial proteinase is undoubtedly useful. It follows from Figure 6.2 that steric and chemical screening of antigenic determinants, which is due to the en-

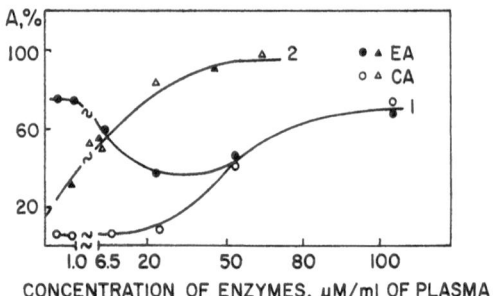

Fig. 6.1. Inhibition of native trypsin (1) and A-trypsin (2)
           by human plasma. EA) Esterase activity; CA) ca-
           seinolytic activity.

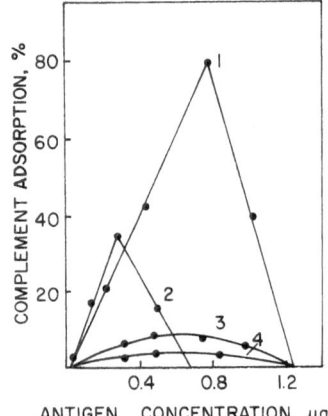

Fig. 6.2. Comparison of antigenic properties of
           native and modified terrilytin in the
           complement-binding reaction: 1) ter-
           rilytin; 2) dextran (D)-terrilytin;
           3) A-terrilytin; 4) VPA-terrilytin.

zyme binding with synthetic and native matrices, leads to a
slackening of its interaction with specific antibodies.

    The above-mentioned protective effect of various poly-
meric matrices provides for a relative increase in the specificity of the modified proteinases' action on fibrin and fibrinogen if fibrinolytic activity is retained. The modification becomes even more effective if a polymer, possessing its

TABLE 6.1. Fibrinolytic Activity of Modified Proteolytic Enzymes

Enzyme	Relative fibrinolytic activity, modified/native
Terrilytin	1.0
IgG-terrilytin	3.3
Terrilytin + heparin	1.2
Trypsin	1.0
IgG-trypsin	11.0
A-trypsin	3.3
VPC-trypsin	4.0

Note: IgG = Immunoglobulin G isolated from the
blood serum of rabbits immunized by fibrin. A =
Human serous albumin. VPC = Copolymer of vinylpyrrolidine with cinnamic acid.

own affinity for fibrin, is used as a matrix. Antibodies obtained from the blood of animals immunized with fibrin [13]
belong to the group of "active" matrices. Fibrinolytic activity of proteinases covalently bound with antibodies to fibrin
increases by 3- to 11-fold (Table 6.1). The increase in enzyme activity with respect to an insoluble substrate points
to the fact that, during modification with IgG immunoglobulins, an additional center of fibrin binding appears in the
enzyme molecule. Adsorption of immunoglobulins on fibrin
leads to a rise in the local concentration of the immuno-
globulin-bound enzyme on the surface of fibrin and to an increase in its fibrinolytic activity. Modification of proteolytic enzymes by the matrices, which are "active" with respect to fibrin, thus makes it possible to provide for selective interaction of proteinases with the "target" (fibrin).

Another approach to regulating the fibrinolytic characteristics of proteolytic enzymes is based on using the unique
capacity of fibrinogen to coagulate because of the splitting
of small peptide fragments (fibrinopeptides) at early stages
of proteolysis. Analysis of the mechanism of proteolytic
fibrinogen degradation under the effect of enzymes with different specificity has already been described elsewhere [14]

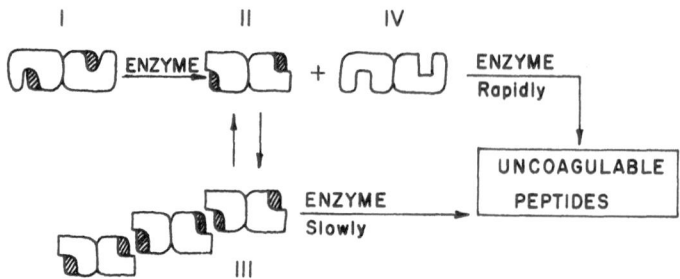

Fig. 6.3. Scheme of fibrinogen hydrolysis (I) under the ef-
           fect of exogenous proteinase. II) Fibrinogen
           fragments capable of aggregate formation (III);
           IV) uncoagulable fibrinogen fragments. Shaded
           areas: polymerization centers.

and can be presented by the scheme given in Figure 6.3. Macromolecular fragments with different characteristics are
found during fibrinogen hydrolysis in the presence of an exogenous proteinase. Some of them (II) are analogous to monomeric fibrin and can form fibrinlike polymeric structures
(III). Other fragments (IV) do not contain polymerization
centers and cannot coagulate. Polymerization of intermediate
products of fibrinogen hydrolysis reduces the total rate of
proteolysis and, under certain conditions, may be manifested
in the formation of insoluble fibrinlike clots during incubation of proteolytic enzymes with plasma.

    The above scheme makes it possible to formulate clearly
the procedures for augmenting proteinase fibrinolytic activity. This may be realized by inhibiting the polymerization
of type-II products by introducing polymerization inhibitors
into the system. Heparin is the most powerful native antiaggregating agent. Fibrinolytic activity of terrilytin
(Table 6.1) is augmented in the presence of heparin, and the
ability of proteinase to inhibit thrombi formation is increased sharply (Figure 6.4). The latter is accounted for
by the acceleration of proteolytic splitting of fibrinogen.
A decrease in fibrinogen concentration leads to the thrombin-
induced prolongation of clot formation time.

    A subtle alteration of proteinase specificity with re-
spect to fibrin may be carried out via chemical modification
by the heparin analogs (polymers negatively charged under

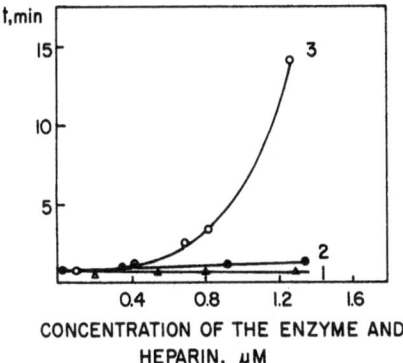

Fig. 6.4. Effect of heparin (1), terrilytin (2),
           and their mixture (3) on the time re-
           quired for fibrin clot formation under
           the action of thrombin.

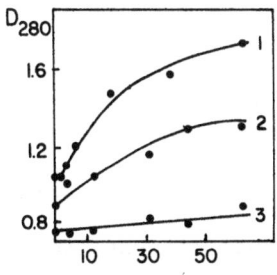

Fig. 6.5. Effect of heparin on the degree of
           hydrolysis of fibrinogen immobilized
           in polyacrylamide gel. The enzymes
           were incubated with the substrate for
           30 min. 1) Terrilytin; 2) trypsin;
           3) A-trypsin.

physiological conditions). A polyanionic carrier facilitates
orientation of the bound enzyme in the proximity of the fibrinogen polymerization centers, primarily stimulates their
splitting, and thus inhibits the formation of intermediate
products of fibrinogen hydrolysis which are capable of developing aggregates. We have demonstrated the feasibility
of this approach in studying the hydrolysis of the model substrate (fibrinogen) immobilized by incorporation into poly-

acrylamide gel, by native and modified enzymes. Figure 6.5
shows the dependence of the degree of hydrolysis on the concentration of the antiaggregating agent heparin. In the
presence of native trypsin and terrilytin, an increase in
heparin content in the solution results in an increase in the
number of peptides split from immobilized fibrinogen.

This result has the same meaning as the data shown in
Figure 6.3: Heparin, by inhibiting the polymerization of
macromolecular products of fibrinogen proteolysis, provides
for an increase in the depth of the substrate hydrolysis. It
should be noted that heparin essentially fails to exhibit its
anticoagulating action when the hydrolysis of immobilized
fibrinogen is affected by trypsin modified by the negatively
charged (under physiological conditions) protein carrier albumin. This finding proves that, during the course of fibrinogen hydrolysis by modified trypsin, the share of intermediate products capable of coagulation is decreased. Measurement of the fibrinolytic and coagulating activity of the
modified trypsin derivatives gave results which agree with
this conclusion. Modification with the negatively charged
carriers (VPC-trypsin, A-trypsin)* is accompanied by a three-
to fourfold increase in trypsin fibrinolytic activity (Table
6.1) and a simultaneous decrease in coagulating activity,
which was evaluated by the plasma recalcification curves in
the presence of proteinases (Figure 6.6).

The stability of modified proteinase under the effect
of denaturants will largely determine the possibility of making standard medicinal preparations. As a result of enzyme
binding with polymers, part of the amino acid residues of the
protein globule changes its properties (charge, affinity for
water) while interacting with functional groups of the carrier. This leads to an impairment in the system of intramolecular cooperative links which secure the stability of the
protein native conformation. The enzyme macromolecule ap-
proximates the denatured state in its energy parameters [15].
On the other hand, the formation of links between the enzyme
and the matrix "freezes" the protein structure which is usually manifested in an increase in enzyme kinetic stability.

The destabilizing effect of modification is displayed
most vividly during an enzyme-matrix multipoint coopera-

---

*VPC = Vinylpyrrolidine copolymer with cinnamic acid. A =
Human serous albumin.

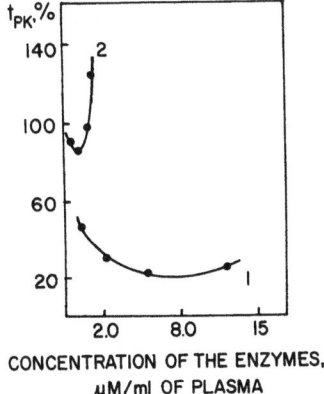

Fig. 6.6. Effect of native trypsin (1) and A-
           trypsin (2) on recalcification time
           of human plasma.

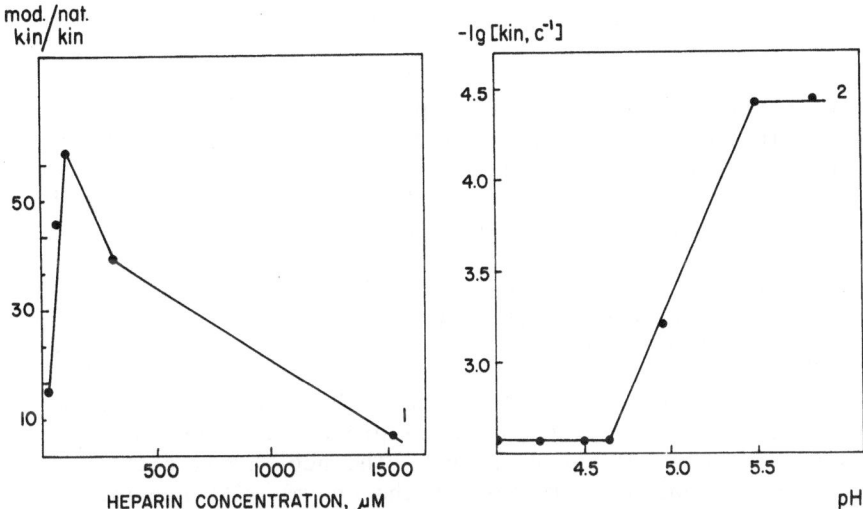

Fig. 6.7. Effect of heparin concentration and pH on the con-
           stant of trypsin (1) and terrilytin (2) inactiva-
           tion rate. 1) pH 4.2, 52°C; 2) molar ratio terri-
           lytin/heparin 1:0.6, 20°C.

tive interaction. We have studied this by analyzing a polyelectrolytic interaction of trypsin and terrilytin with heparin [16, 17]. It was demonstrated that proteinases combine

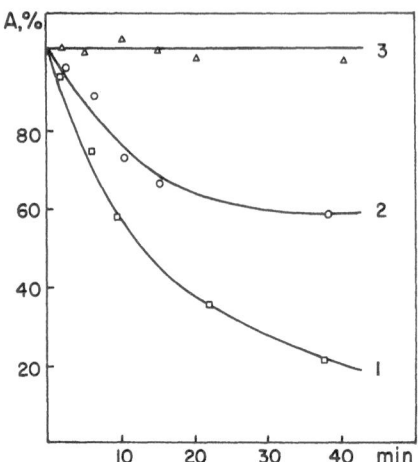

Fig. 6.8.   Inactivation kinetics of native terrilytin (1),
VPA-terrilytin (2), and VPAS-terrilytin and VPAM-
terrilytin (3) in the presence of heparin (49 µM) at
pH 4.5, 20°C.

with heparin into dissociating polyelectrolytic complexes of
variable composition. The degree of enzyme destabilization
within such a complex is largely determined by external conditions and may be widely regulated by altering pH, heparin
concentration, and ionic strength (Figure 6.7).

Chemical modification is an effective way of protecting
the proteinase from the polyelectrolytic interaction with
heparin, which has an "unfolding" effect on its structure.
Modified derivatives of terrilytin are inactivated in the
presence of heparin at a lower rate than the native enzyme
(Figure 6.8). The effect is more pronounced when the matrix
carries an electrostatic charge. The negatively charged matrix blocks complex formation of the enzyme with heparin
(VPAS-terrilytin).* If a positively charged matrix (VPAM-ter-
rilytin)* is used, functional groups of the carrier are mainly associated with heparin while the proteinase remains intact [17].

---

*VPAS = Copolymer of vinylpyrrolidine with acrolein (VPA) ad-
ditionally modified with negatively charged bisulfate groups.
VPAM = VPA additionally modified with positively charged
amine groups.

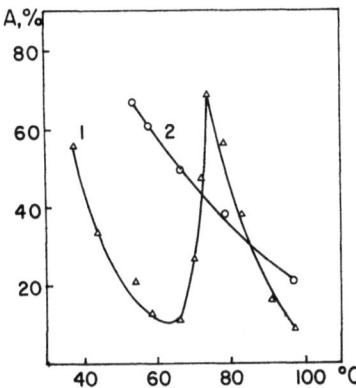

Fig. 6.9. Decrease in the activity of trypsin (1) and VPA-tyrosine (2) following incubation at different temperatures (pH 4.5) for a period of time sufficient to set equilibrium.

Comparison of the inactivation rates of native and modified trypsin, both during thermodenaturation and autolysis, shows that the polymeric proteinase derivatives are characterized by increased stability compared to the native enzyme. Detailed investigation of the causes of this phenomenon [18] show that binding with the carrier leads to a change in the mechanism of inactivation. In the experiment presented in Figure 6.9, native trypsin forms dimers and both the monomeric and dimeric forms are subjected to reversible denaturation. Chemical modification inhibits trypsin dimerization. The modified trypsin is denatured in monomeric form, which accounts for its increased stability at 40-70°C.

At optimum pH, autolytic splitting of a protein globule is the limiting stage in the inactivation of native trypsin. Amino acid residues of lysine, which develop peptide links when subjected to tryptic hydrolysis, undergo chemical transformation as a result of binding with the carrier. Also, a polymeric matrix limits protein accessibility to autolysis.

These factors result in a change in the inactivation mechanism: At optimum pH, the modified forms of trypsin are not autolyzed but are reversibly denatured, which accounts for the decrease in their inactivation rate (Figure 6.10).

174   B. V. MOSKVICHEV ET AL.

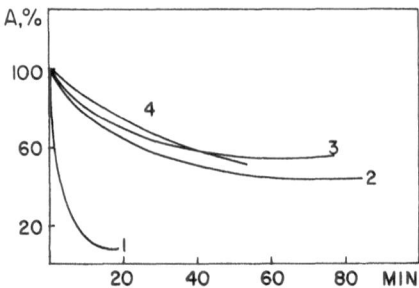

Fig. 6.10. Inactivation kinetics of native and
modified trypsin under conditions of
autolysis (pH 8.0): 1) trypsin, 47°C;
2) A-trypsin, 50°C; 3) D-trypsin,
65°C; 4) VPC-trypsin, 47°C.

The results of comparative in vitro studies of native
and modified enzymes indicate that specific characteristics
of proteinases may be so modified that their therapeutic fibrinolytic effect is enhanced. Experimental studies in various model systems substantiate the selection of such a carrier for modification which helps to (a) protect the enzyme
from inactivation under physiological conditions, (b) stimulate the prevailing concentration of a high-molecular-weight
proteinase derivative on the surface of a thrombus, and (c)
affect the efficacy of fibrin polymeric chain degradation by
regulating the sequence of splitting of the internal peptide
links.

However, it should be remembered that, after administration of fibrinolytics into an organism, factors of a higher
order become important. Their function is regulated by the
central nervous system and serves to maintain general hemostasis. Therefore, it is of primary importance to corroborate the advantages of the modified proteolytic enzymes. It
was demonstrated that the modified proteinases are characterized by a lower acute toxicity than the native enzymes. With
trypsin immobilization on serous human albumin, a 10- to 15-
fold increase in LD_{50} was observed. Multiple administration
of the proteolytic enzymes immobilized on water-soluble matrices does not cause vascular damage or macro- and microscopic changes in animal viscera. A decrease in toxicity of the
proteinases after binding with polymers reflects the reduction in their side interactions with plasma proteins and con-
forms well with the conclusions of in vitro studies [19].

TABLE 6.2. Thrombolytic Activity of the Modified Fibrinolytic and Fibrinolytic—Heparin Complex in Experiments on a Model Thrombus of the Rabbit Jugular
Vein

Enzyme	Dose, FU/kg	No. of animals	Efficacy of therapy (lysis of thrombus)		
			complete	incomplete	no lysis
Native fibrinolytic	800	4	1	0	3
Modified fibrinolytic	400	7	5	1	1
	800	3	3	0	0
Fibrinolytic + heparin	800	3	2	0	1
Modified fibrinolytic + heparin	400	7	7	0	0

In order to study the action of immobilized proteinases,
a model thrombus of the rabbit jugular vein was treated with
native and modified fibrinolytic (Table 6.2). Lysis of the
thrombus was corroborated by angiographic and flowmetric control of the restoration of vessel patency. A single stream-
line injection of native enzyme in the doses selected proved
to have low efficiency. Biochemical study of the bloodclot-
ting system by a number of generally accepted tests (plasma
recalcification time, fibrinogen content, prothrombin con-
sumption, time of lysis of the euglobulin fraction of plasma,
etc.) showed that the native fibrinolytic enzyme induces the
activation of coagulation during the first hours after administration. Activation of the coagulating system is almost
always observed during fibrinolytic therapy. This is a compensatory reaction of the macroorganism in response to the
shift in hemostatic balance [10].

The data given in Table 6.2 demonstrate that the modified
fibrinolytic enzyme has a more pronounced thrombolytic action
than the native enzyme although the dosage of the former was
half that of the latter. A short-term activation of the coagulating system occurred, while the fibrinolytic action was
manifest for 1-2 days.

Inhibition of thrombi formation is usually achieved via
fibrinolytic therapy with simultaneous administration of anti-

coagulants. Treatment of the experimental thrombosis of the
rabbit jugular vein with a fibrinolytic—heparin complex was
more effective than that with the fibrinolytic agent only
(Table 6.2). However, the result was accompanied by a significant increase in bleeding and a series of pathological
changes in the blood coagulating system (fall of fibrinogen
content to zero, sharp delay of thrombin time). An increase
in the dose of the complex to 4000 fibrinolytic units per
kilogram (FU/kg) leads to death of the animals caused by fibrinolysis. Application of the modified complex (fibrinolyt-
tic—heparin) yields good therapeutic results. Lysis of the
thrombus was observed in all of the animals studied. The
fibrinolytic action of the preparation is prolonged (1-2 days
after administration), is not accompanied by sharp changes
in the blood coagulation system, and much more seldom leads
to hemorrhagic complications.

Based on the analysis of the effect of a polymer microenvironment on enzymatic characteristics of proteolytic enzymes, a number of general principles can be formulated.
Drawing from these principles, one can directionally regulate
the selectivity of proteinase interaction with various components of the macroorganism. Immobilization on water-soluble
matrices, carried out with regard for the mechanisms by which
proteolytic enzymes interact with blood proteins, and utiliz-
ing the characteristics of chemical structure and the specific properties of the carrier, allows one to vary the lifetime
of proteinases widely under physiological conditions and to
decrease their activating effect on hemostasis.

It should be noted that chemical modification helps to
substantially enhance enzyme stability in the presence of
heparin. This opens up the possibilitty of developing complex preparations which would have both fibrinolytic and anticoagulating activity. Such a research direction is valuable
given that the complexity of the physiological system regulating the fluid state of the blood requires the combined application of fibrinolytics and anticoagulants.

Biological studies demonstrate that modified proteolytic
enzymes exhibit a high thrombolytic activity in doses lower
than those of native enzymes. The application of polymeric
derivatives of proteinases is characterized by a decreased
probability of hemorrhagic complications. Such preparations
also have a prolonged fibrinolytic action. Finally, the technology of obtaining the water-soluble polymeric enzyme form is

simple insofar as all of the operations are conducted under
mild conditions in a homogeneous water medium. All of these
assets indicate that immobilization of enzymes on water-soluble carriers is a promising method for developing highly effective fibrinolytic preparations.

REFERENCES

1. W. H. E. Roschlau and D. A. J. Ives, "Review of the bio-
    chemistry and coagulation physiology of Brinolase (fib-
    rinolytic enzyme from Aspergillus oryzae)," Folia Haem-
    atol., 101, 22-37 (1974).
2. E. P. Frisch, "Clinical review of Brinolase, a protease
    from Aspergillus oryzae," Folia Haematol., 101, 63-82
    (1974).
3. R. W. Colman, "Proteolytic enzymes in clinical medicine,"
    Clin. Pharmacol. Ther., 6, 3-14 (1965).
4. A. A. Imshenetsky, S. Z. Brotskaya, and V. V. Korshunov,
    "On the effect of mold proteinases on blood thrombi,"
    Dokl. Akad. Nauk SSSR, 163, 737-740 (1965).
5. G.V. Samsonov, L. K. Shataeva, and O. V. Orlievskaya,
    "Physicochemical and enzymatic characteristics of the
    protease from Aspergillus terricola," Dokl. Akad. Nauk,
    SSR, 296, 497-499 (1972).
6. T. B. Tennikova, E. F. Panarin, O. A. Mirgorodskaya,
    G. V. Samsonov, and B. V. Moskvichev, "Immobilization
    of the proteolytic enzyme terrilytin on a water-soluble
    polymeric matrix," Khim. Pharmacol.Zh., 11, 86-90 (1977).
7. G. M. Lindenbaum and O. A. Mirgorodskaya, "Chemical
    modification of trypsin by water-soluble dextrans,"
    Prikl. Biokhim. Mikrobiol., 14, 719-728 (1978).
8. T. B. Tennikova, B. V. Moskvichev, and G. V. Samsonov,
    "Study of physicochemical properties of terrilytin mod-
    ified by a copolymer on the basis of vinylpyrrolidone,"
    Biokhimiya, 45, 438-448 (1980).
9. I. M. Tereshin, T. I. Bogacheva, and B. V. Moskvichev,
    "Immobilized enzymes in medicine and the medical indus-
    try," in: Topical Problems of Hemostasiology," B. V.
    Petrovsky, E. I. Chazov, and S. V. Andreeva (eds.), Nauka,
    Moscow (1981), pp. 337-345.
10. E. I. Chazov and K. I. Lakin, Anticoagulants and Fib-
    rinolytic Preparations, Meditsina, Moscow (1979), p. 311.
11. J. Bieth and M. Aubry, "The interaction of human cation-
    ic trypsin and chymotrypsin II with serum inhibitors,"
    in: Proteinase Inhibitors, Bayer Symposium V, Springer-
    Verlag, Berlin (1974), pp. 53-62.

178   B. V. MOSKVICHEV ET AL.

12. T. I. Bogacheva, O. A. Mirgorodskaya, and B. V. Moskvichev, "Certain physicochemical properties of modified
trypsin," Biokhimiya, 42, 609-615 (1977).
13. T. I. Bogacheva, I. V. Moskvichev, V. N. Rutkovskaya,
B. V. Moskvichev, and I. M. Tereshin, "Covalent binding
of proteolytic enzymes with antibodies," Bioorg. Khim.,
6, 623-626 (1980).
14. T. I. Bogacheva, O. A. Mirgorodskaya, B. V. Moskvichev,
and I. M. Tereshin, "Study of the interaction of modified proteolytic enzymes with fibrinogen and fibrin,"
Biokhimiya, 44, 2134-1243 (1979).
15. L. V. Kozlov, "Thermodynamic principles of immobilized
enzyme immobilization," Bioorg. Khim., 6, 1243-1254
(1980).
16. O. A. Mirgorodskaya, T. B. Tennikova, and B. V. Moskvichev, "Certain characteristics of trypsin interaction
with heparin," Biokhimiya, 43, 1924-1928 (1978).
17. O. A. Mirgorodskaya, G. P. Ivanova, T. B. Tennikova,
and B. V. Moskvichev, "Study of terrilytin–heparin interaction," Bioorg. Khim., 4, 972-977 (1978).
18. G. P. Ivanova, O. A. Mirgorodskaya, B. V. Moskvichev,
and L. V. Kozlov, "Thermodenaturation mechanism of native trypsin and trypsin modified by a neutral soluble
polymer," Biokhimiya, 46, 1815-1822 (1981).
19. T. I. Bogacheva, O. A. Mirgorodskaya, G. E. Grinberg,
G. A. Mikhailets, B. V. Moskvichev, and I. M. Tereshin,
"Effect of chemical modification on proteinase interaction with inhibitors, their clotting activity, and
acute toxicity," Biokhimiya, 46, 863-871 (1981).

FIBRINOLYTICALLY ACTIVE POLYMERS

D. M. Zubairov, O. D. Zinkevich,* and

M. G. Asadullin

Kazan Medical Institute and Kazan Research Institute of Epidemiology and Microbiology,* Kazan

Advances in cardiovascular surgery and the use of extra-corporeal circulation are considerably hampered by blood clotting on the surface of artificial devices primarily made of synthetic polymers. These devices (cardiac prostheses, prosthetic valves, vascular grafts, dialyzer membranes, hemo-sorbents, etc.) are designed for either temporary contact with the blood or for permanent implantation into the body.

Researchers have not yet found a material that is suffi-ciently athrombogenic to meet medical requirements. Athrombo-geneity is only one of the criteria for bicompatibility of blood-contacting materials. These materials also must not cause mechanical trauma or necrotic changes in surrounding tissues, hemolysis, protein denaturation, or a shift in elec-trolytic balance at large volumes of ion exchange, or activa-tion of the complement system. Such materials should also be devoid of carcinogenic, allergic, and antigenic action. While degrading in the organism, they must not form toxic products. Among all of these qualities, athrombogeneity is the most difficult to achieve.

It has long been considered that normal endothelium can-not moisten, and thus does not initiate blood coagulation. Therefore, in order to initiate natural athrombogeneity the surface of devices intended for contact with the blood was

179

made as hydrophobic as possible. Our studies, however, have
shown that undamaged vascular endothelium is able to moisten
[1]. And although the hydrophobic surface of polydextran is
one of the most inert surfaces in initiating blood clotting,
materials with low in vitro thrombogeneity do not guarantee
the prevention of clot formation when incorporated into systemic circulation in vivo and ex vivo.

The hemocoagulation system in the general circulation
of the human organism is regulated by the nervous and humoral
systems in response to an immediate danger of blood loss [3].
When this happens, the components of the hemostasis system
are permanently activated and utilized. Fibrin and soluble
complexes of fibrin monomer on the surface of normal endothelial cells are subjected to extracellular lysis by plasmin
and intracellular proteolysis by lysosomal enzymes.

Thus, it is not sufficient to use polymers which minimal-
ly activate the contact phase of blood clotting (Hageman factor, prekallikrein, high-molecular-weight kininogen, and factor
XI) and cause minimal platelet adhesion in order to obtain a
maximal decrease in thrombi formation on a synthetic surface.
It is also important that the surface be exceedingly smooth
and possess fibrinolytic properties.

For these reasons, the development of a self-cleaning
surface which, in our view, can be attained via proteolytic
enzyme immobilization, is a promising way of solving the
present problems in cardiovascular surgery, extracorporeal
circulation, prolonged infusion through vascular catheters,
etc. Given the requirements, it is not advisable to use enzymes with broad specificity, such as trypsin, chymotrypsin,
or papain, for providing the material with fibrinolytic qualities. They may be used, however, for model studies of enzyme
behavior on the surface of a modified polymers.

Craddock et al. [4] have shown that cellophane membranes
used for hemodialysis can adsorb factor B of the properdin
system and activate the complement system which, in turn,
leads to the aggregation of polymorphonuclear leukocytes and
to their adhesion on cellophane, and to the discharge of lysosomal granules into the vessel bed. Biologically active substances synthesized during the interactions of the given type
lead to the formation of platelet aggregates which, together
with leukocytes, can cause capillary clotting, ischemia, and
infarction in various organs.

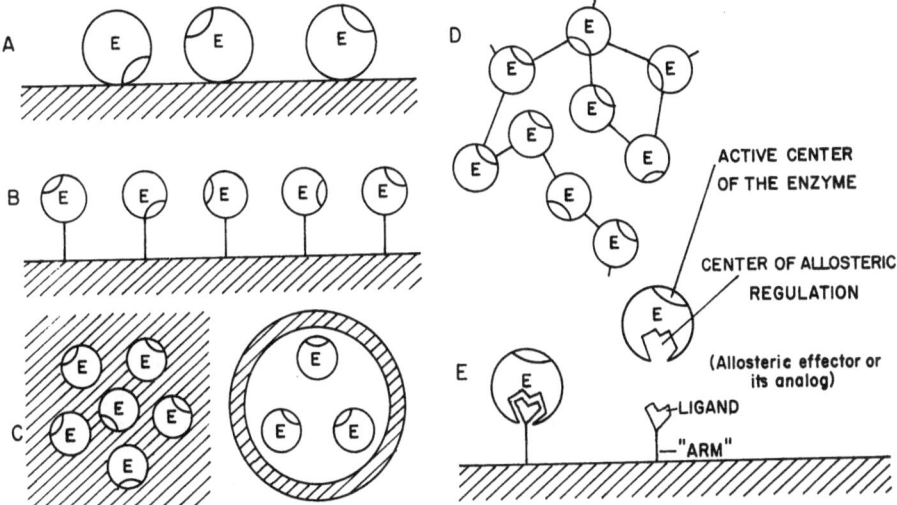

Fig. 7.1.   Methods of enzyme fixation to polymers: A) adsorption on a solid surface; B) covalent binding;
C) introduction of gel into the matrix and encapsulation; D) cross-linking with bifunctional reagents; E) affinity binding.

Numerous methods of enzyme fixation on a variety of polymers has been described in the literature [5].   They may be
classified into five main groups (Figure 7.1).   Hydrophilic
intensively swelling matrices, such as Sepharose, Sephadex,
cellulose, starch, and polyacrylamide, are most widely used
for the immobilization of proteolytic enzymes.   Porous glass,
nylon, and polystyrene and its derivatives are used less frequently, and there are no data on the use of polyethylene or
Teflon for immobilization of any enzyme.   A detailed account
of the chemism of most enzyme immobilization reactions is
given in the reference work by Lurje [6].

Hybrid methods of binding also exist, involving several
ways of immobilization simultaneously, e.g., adsorption with
subsequent cross-linking by a bifunctional reagent.

Many authors have utilized adsorption to prepare insoluble derivatives of proteolytic enzymes.   The shortcoming of
this method is the low energy of protein binding with the carrier and, as a result, transition of the enzymes into the
solution when its ionic strength or pH is changed.   A local

pH alteration may be caused by the reaction catalyzed by this enzyme.

The technique of enzyme incorporation into gel latex and encapsulation is very seldom used for proteases [7], since the size of the enzyme and substrate molecules is proportionate. Otherwise, the increase of latex pores would render no effect on enzyme immobilization.

The procedure of covalent binding with the carrier [8-10] is most frequently used to obtain immobilized proteases. Covalently fixed enzymes are bound firmly with the matrix and do not migrate into the solution during sustained and multiple utilization in various processes. Likely due to this quality, this method has been used for immobilization of chymotrypsin alone in more than 120 studies.

Cross-linking of molecules is a technique which rules out the effect of the carrier on the immobilized enzyme, but provides for a secure bond, which is characteristic of covalent immobilization. It is essentially a covalent technique. The role of the matrix is played by the enzyme molecules themselves which pass into an insoluble state [11].

Affinity binding was first used in our experiments [12] to link kallikrein with epinephrine-Sepharose.

Various functional groups of amino acid residues contained in the enzymes may participate in covalent binding: the indole group of tryptophan, the gamma-carboxylic group of glutamine acid, the imidazole group of histidine, the ε-amino group of lysine, the hydroxyl group of serine, the cysteine SH group, the phenolic hydroxyl group of tyrosine, and the carboxylic and amino groups of amino-terminal acids. Some of these groups may be essential for preserving the catalytic activity of the enzyme and, consequently, it is important that such a group, or even several of them, remain intact in the molecule of the immobilized enzyme. Such a role is very often played by the disulfide cysteine link, which forms intramolecular bridges responsible for preserving the tertiary enzyme structure, or by the imidazole ring, which is an active center constituent in certain enzymes.

As already noted, minimal thrombogeneity, sometimes termed athrombogeneity, is one of the requirements of the material intended for contact with the circulatory blood.

When selecting the polymers which satisfy this requirement,
one should take into account their effect (a) on the contact
phase of blood coagulation [13] and (b) on platelet adhesion.
The first process can be evaluated specifically by the hydrolysis of N-benzoyl-arginine ethyl ester, N-benzoyl-L-arginine
methyl ester, p-tosyl-L-arginine methyl ester, and N-D-Pro-
Phen-Arg-pNA under the effect of sorbed and activated enzymes
of the blood coagulation contact phase. The indicator for
platelet adhesion is a catalytic splitting of inorganic phosphate from adenosine monophosphate under the effect of 5'-
nucleotidase of platelets adherent to the surface.

As shown in our experiments [2], adsorption and activation of the contact factors of blood coagulation takes place
on polymers with both hydrophobic and hydrophilic characteristics, and carrying both negative and positive charges, but
does not occur on a hydrophilic surface devoid of ionogenic
groups. An increase in the distribution density of negative
charges on polymers to the value proximate to the distance
between the peptide links in proteins (0.4 nm) augments the
thrombogeneity of the surface. Ionic and hydrogenous links
play an essential role in the adsorption of the Hageman factor
and the activation of the blood coagulation contact phase.
Then the Hageman factor serves as a proton donor. Therefore,
in selecting materials, those carrying ionogenic groups on
the surface (especially those with a dense distribution)
should be avoided.

Today, chemically inert materials (Teflon, Lavsan, Kapron, polyethylene, silicone rubber) are used to manufacture
artificial heart valves, cardiac prostheses, and vascular
catheters. The chemical inertness of the material makes surface immobilization of fibrinolytic enzymes with retained
catalytic activity a difficult task. Because of this, not
all of a wide variety of procedures for chemical and physical
immobilization are appropriate. The surface of these rather
inert compounds has to be modified chemically by inoculating
the group that is capable of reaction. If these groups are
bonded completely during a subsequent interaction with the
enzyme, it may be assumed that the initial level of hemocompatibility of these surfaces will not be impaired. However,
if a part of the group does not react with the enzyme and remains free, the danger of impaired hemocompatibility arises.
Hence, the remaining functional groups must be additionally
neutralized with low-molecular-weight reagents which do not
inhibit enzymatic action or increase thrombogeneity.

In the utilization of immobilized enzymes, it is very
important that such derivatives have the capacity to retain
their initial activity for a long time during multiple use
in enzymatic reactions. In general, the matrix on which an
enzyme is immobilized can either increase or decrease its stability. Similar to organic solvents, the carriers with hydrophobic cores are able to denature enzyme molecules. Positively or negatively charged hydrophilic matrices can either aug-
ment or reduce the stability of immobilized enzyme derivatives
via electrostatic interaction between the molecules of an enzyme and a carrier. As shown in experimental studies [14],
if the enzymes immobilized on hydrophilic matrices are kept
at 4°C, no decrease in their activity is registered for 7
months.

Insoluble derivatives of proteolytic enzymes on hydrophobic polymers do not alter their activity when kept in the
form of a water suspension, but are completely inactivated
during lyophilization [15].

The thermal stability of enzymes immobilized by covalent
binding and adsorption is reduced compared to that of soluble
forms [16], although some authors [9] report that the thermal stability of immobilized enzymes is much higher. An opinion was expressed in a review article [5] that the reduced
stability of adsorbed or covalently bound enzymes depends
on a decreased probability of the enzyme molecule returning
to an active conformation following the thermal changes. However, this does not take into account that, because of the
loss of one, two, or more degrees of freedom, the probability
of such a change is lower compared to a soluble enzyme.

The stability of enzymes in relation to the alteration
of medium activity depends on the type of polymer used for
immobilization. Resistance to change in acidic pH [17] is
improved if the enzyme is bound on a polyanionic matrix. Immobilization of enzymes on polycationic matrices increases
stability in an acidic medium and decreases stability in an
alkaline medium. According to Levin et al. [14], this effect
is accounted for by the microenvironment of the enzyme on the
matrix, since the pH and stability of enzymes immobilized on
neutral matrices remain unchanged. The stability of immobilized enzymes to a change in the active reaction of the medium
depends not only on the type of polymer used, but also on the
chemical enzyme modification caused by covalent binding, i.e.,
on the kind of charged groups of the protein enzyme molecule
which participate in immobilization [5].

The effect of enzyme binding on the enzyme's kinetic parameters results from (a) limitation of substrate diffusion to the enzyme, (b) steric obstacles in the way of substrate interaction with the enzyme, (c) chemical enzyme modification, and (d) microenvironment of the enzyme on the matrix. Study of the kinetics of chymotrypsin bound with Sepharose [18] showed that the K_M of the bound enzyme is ten times higher than that of the native one. After solubilization of the insoluble derivative under the action of dextranase, the K_M value equals that of the native enzyme.

As in heterogeneous catalysis, the rate of substrate diffusion to the catalyst surface plays a significant role in determining the kinetics of the reaction. Insoluble enzyme particles in water suspension are surrounded by an unmixable layer of the solvent, the width of which is conditioned by the rate of stirring. This circumstance may play a definite role in the functioning of the enzyme that is immobilized on the surface of an artificial circulation apparatus or vascular grafts. The substrate concentration gradient is permanent across the layer. To saturate active centers of the carrier-bound enzyme by the substrate, a higher concentration of the substrate is required compared to the enzyme saturation in the solution. In reality, it is expressed in an increase of the Michaelis constant. Full activity of the enzyme is manifested only when the local concentration of the substrate exceeds the K_M value for the native enzyme [16].

Specific activity of immobilized proteases with respect to proteins is considerably lower than that of the corresponding soluble derivatives. In most cases a decrease in proteolytic activity with respect to high-molecular-weight substrates can be accounted for by steric obstacles to the substrate—enzyme interaction caused by the matrix. A fall in the rate of protein hydrolysis by immobilized proteolytic enzymes is often accompanied by a decrease in the number of peptide links sensitive to hydrolysis [19]. The effect of covalent enzyme binding with a polymer is also mediated via a conformational change of the enzyme molecule, its active center, or an area adjoining the active center. These changes are analogous to those caused by low-molecular-weight reagents. These effects and phenomena should be taken into account when selecting a polymer and an enzyme immobilization technique that will impact the hemocompatibility characteristics of the surface.

TABLE 7.1. Hemocoagulation Characteristics of Proteases Immobilized on Polyaminestyrene Resin

Enzyme	Prothrombin activation (sec)	Fibrinogen clotting (sec)	Plasminogen activation (mm/h)	Fibrin lysis (mm/h)	BAEE* and TEE† hydrolysis (cat/M²)
Trypsin	–	–	0.6	0.3	$1.01 \cdot 10^{-4}$
Thrombin	5	25	–	–	$5.07 \cdot 10^{-4}$
Plasmin	–	–	0.3	0.2	$0.49 \cdot 10^{-4}$
Papain	–	–	0.3	–	$0.82 \cdot 10^{-4}$
Chymotrypsin	–	–	0.4	–	$3.1 \cdot 10^{-4}$
Urokinase	–	–	0.5	–	–
Control	–	–	–	–	–

*BAEE = N-benzoyl-L-arginine ethyl ester.
†TEE = L-tyrosine ethyl ester.

Our preliminary studies [10] have shown that enzymes
(trypsin, thrombin, plasmin, papain, chymotrypsin, urokinase)
immobilized on polyaminestyrene resin retain their specificity (Table 7.1). Fibrinolytic and autocatalytic activity
was found in immobilized plasmin. Direct fibrinolytic action
is achieved only during an immediate contact of the surface-
bound immobilized enzymes with fibrin. The absence of enzyme
diffusion into the underlying layers of fibrin points to the
firmness of covalent binding of the proteases with the carrier. Two of the other enzymes (papain and chymotrypsin)
lost the ability to cause direct fibrin proteolysis after
covalent binding to the polymer surface. They retained only
the fibrinolytic action mediated by plasminogen activation.
The same type of action was detected in immobilized urokinase
but it is specific for this enzyme.

The experimental data support the conclusion that only
trypsin and plasmin, of all the enzymes studied, retain direct
fibrinolytic action after immobilization on the surface of a
polyaminestyrene polymer.

Proceeding from this conclusion, we decided to impart
to vascular prostheses the ability to remove the fibrin deposits from their surface. To immobilize the enzymes on
crimped Kapron grafts we used a procedure for which the chemical reaction is given below:

Plasmin, papain, trypsin, streptokinase, and chymotrypsin
were immobilized on Kapron with glutaraldehyde. The prostheses were treated with 3.65 M hydrochloric acid at 40°C for
30 min with subsequent rinsing in distilled water and sodium-
borate buffer (0.1 M, pH 8.0). The reaction between the matrix and glutaraldehyde was carried out in the same buffer.
The prostheses were successively washed of the glutaraldehyde

excess with distilled water, 0.2 M sodium–borate buffer, and
again with distilled water. Immobilization was carried out
using the enzyme solution (1% protein concentration) in 0.05
M KH_2PO_4 (pH 7.0). The prostheses were washed of noncovalently bound proteins with 1 M solution of cesium chloride in 0.1
M sodium–borate buffer (pH 8.0) to zero extinction at a 280 nm
wavelength. They were then washed with sterile physiological solution, placed into polyethylene containers, covered with donor blood, hermetically packed, and, after obturat-
ing clot formation, incubated in a thermostat at 37°C. Following 24 h, the containers were opened; using weight analysis, the percentage of clot lysis was calculated compared to
the control grafts devoid of immobilized enzymes. The examined Kapron prostheses with immobilized papain, streptokinase,
plasmin, trypsin, and chymotrypsin had the capacity to lyse
human blood clots by $60 \pm 0.6\%$ ($p < 0.01$), $78 \pm 1.3\%$ ($p <$
0.001), $57.5 \pm 4.1\%$ ($p < 0.001$), $39.4 \pm 6.7\%$ ($p < 0.01$), and
$12 \pm 2.6\%$ ($p < 0.05$), respectively.

Covalent trypsin binding with Lavsan was carried out according to the scheme below:

To increase the amount of free carboxylic groups, partial
surface hydrolysis with 3 M HCl at 40°C for 30 min was performed. The polymer powder was washed of the acid and vacuum-
dried over alkali to constant weight. The dried Lavsan
was then placed into a retort and treated with a mixture of
anhydrous thionyl chloride and absolute methanol while boil-

ing with reverse cooling for 2 h. The Lavsan was then transferred to a glass filter, washed with absolute methanol, and
vacuum-dried to a constant weight.

The dried powder (100 mg) was covered with anhydrous
hydrazine hydrate and incubated for 2 h at 45°C. The product was filtered through a glass funnel to remove excess
hydrazine hydrate, vacuum-dried over sulphuric acid, and subsequently used for the activation with glutaraldehyde. Fifty
milligrams of the obtained polymer were treated with 5 ml
of 12.5% glutaraldehyde at 37°C for 1 h. Excess aldehyde
was removed by rinsing on a glass filter with 0.05 M potassium-phosphate buffer, pH 6.8. Trypsin binding with activated Lavsan was carried out in the same buffer using enzyme
solution (1% protein concentration). Lavsan was washed of
the noncovalently bound enzyme with 0.05 potassium-phosphate
buffer (pH 6.8) with sodium chloride, and then with isotonic
sodium chloride to zero extinction at 280 nm. Before use,
the ready preparation was kept in the same solution at 0-4°C.

Plasminogen activation was studied by the lysis of a 1%
fibrin—plasminogen column following accretion of the immobilized enzyme on its surface. Direct fibrinolytic action was
similarly studied but, in this case, a 1% fibrin—plasminogen
column was first warmed at 80°C for 45 min to inactivate the
plasminogen. In the control experiments we used a Lavsan derivative without trypsin immobilization prior to the reaction
with glutaraldehyde.

Immediate fibrinolytic action was 1.28 ± 0.1 mm/h (p <
0.001) and, together with plasminogen activation, amounted
to 2.64 ± 0.2 mm/h (p < 0.001). In the control Lavsan had
neither fibrinolytic nor plasminogen-activating activity.

Besides immediate immobilization on the surface of polymers, procedures for covalent streptokinase-spacer binding
to a silicone rubber film [20] have been suggested. The chemical activity is shown below. Specifically, urokinase was
immobilized on nylon [21]. Polyethylenimine and copolymer of
methylvinyl ester of maleic aldehyde were used to increase the
amounts of immobilized urokinase.

Immobilization of urokinase and streptokinase with a
spacer makes it possible to obtain high fibrinolytic activity
on the surface of polymers. Direct fibrinolytic action of
trypsin immobilized on Lavsan is only part of the total fib-

$$\overset{|}{\underset{|}{\text{—OH}}} \quad \xrightarrow{\text{CNBr}} \quad \overset{|}{\underset{|}{\begin{array}{c} \text{—O} \\ \text{—O} \end{array}}}\!\!\!\!>\!\!\text{C=NH} \quad \xrightarrow{\overset{O}{\overset{\|}{\text{HO—C—(CH}_2)_5\text{—NH}_2}}}$$

$$\overset{|}{\underset{|}{\begin{array}{c} \text{—O} \\ \text{—O} \end{array}}}\!\!\!\!>\!\!\text{C=N—(CH}_2)_5\!\!\overset{O}{\overset{\|}{\text{—C—OH}}} \xrightarrow[\text{H}_2\text{N—enzyme}]{\text{R—N=C=N—R}} \overset{|}{\underset{|}{\begin{array}{c} \text{—O} \\ \text{—O} \end{array}}}\!\!\!\!>\!\!\text{C=N—(CH}_2)_5\!\!\overset{O}{\overset{\|}{\underset{\underset{\text{H}}{|}}{\text{—C—N}}}}\!\!\text{—enzyme}$$

rinolytic potential of the immobilized enzyme. Plasmin activity, due to plasminogen transformation into an active enzyme under the action of immobilized proteases, is responsible for nearly half of the fibrinolytic effect. Fibrinolytic action of immobilized urokinase and streptokinase preparations is conditioned only by this mediating influence.

Other than the immobilization of fibrinolytic enzymes on the surface of Kapron and Lavsan, we have experimented with commercial polyethylene subclavian catheters. After immobilization of a specific protease on their surface, the catheters acquire fibrinolytic properties. When such catheters are immersed into fibrin—plasminogen—agar gel, fibrin lysis takes place around its surface (Figure 7.2).

Factors affecting the fibrinolytic activity of the catheters were studied. The effect of air storage duration (at room temperature) on the fibrinolytic activity of proteases immobilized on the catheter surface is given in Table 7.2. Table 7.2 shows that the catheters exhibit the highest fibrinolytic activity within a week and retain a high level of activity subsequently. Thus, enzymes immobilized on a polyethylene surface and kept under the above conditions do not lose their fibrinolytic characteristics. Consequently, such catheters can be delivered to hospitals after fibrinolytic enzyme immobilization without special precautions.

Because catheters introduced into a vessel sometimes remain there for several days, we also studied the fibrinolytic activity of enzymes immobilized on their surface during the interaction with human blood serum (Table 7.3). This table shows that the fibrinolytic activity of immobilized proteases

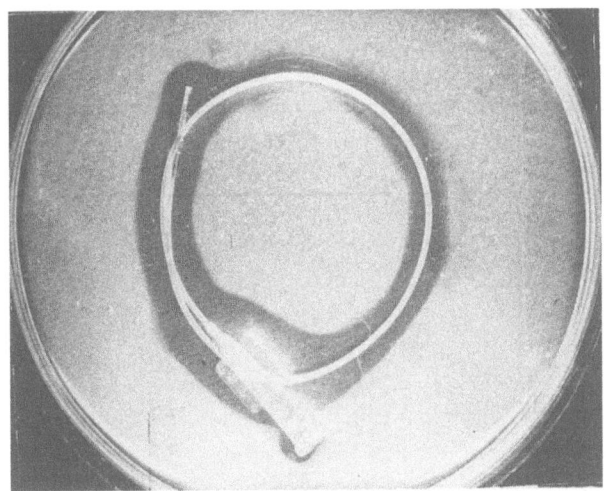

Fig. 7.2. Fibrinolysis around a polyethylene subclavian
           catheter modified by the fibrinolytic enzyme and
           placed into a fibrin—plasminogen—agar medium ac-
           cording to Astrup and Mullertz [22].

TABLE 7.2. Fibrin—Plasminogen—Agar Gel Lysis by
            Catheters

Storage duration (days)	Fibrinolysis at 37°C after 16 h (min)
0	22
7	20
15	16
25	11

is decreased gradually as during air storage. The rate of
decrease in fibrinolytic activity in the blood serum is
higher, however, which can be accounted for by the action of
natural proteinase inhibitors (antiplasmin, α_2-macroglobulin,
α_1-antitrypsin, α_1-antichymotrypsin, Cl-inactivator) contained in human serum. One should note that the data in Table
7.3 demonstrate that the fibrinolytic activity of the catheter is retained for a sufficiently long time, comparable to
the usual period that catheters are maintained in human vessels.

TABLE 7.3.   Effect of Time of Human Serum Action
             on Fibrinolytic Activity of Modified
             Catheters

Time of contact with human serum	Fibrinolytic activity after 16 h, 37°C (mm)
30 min	20
2 h	20
12 h	20
24 h	20
2 days	18
4 days	17
6 days	15
8 days	12
10 days	10
11 days	7
12 days	4
13 days	2
14 days	0

Graphite—polyvinylchoride rings with immobilized urokinase retain local fibrinolytic activity [23] 10 days after implantation into the dog aorta, although no increase in fibrinolytic activity is revealed in the general circulation.

All of the data obtained demonstrate that, when approp-
riately modified, fibrinolytically active agents (plasmin,
urokinase, streptokinase) can be covalently immobilized on
the surface of certain inert polymers which are presently
used to manufacture blood-contacting devices. These agents
can directly cause (immobilized plasmin) or mediate (streptokinase and urokinase) the lysis of fibrin fibers accumulating
on the surface of foreign materials (Figure 7.3). A fibrin
net is accumulated on a foreign surface because of the activation of the blood coagulating system by the contact, and it
falls out of the bloodstream due to continuous physiological
microclotting. Fibrin is also lysed either directly by immobilized plasmin (to a lesser extent) or by activated plas-

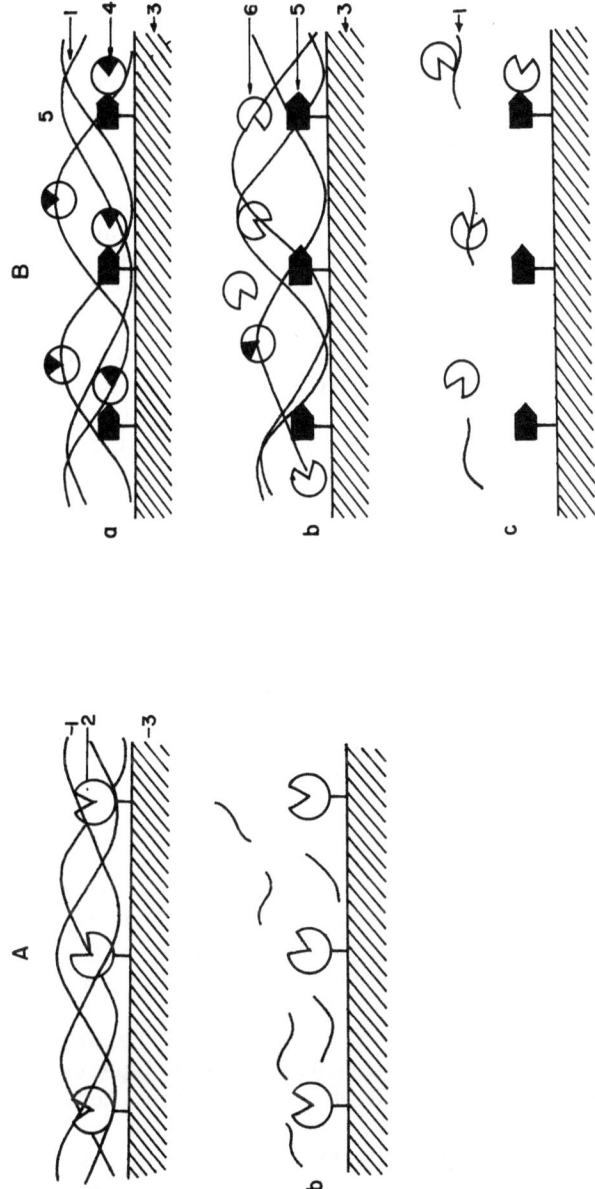

Fig. 7.3.   Scheme of direct and mediated enzymatic fibrinolysis on a polymer. A)
Direct fibrinolysis by immobilized plasmin; B) mediated fibrinolysis
by immobilized urokinase or streptokinase. 1) Fibrin and its degradation products; 2) immobilized plasmin; 3) polymer carrier; 4) nonimmobilized plasminogen; 5) immobilized plasminogen activator; 6) nonimmobilized plasmin. a) Accumulation of a fibrin net on the polymer;
b) plasminogen activation by an immobilized activator; c) cleaning of
the polymer surface.

minogen activator (nonimmobilized plasmin) present in the
complex with the accumulated fibrin and circulating freely in
the blood (to a greater extent). This process continues without interruption until inactivation of the enzyme.

The amount of immobilized, fibrinolytically active enzyme per polymer surface unit should be assessed based on the
purpose of the device. Since artificial heart valves, vascular grafts, and catheters have a small area, the density
of enzyme saturation may be considerable. In contrast, a
vast blood contact with the surface of dialyzer membrane oxy-
genators and hemosorbents requires a low density of fibrinolytically active enzymes, since excessive fibrinolysis may
lead to a disturbance in the hemostasis of the patient. Preservation of fibrinolytic activity for several days answers,
for practical purposes, the major medical requirements for
catheters, extracorporeal circulation, and vascular grafts.
Further experimental and clinical testing of fibrinolytically
active polymers under in vivo conditions appears to have high
priority.

REFERENCES

1. D. M. Zubairov, A. V. Repeikov, and V. N. Tomerbaev,
   "On vascular endothelium moistening," Fiziol. Zh. SSSR,
   49, 85-91 (1963).
2. D. M. Zubairov, Z. Z. Asadullina, and L. G. Popova,
   "Hageman interaction with surfaces," Probl. Hematol.
   Pereliv. Krovi, 10, 49-53 (1971).
3. D. M. Zubairov, Biochemistry of Blood Clotting, Meditsina, Moscow (1978).
4. Ph. R. Craddock, J. Fehr, A. P. Dalmasso, K. L. Brigham,
   and H. S. Jacob, "Hemodialysis leukopenia. Pulmonary
   vascular leukostasis resulting from complement activation by dialyzer cellophane membranes," J. Clin. Invest.,
   59, 879-888 (1977).
5. R. Goldman, L. Goldstein, and E. Katchalski, "Water-
   insoluble enzyme membrane," in: Biochemical Aspects of
   Reactions on Solid Supports, G. R. Stark (ed.), Academic
   Press, New York (1971).
6. A. A. Lurje, Chromatographic Materials, Khimiya, Moscow
   (1978).
7. P. Bernfeld and J. Wan, "Antigens and enzyme made insoluble by entrapping them into lattices of synthetic polymers," Science, 142, 678-679 (1963).

8. G. Kay and F. M. Crook, "Coupling of enzyme to cellulose
    using chloro-s-triazines," Nature, 216, 514-515 (1962).
9. R. Axen and S. Ernback, "Chemical fixation of enzymes
    to cyanogen halide-activated polysaccharide carriers,"
    Eur. J. Biochem., 18, 352-360 (1971).
10. D. M. Zubairov and O. D. Zinkevitch, "Hemocoagulating
    properties of immobilized proteases," Vopr. Med. Khim.,
    2, 187-191 (1976).
11. S. Avramees and T. Ternynck, "Cross-linking of protein
    with glutaraldehyde and its use for preparation of im-
    munosorbents," Immunochemistry, 6, 53-66 (1969).
12. L. G. Popova, D. M. Zubairov, O. D. Zinkevitch, and G. Y.
    Konchurina, "Effect of immobilized norepinephrine on
    arginine-esterase of human blood plasma activated by
    factor XII," Biokhimiya, 42, 403-407 (1977).
13. I. A. Andrushko, Z. Z. Asadullina, D. M. Zubairov, and
    N. Y. Dobrova, "A technique for studying polymer thrombo-
    resistance," in: Abstracts, Vth USSR Symposium on Syn-
    thetic Polymers for Medical Purpose, Ministry of Health
    of the USSR, Riga (1981), pp. 104-105.
14. I. Levin, M. Pecht, L. Goldstein, and E. Katchalski, "A
    water-insoluble polyanionic derivative of trypsin. 1.
    Preparation and properties," Biochemistry, 3, 1905-1913
    (1964).
15. L. Goldstein, "Water-insoluble derivatives of proteolyt-
    ic enzymes," Methods Enzymol., 19, 935-962 (1970).
16. R. Goldman, O. Kedem, and E. Katchalski, "Papain colloid
    in membranes. II. Analysis of the kinetic behavior of
    enzymes immobilized in artificial membranes," Biochem-
    istry, 7, 466-500 (1968).
17. L. Goldstein, I. Levin, and E. Katchalski, "A water-
    insoluble polyanionic derivative of trypsin. II. Effect
    of polyelectrolyte carrier on the kinetic behavior of
    the bound trypsin," Biochemistry, 3, 1913-1919 (1964).
18. R. Axen, P. A. Myrin, and J. C. Janson, "Chemical fixa-
    tion of chymotrypsin to water-insoluble crosslinked dex-
    tran (Sephadex) and solubilization of the enzyme deriv-
    atives by means of dextranase," Biopolymers, 9, 401-403
    (1970).
19. I. H. Silman and E. Katchalski, "Water-insoluble deriv-
    atives of enzyme, antigens, and antibodies," Annu. Rev.
    Biochem., 35, 873-908 (1966).
20. A. S. Hoffman, G. Schmer, C. Harris, and W. G. Kraft,
    "Covalent binding of biomolecules to radiation-grafted
    hydrogels on inert polymer surfaces," Trans. Am. Soc.
    Artif. Intern. Organs, 8, 10-15 (1972).

21. A. Sugitachi, K. Takagi, S. Imaoko, and G. Kosaki, "Immobilization of plasminogen activator, urokinase, on nylon," Thromb. Haemostas., 39, 426-436 (1978).
22. T. Astrup and S. Mullertz, "The fibrin plate method for estimating fibrinolytic activity," Arch. Biochem. Biophys., 40, 346-352 (1952).
23. B. K. Kusserov, R. W. Larrov, and J. E. Nichols, "The surface-bonded covalently cross-linked urokinase synthetic surface. In vitro and chronic in vivo studies," Trans. Am. Soc. Artif. Intern. Organs, 9, 8-12 (1973).

MICROBIOLOGICAL SYNTHESIS OF PROTEOLYTIC ENZYMES POSSESSING FIBRINOLYTIC ACTIVITY

N. S. Egorov

Chair of Microbiology
M. Lomonosov Moscow State University, Moscow

ABSTRACT

Enzymes of microbic origin which possess thrombolytic activity in human blood are studied. The physiology of microorganism producers is investigated and the optimum conditions for biosynthesis of active fibrinolytic enzymes are determined. The role of induction and catabolic repression, and the specific and nonspecific effects of amino acids on protease formation are studied in detail. New methods for obtaining mutants, i.e., enzyme supersynthetics, are described. An ecological approach (using a combined cultivating technique), which provides for effective synthesis of highly active proteases with fibrinolytic activity, is applied.

GENERAL CHARACTERISTICS OF PROTEASES WITH FIBRINOLYTIC ACTIVITY

Proteolytic enzymes are attractive to researchers not only because of their widespread occurrence and significance in life, but also because of their ever-increasing potential for application in society. Proteases are used in leather, film, and photographic industries, in brewing, in baking bread, in everyday chemistry and agriculture, and in the food industry for tenderizing meat products and processing meat and fish

197

wastes. They are also used in the microbiological industry
to prepare culture media, in theoretical biology, and mainly
in medicine for the treatment of suppurations, infected
wounds, burns and inflammations, some tumors, and thrombolysis.

Experimental study of thrombolytic drugs has advanced
considerably in recent years. Possible ways of treating disturbances in hemostasis in various types of pathology have
been devised. However, the range of drugs available which
physicians need in clinical settings is still not large, while
the number of diseases accompanied by thrombogenesis in the
main vessels and in the microcirculation increases.

For this reason, the ability of microorganisms to lyse
thrombi in human blood arouses the interest of scientists
around the world in microbes that are notable for a high
growth rate, short developmental cycle, and ability to grow
in simple and advantageous culture media.

In contrast to the microbic activators of the fibrinolytic system, microbic proteolytic enzymes which lyse thrombi in
human blood via a direct hydrolytic effect on fibrin possess
neither pyrogenicity nor antigenicity and are incapable of
activating the blood coagulation system. These properties
and the positive results from medical and biological trials
on preparations of microbic proteases with thrombolytic action demonstrate that such compounds are promising in the
treatment of thromboses.

MICROORGANISMS AND FIBRINOLYTIC PROTEASE PRODUCTS

In the Soviet Union, the first studies of proteases capable of lysing thrombi in human blood were conducted in 1962
in the Department of Microbiology, Moscow State University,
during examination of a series of microscopic fungi [1-3].
The possibility of forming thrombolysing proteases by groups
of microorganisms not known to have this capability was studied later. Examination of 800 cultures of microorganisms in
different media showed that the ability to synthesize fibrinolytic enzymes is spread widely among microbes (Table 8.1).
These enzymes can be formed by the following bacteria: representatives of the Bacillus and Clostridium genera [4], mesophilic and thermophilic Streptomyces [5-8], proactinomyces
Nocardia [6, 9, 10], Mycobacterium [6, 11], mold fungi from

TABLE 8.1. Formation of Proteolytic Enzymes Possessing Fibrinolytic Activity by Various
Groups of Microorganisms

Microorganism group	Strains possessing fibrinolytic activity (%)
Bacteria	65
Streptomyces mesophilic	97
thermophilic	78
Nocardia	61
Mycobacterium	36
Mycococci	20
Microscopic fungi	76

Penicillium and Aspergillus genera [1, 12, 13], and defective
fungi from Alternaria, Cladosporium, and Fusarium [14] (Table
8.1). Fungi and Streptomyces were studied most thoroughly.

Due to their ability to synthesize proteases, molds from
the genus Aspergillus have been studied extensively at various laboratories. Examination of 184 Aspergillus cultures
relating to 24 various species demonstrate that this fibrino-
lysing ability is intrinsic to many of the species of fungi
studied (82% of the total amount). Fibrinolytically active
cultures are most common for the species Asp. carneus, Asp.
fumigatus, Asp. candidus, Asp. terreus, and Asp. nidulans,
which have been the least studied in this respect. For the
most part, the fibrinolytic activity of mold fungi of the
genus Aspergillus correlated with the thrombolytic activity.

Defective fungi of Fusarium, Alternaria, and Cladosporium are interesting ecological groups. Among them, more than
80% of the cultures possess fibrinolytic activity. The fact
that most of them can form proteolytic enzymes is certainly
associated with their specific existence in nature and the
need to grow on other living organisms.

In examining Streptomyces, which have been studied previously primarily as antibiotic producers, it was found that
most of the microorganisms form proteases that produce fib-
rino- and thrombolytic effects. Thermophilic Streptomyces,

which are known to grow at high temperatures, have a short
evolutionary cycle, have high activity, and are thermostable
enzymatic systems, are of special interest [7, 15, 16].

Mesophilic Streptomyces are also of interest, especially those which form both antibiotics and thrombolytic enzymes,
thus permitting the cultivation of two physiologically active
compounds with the expense of only one culture.

Producers of highly active proteases of fibrinolytic action range from 20 to 60% among Nocardia, Mycobacterium, and
mycococci.

Until the mid-1960s, proteolytic enzymes of putrefactive
bacteria were thought to be either devoid of thrombolytic
activity or capable of activating the fibrinolytic system of
human blood [17, 18].

Culturing of putrefactive bacteria from the Bacillus
subtilis and mesentericus group on various media compositions
demonstrated to us in 1968 [4] that 80% of the bacterial
strains studied are characterized by their ability to synthesize fibrinolytic enzymes which can lyse experimentally induced thrombi. This observation emphasized the importance
of creating optimum conditions for the culturing of a microorganism in order to determine its properties.

Of more than 800 microorganisms studied, 601 cultures
(more than 73%) were found to possess lytic activity toward
fibrin and thrombin in human blood. However, it should be
noted that the ability to form proteolytic enzymes having
fibrinolytic action is, to a large extent, expressed only
during the growth of these microorganisms on specific media
(giving consideration to the composition of these media).

The growth of organisms is slowed down in synthetic
media, as shown by most of the cultures studied, such as Bac.
mesentericus, Bac. subtilis, Asp. oryzae, Asp. fumigatus, Nocardia sp. 1, and Rhodococcus rubropertinctus. Culturing of these
producers on complex natural media under refined conditions
is accompanied by a decrease in the fibrinolytic activity of
the culture fluid.

This dependence of the amount of proteases formed on the
composition of the media indicates that physiologic ways of
increasing the proteolytic activity of microorganisms are very
promising.

MECHANISM OF EFFECT OF PROTEOLYTIC ENZYMES
ON FIBRIN IN BLOOD

There are data [19] which indicate that the fibrinolytic activity of microorganisms is dependent on the formation of β-type hematoxins. Thus, it has been concluded that fibrinolytic enzymes are toxic by nature. It is also known that erythrocyte hemolysis may be induced by substances of various origins and natures, such as proteases, saponins, ricin, lecithinases, and lipolytic substances such as alcohol, ether, and chloroform. Even in healthy human organisms, moreover, there is an equilibrium between the erythrocytolytic enzymes and their inhibitors. Our studies show no direct dependence between fibrinolytic activity and the size of the hemolysis zone. Erythrocyte hemolysis caused by the culture fluid of putrefactive bacteria, Nocardia, Streptomyces, and Mycobacterium sometimes occurs when they lack fibrinolytic activity. In some microorganisms that possess fibrinolytic activity, hemolysis was not observed. In addition, in thoroughly boiled solutions of some Streptomyces fibrinolytic preparations, hemolytic properties were preserved.

These data do not permit us to identify microbic proteases with hemotoxins and, so, saprophytic microorganism proteases are regarded as promising thrombolytic agents. Either activators catalyzing the conversion of blood protofibrinolysin to its active thrombolysing form (fibrinolysin) or proteases of direct action, capable of splitting arginylglycine bonds of fibrin and thereby converting it to a soluble form, can lyse intravascular thrombi. We have studied many microorganisms and have found that microbic enzymes, in general, produce a direct proteolytic effect on fibrin in human blood by breaking the arginylglycine bonds. Some microorganisms (Penicillium lilacinum Thom., Asp. fumigatus, Asp. candidus, Streptomyces spheroides), however, form fibrinolytic enzymes which have a weak ability to activate the conversion of blood profibrinolysin to fibrinolysin.

Both the coagulation of blood and the lysis of clots formed in the human organism are closely coordinated, with the main role played by protease inhibitors occurring in blood plasma [20]. For this reason, we decided that it was significant to determine to what extent the proteases under study were sensitive to the inhibitors mentioned. The selected, highly active strains of microorganisms synthesized the enzymes sensitive to plasma inhibitors in in vitro experiments.

The degree of sensitivity depended, first, on the species of the producer. Thus the enzymes formed by Asp. candidus, Asp. ochraceus, and some other microorganisms show weak sensitivity to the inhibitors. Second, fibrin concentration was also prominent. Microbic protease preparations and proteases of animal origin (trypsin and blood protease fibrinolysin) at high concentrations (about 2 mg protein per 1 ml) did not decrease their enzymatic activity in the presence of plasma inhibitors. Complete inhibition of trypsin and fibrinolysin activity occurred upon a reduction of the enzyme concentration to 0.3 mg protein/ml, whereas some proteases (from Asp. candidus) performed this activity even at 0.03 mg protein/ml. Thus, microbic proteases seem to be more promising for in vivo use than proteases of animal origin.

We also found that heparin blocks the activity of microbic proteases in order to protect them from blood plasma inhibitors [21]. A combined administration of microbic proteases and heparin in the blood enables one to, first, increase thrombolysis and, second, decrease the dose of protease injected. The thrombolysing activity of aspergillin, obtained from the culture fluid of Asp. oryzae, Moscow University strain, averaged 24 h. With a combined administration of aspergillin and heparin, thrombolysis began 1-3 h after the injection. A threefold decrease in the dose of the preparation from the Nocardia culture fluid introduced and its combined administration with heparin resulted in 2 h of thrombolysis of a rat's jugular vein without the death of the animal.

We also showed that the sensitivity to blood inhibitors may be decreased (depending on the composition of the medium) due to a change in the relationship among the components of the protease complexes formed by the microorganisms.

Thus, the choice of microorganism strain sensitive to blood inhibitor, the creation of appropriate culturing conditions, and the combined administration of microbic enzyme preparations with heparin permit the use of small doses of proteases of microbic origin and, thereby, decreases the rejection of heterologous protein by the immune system.

METHODS FOR INCREASING PROTEASE FORMATION
BY MICROORGANISMS

The need to produce proteolytic enzymes on a large scale accentuates the issue of increasing the yield of microbic

protease. One method for approaching this problem is to obtain mutants which demonstrate increased synthesis of enzymes.
To this end, diethylsulfate vapor has been used to obtain
the strain Streptomyces odorifer 492, which, by its activity
to fibrinolysis, exceeded the initial culture two- to threefold [22]. However, the use of such effective mutagenic factors often leads to the occurrence of deficient, weakly growing microorganisms which require complex and expensive media
for growth.

Therefore, in addition to conventional techniques of mu-
tagenesis, we have used our newly developed method for obtaining mutants under the influence of electromagnetic waves within the millimeter range [23]. These waves produce no thermal
effect on the organism and leave the cellular structure intact. However, they cause a change in polar orientation of
the molecules, which leads to the new indications of reproduction. Our experiments demonstrate that a milliwave irradiation of Asp. oryzae, Moscow University strain, results in a
2- to 2.5-fold increase in the synthesis of fibrinolytic protease, accompanied by an increase in the productibility of
cells and in sporulation. Both the indication of protease
formation and the qualitative composition of the protease complex synthesized by the organism depend greatly on electromagnetic field parameters such as the wavelength and short-
term irradiation in particular [24].

Combined Cultivating Technique

We were also the first to create another technique for
increasing the yield of microbic proteases which possess fib-
rino- and thrombolytic activity, i.e., a combined culturing
of two microbic species. This physiologic method is simple
and requires no additional expenditures for culturing.

In our experiments we studied the peculiarities of thrombolytic enzyme biosynthesis in mixed/combined cultures composed of Streptomyces, fungi, and corynebacteria. By mixing
the cultures, we tried to create the conditions under which
these microorganisms exist naturally and have to synthesize
such biologically active compounds as thrombolysing proteases.
The components of microbic associations were selected according to protease formation and, therefore, mixed cultures consisting of either two active producers of thrombolytic enzymes
and two inactive microorganisms, or a protease producer together with an inactive partner (Table 8.2), were selected.

TABLE 8.2. Biosynthesis of Proteolytic Enzyme with Combined Cultivation of Micro-organisms

Test variant	pH (end)	Biomass (g/liter)	Fibrinolytic activity (FU/ml)
Streptomyces spheroides 35	6.8	4.9	560
Streptomyces rimosus	7.8	6.6	780
Str. spheroides + Str. rimosus	7.5	6.38	1400
Fusarium sp.	8.3	5.35	749
Aspergillus kanagawaensis	7.0	4.84	480
Fus. sp. + Asp. kanagawaensis	8.2	4.95	213
Aspergillus oryzae 168	7.3	4.99	800
Asp. oryzae 168 + Asp. kanagawaensis	7.0	5.2	999
Nocardia sp. 1	6.4	2.09	1600
Arthrobacter ramosus	5.8	5.8	270
Corynebacterium ulcerans	6.3	1.85	649
Nocardia sp. + Arth. ramosus	6.6	1.37	2100
Nocardia sp. + Coryn. ulcerans	6.2	1.09	2561

Streptomyces violaceus	7.4	6.14	0
Fusarium oxysporum	8.0	6.0	0
Aspergillus wentii	6.1	2.8	0
Fus. oxysporum + Asp. wentii	7.5	6.38	0
Arthrobacter citreus	6.2	1.3	0
Str. spheroides 35 + Str. violaceus	7.3	6.06	540
Str. rimosus + Str. violaceus	7.6	6.15	1840
Asp. kanagawaensis + Asp. wentii	7.2	5.0	952
Nocardia sp. 1 + Arth. citreus	6.3	3.43	2894
Asp. oryzae 168 + Fus. oxysporum	8.5	6.52	130

Two active producers of thrombolytic proteases (fungi, Streptomyces, or corynebacteria) grew well, as a rule, upon combined cultivation. However, the enzymatic activity of the combined cultures was very much species-dependent. In mixture, Streptomyces demonstrated total activity while fungi exhibited weaker activity compared to the initial cultures. The enzymatic activity of corynebacteria was somewhat increased.

In a combined cultivation of two microorganisms incapable of synthesizing the proteases secreted, enzymatic activity was exhibited in neither of the test variants and the growth of the microorganisms was normal. The most interesting results and the highest production of proteases were obtained by combining the active enzyme producer with the related inactive microorganism. The enzymatic activity of these mixtures increased two- to threefold and was not species-dependent.

In order to obtain productive associations, we used representatives from the same systemic and ecological groups. An increase in the formation of proteases performing a protective function was observed in combinations of microorganisms from the same ecological group, e.g., Asp. kanagawaensis + Asp. wentii, Str. rimosus + Str. violaceus or Str. sp. 5G, and Nocardia sp. 1 + Arthrobacter citreus.

A different relationship was found by the author [25] among the representatives of diverse ecological groups upon the cultivation of mold and phytopathogenic fungi. The development of a culture composed of Asp. oryzae 168, which is a highly active fibrinolytic enzyme producer, and Fusarium oxysporum, which is incapable of synthesizing these hydrolases (zero strain), showed that these fungi depress each other so that 80% of the fibrinolytic activity in the culture fluid is lost. To achieve intensive synthesis of thrombolytic enzymes in mixed cultivations of active and inactive microorganisms, the conditions must be created under which the growth of the inactive organisms is inhibited while the development of the producer remains normal [26]. Thus, the greatest amount of proteases under study is obtained either in the presence of 75-80% of the fungus producer and 25-20% of the inactive fungus or in the presence of 90% of the Streptomyces producer and 10% of the inactive partner. In the association with corynebacteria, one or two colonies of the inactive component Arthrobacter citreus appear among the multiple colonies of the producer at the phase of stationary growth [26].

TABLE 8.3. Conditions for Intensifying Exoprotease Biosynthesis in Combined Cultures of Microorganisms upon Simultaneous Incorporation of Both Cultures

Microorganism association	Quantitative ratio of cultures at the moment of incorporation	Medium for cultivation	Relationship of cultures during association development	Fibrinolytic activity (in % of control)
Streptomyces rimosus + Streptomyces violaceus	1.25% + 5%	Soybean medium for both cultures	90% Str. rimosus and 10% Str. violaceus	250
Streptomyces rimosus + Streptomyces sp. 5G	5% + 1.25%	Soybean medium for both cultures	5-10 colonies Str. sp. 5G in the complete growth of Str. rimosus	200
Aspergillus kanagawaensis + Aspergillus wentii	5% + 5%	Wort (4 Blg.) for Asp. kanagawaensis; synthetic medium for Asp. wentii	75-80% Asp. kanagawaensis and 25-20% Asp. wentii	200
Nocardia sp. 1 + Arthrobacter citreus BKM-654	5% + 1.25-5%	Beef-extract broth + 3% glucose for both cultures	1-2 colonies Arth. citreus in the complete growth of Nocardia sp. 1	180

The inactive organism is inhibited by the producer upon
incorporation into the media of various amounts of components at different times, upon the passage of the partner on
certain media before mixing, and because of the relationship
among the organisms during combined cultivation (Table 8.3)
[27]. The domination of one of the cultures in the association intensifies synthesis of the enzyme. This was experimentally proved to result from the interactions of the organisms affecting each other through the products of metabolism.

Study of the metabolic interactions in cultures during
combined cultivation showed that the development of a fungus
association is accompanied by the secretion from the inactive
component into the culture fluid of a substance which was excreted and identified as a hydroxyanthraquinonic pigment.
This compound increased the production of fibrinolytic enzymes by 2.8-fold, while productivity of the active mycelium
component increased 3.7-fold. De novo stimulation of protease synthesis in a pair of fungi was found [28] to proceed
at the level of transcription.

In corynebacteria combined culture (Nocardia sp. strain
1 + Arthrobacter citreus), proteolytic enzyme production was
shown to increase by 95% in Nocardia sp. The stimulator
formed by Arth. citreus was found to be a polysaccharide. It
was isolated, purified, and shown [29] to be composed of
three monosaccharides: glucose, galactose, and mannose.

It should be noted that the degree of increase in the
biological effect and its direction in the metabolites stimulating the biosynthesis of thrombolytic enzymes in microbic
associations depend greatly on the concentration of the metabolites. At high concentrations, they completely block synthesis of the enzyme and significantly inhibit growth of the
producer. Such dependence is typical for all of the combined
cultures of microorganisms studied.

Detailed study of the morphology, physiology, and kinetics of the growth of protease producers in both monocultures·
and associations showed considerable morphofunctional changes
and changes in the growth rates and enzyme biosynthesis in
the active components in mixed cultures. These changes were
also found to depend on the effect of stimulating factors,
i.e., the regulators of exoprotease biosynthesis are also involved in regulating the life cycle of microorganism producers [30].

The growth and development of active association components are somewhat due to the incorporation of stimulators. A
slowing down of the growth lag (by 6-12 h) has already been
observed in the germination of conidia (in fungi) and in
sporulation (in Streptomyces) of the inoculate. In protease
producers of these cultures, the formation and ripening of
mycelium of the second hypha generation and intracellular
sporulation lagged, and the rate of cell cytolysis decreased.
That the growth and development of microorganism producers
proceed with a lag was also shown by analyzing the growth
rates and enzyme synthesis in the presence of stimulating
factors.

In all of the pairs of microorganisms under study, the
stimulators induced marked changes in the morphology of cultures. For instance, the development of diffuse mycelium
(small, single parts of fungus mycelium) was observed in a
fungus producer affected by hydroxyanthraquinone. However, in
monoculture, the colonial type of mycelium (mycelium in the
form of pellets) was formed.

The polysaccharide bioregulator from Arth. citreus also
favored the increase of polymorphism in Nocardia sp. strain 1
during the entire developmental cycle.

In the Str. rimosus + Str. sp. 5G association, the protease producer developed as in the monoculture, i.e., in the
form of microcolonies, but with a more diffuse periphery. In
addition, we noted an unordered growth process expressed inside the hypha. It is of interest that both cytologic and
morphologic changes were observed in the inactive component
Str. sp. 5G upon combined cultivation. This actinomycete was
characterized by the following: loss of the capability to
form a capsule, delay in sporulation, hypersynthesis of pep-
tidoglycan, and callosity of the vascular wall — all of which
developed in parallel with hypertropy of the mesosomes.

Thus, we showed that the partners affect each other in
associations of microorganisms through various products of
metabolism excreted to the surrounding medium. The metabolites studied so far differ in both their chemical composition
and the way they affect another organism. These substances
may influence certain stages of biogenesis of physiologically active compounds and may be specific stimulators or inhibitors of the growth and development of microorganisms.

REGULATION OF PROTEASE BIOSYNTHESIS

Targeted biosynthesis of enzymes can proceed only after the regulatory mechanisms of their biosynthesis are studied thoroughly. These mechanisms permit microbic cells to use existing nutrients in the best way possible, following the principle of maximum economy. This specifies the ecological advantages of the species and makes it competitive in relation to other microorganisms and living organisms [31, 32].

Two processes — induction and repression — form the basis for the regulation of metabolism. Proteolytic enzymes, degrading proteins, and polypeptides are not only referred to as degradative enzymes, but they are also capable of amino acid formation for the purposes of biosynthesis. Their synthesis is therefore likely to be controlled by induction, catabolic repression, ammonium ion repression, change in membrane transport, level of active enzyme in the medium, and specific action of amino acids.

Induction of fibrinolytic enzyme synthesis in the presence of proteins has been found in a series of microorganisms. However, the set of inducing proteins and this type of regulation depend greatly on the features of the microorganism producer. For instance, casein and edestine are inducers for Asp. oryzae, peptin for Str. streptomycini, fibrinogen for the Streptomyces species, casein for Nocardia sp. 1, and albumin for Serratia marcescens VI. For most producers, incorporation of negligible amounts of protein (to 0.5%) into a synthetic medium increases the formation of proteases possessing fibrinolytic activity. The increase in protein concentration in the medium accentuates growth and decreases the level of enzymes. Experiments with specific inhibitors of protein synthesis demonstrate that proteins induce synthesis of fibrinolytic enzymes other than the release of protease molecules previously formed in the cell [33, 34].

The inducer must first penetrate into the cell, which is impossible in the case of such high-molecular-weight substances as proteins. Therefore, either information on the presence of protein in the media should be carried inside the cell via receptor sites located on the cell surface or, which is more likely, the basal enzyme degrades the protein to peptides and amino acids, which are true inducers. For example, in Serratia marcescens VI, leucine, i.e., a primary product of albumin enzymatic hydrolysis by this protease, appeared to induce protease formation on the medium with albumin [35]. Such amino acids as arginine, ornithine, and glutamine are inducers of the synthesis of proteases with fibrinolytic activity in Streptomyces thermovulgaris. Alpha-alanine stimulates the formation of proteases in Nocardia sp. 1.

Protease synthesis, however, can be induced in the presence of proteins only in some of the producers studied. For example, protein, as a single source of carbon, nitrogen, and sulfur in the medium for Asp. candidus and Asp. spheroides growth, does not induce but rather inhibits the biosynthesis of exoproteases by these organisms. Amino acids that are formed under the effect of protease on protein seem to make up for the nitrogen, carbon, and sulfur of the cell and to induce repression by metabolites [36, 37].

The lack of increase in the synthesis of exoenzymes on media with proteins permitted some conclusions about the constitutive nature of protease biosynthesis in Bacillus thuringiensis [38].

The biosynthesis of fibrinolytic enzymes by various microorganisms is also under the control of catabolic repression. Use in the media of readily available carbon sources leads, generally, to a decrease in the amount of proteases synthesized by the producer. For example, the existence in the medium of more than 1% glucose concentration inhibits the formation of enzymes in Asp. oryzae, Rhodococcus rubropertinctus, Bac. subtilis, and Bac. mesentericus [2, 39, 40]. The effect of catabolic repression may depend on both the glucose concentration in the medium and the nitrogen and phosphorus sources. Provided that the latter concentrations are limited, the inhibitory effect of sugars is decreased [41].

These easily assimilable sources, however, are likely to affect both catabolic repression and the regulation of cell membrane permeability. We showed [35] that the sensitivity to glucose of fibrinolytic protease synthesis in Serratia marcescens IV is not linked to the catabolic repression, but is specified by the inhibition of inducer transport into the cell, which is induced by oxalic and fumaric acids formed during glucose metabolism.

We have experimentally proved the hypothesis of some authors that glucose is capable of producing multiple effects depending on the time of its incorporation into the medium.

For example, while studying the effect of various carbon
sources on fibrinolytic enzyme formation in Bac. thuringiensis, it was found that the synthesis of this enzyme is exposed
to catabolic repression in an exponential growth phase. This
process is stimulated at the cell sporulation stage, mainly
because of the supply of an energy source to the synthesis
process. Interestingly, cyclic adenosine monophosphate (cAMP)
eliminates the inhibitory effect of the carbon sources on enzyme formation by the cells of the exponential growth phase.
However, AMP and adenine possess the same activity, which
causes the highest effect in a medium deprived of readily
metabolized carbon sources. Unlike cAMP and its analogs,
which have a nonspecific effect on protease synthesis, cyclic
guanosine monophosphate (cGMP) exhibits a specific effect,
i.e., it inhibits protease formation in the presence of additional carbon sources, increasing their intake by the cell.
cGMP stimulates enzyme synthesis by exponential cells in a
medium which lacks readily metabolized carbon sources. GMP
and guanine produce similar effects.

Thus, when constructive metabolism exchange is possible
in Bac. thuringiensis, the presence of nucleotides saves both
the energetic and constructive potential of the cell which
may be directed at exoprotein synthesis. This concerns spor-
ulating cells to a lesser degree when the construction stops.
Therefore, nucleotides occurring in the medium do not affect
exoprotease synthesis, especially in the presence of an additional source of energy [42].

Specific inhibition by ammonium ions of the synthesis of
most enzymes which are induced in nitrogen metabolism is widespread [43]. We observed the effect of nitrogen metabolic
repression in various microorganisms. It was shown that the
biomass of Serratia marcescens increased with an increase in
ammonium ion concentration in the medium. In the same situation, intracellular protease synthesis is inhibited and is
terminated at 0.3% concentration [44]. The addition of 0.1%
ammonium sulfate to the medium leads to a sharp increase in
biomass and a decrease in enzyme formation in Str. thermovulgaris [45]. The substitution of an ammonium source of nitrogen by a nitrate source causes a considerable increase in proteases in Fusarium graminearum and Asp. oryzae [2, 14]. Removal of the nitrogen source from the substrate stimulates
proteolytic enzyme formation in Str. spheroides [36]. Exoprotease synthesis by the cells of B. thuringiensis in its
exponential growth phase is sensitive to nitrogen metabolic

repression, while enzyme formation by sporulation is not repressed in the presence of high concentrations of ammonium
ions [38].

In studying the regulation of proteolytic enzyme synthesis in Asp. candidus, we found a different type of repression which we designated "metabolic repression." This type
of repression demands that readily metabolized carbon, nitrogen, and sulfur sources be present side by side in the medium
in order to inhibit protease formation in the organism [37].

As mentioned above, individual amino acids are actual
inducers, and thus can induce the biosynthesis of proteolytic
enzyme. We also observed multiple repressions of protease
synthesis by amino acids. In Str. spheroides the repression
occurred in the presence of D-alanine, in Nocardia sp.1, in the
presence of tyrosine, and in Asp. candidus, in the presence
of arginine, histidine, lysine, and ornithine [41].

In many cases, the repression by amino acids is restrict-
ed to the repression of ammonium ions. For example, asparaginic acid and asparagine inhibit, to a large extent, the
synthesis of alkaline and neutral proteases of Bac. mesentericus [40]. However, substitution of these amino acids by
their organic analogs, i.e., fumaric acid and malic acid, removes the repression and leads even to an increase in enzyme
synthesis.

Another possible explanation of the repressive effect
of amino acids follows. Proteolytic enzymes perform a double
function: Involved in catabolism, they build up stores of
amino acids for the purposes of biosynthesis. Their synthesis is therefore regulated through the repression by the end-
product, as in the case of anabolic enzymes. Since the mixture of amino acids is usually formed as a result of proteolysis, proteolytic enzyme synthesis is often controlled by
these amino acids. If, for example, a mixture containing 18
amino acids is added, this will lead to a total inhibition
of extracellular protease synthesis in Serratia marcescens VI
[44]. Protease synthesis in Str. thermovulgaris is repressed
completely due to the addition of a combination of five amino
acids: glycine, serine, proline, tyrosine, and glutamine [46].

Individual amino acids are also capable of repressing
enzyme synthesis to a degree, and the same amino acid is capable of both repressing protease formation from the enzyme

complex synthesized and of inducing the formation of another
protease. For example, asparaginic acid completely represses
the synthesis of neutral and acidic proteases in Str. spheroides and induces the synthesis of protease specific to fibrin [47].

Amino acids entering into the cell were found to make
up for a deficiency in carbon, nitrogen, and, probably, sulfur
sources, and to restore metabolic repression. Free intracellular amino acids were also shown [48] to be a component
in the chain of signals transmitting information on the presence of carbon, nitrogen, and sulfur sources in the medium;
in other words, they represent a physiological signal which
can "switch on/induce" and "switch off/inhibit" the system
of extracellular protease synthesis.

MODEL OF REGULATORY MECHANISM
OF EXTRACELLULAR PROTEASE SYNTHESIS

Based on our experimental results and some data in the
literature, we propose a hypothetical model of the mechanism
which regulates the biosynthesis of extracellular proteases
in microorganisms (Figure 8.1). Our model describes this regulatory mechanism as follows. The presence in the medium of
readily metabolized carbon, nitrogen, and sulfur sources, as
in Asp. candidus, maintains a high amino acid pool. Under
these conditions, the cell synthesizes intensively the non-
translated RNA and i-RNA for the intracellular proteins. Thus,
the concentration of "free" factors of initiation and RNA-
polymerase which can be involved in the synthesis of exoprotease i-RNA is negligible. This accounts for the low frequency
of exoprotease i-RNA transcription and, consequently, a low
level of extracellular protease formation at the beginning
of the exponential growth phase.

It was shown that the exhaustion of one of the substrates
during the growth of fungus leads to a decrease in the free
amino acid pool in the cell which, in turn, considerably decreases RNA synthesis. Consequently, the initiation factors
of RNA and RNA-polymerase transcription are released. The
frequency of initiation and of exoprotease RNA increases, together with the synthesis of enzyme.

Proteases degrade proteins in the medium to peptides and
amino acids which are capable of entering the cell and serv-

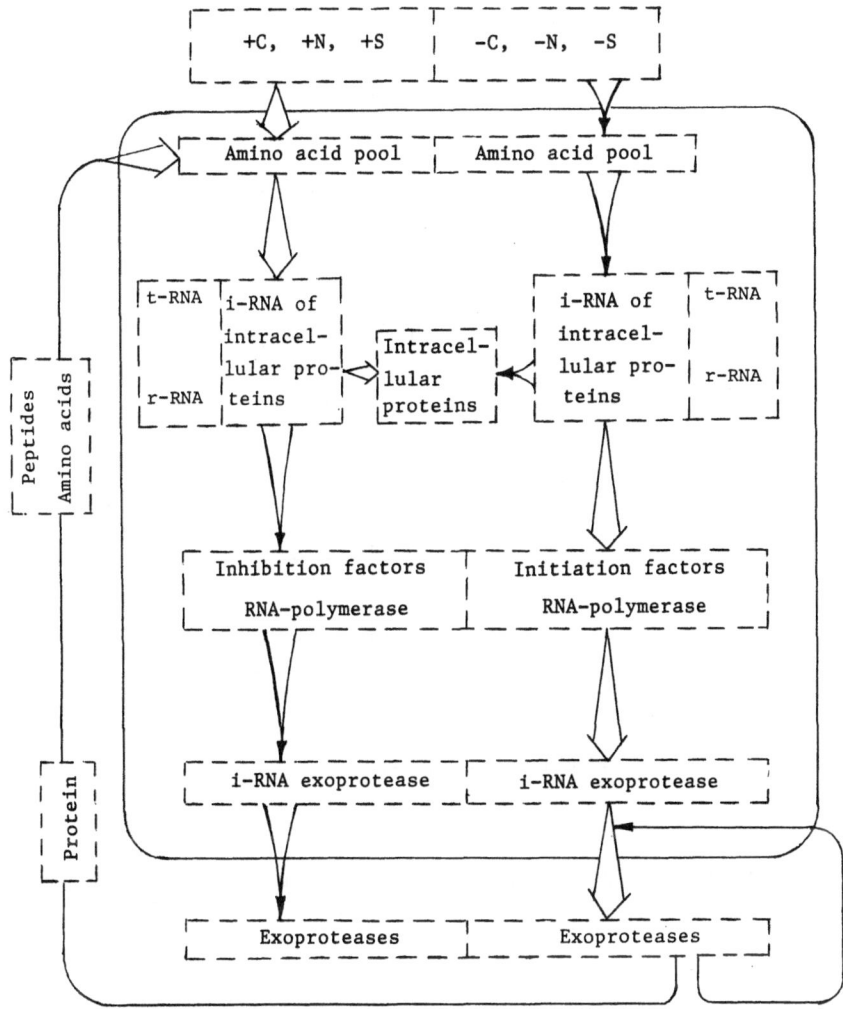

Fig. 8.1.   Control mechanism of intracellular protease bio-
            synthesis in microorganisms.

ing as a limiting substrate. Provided that a sufficient
amount of amino acid enters the cell to make up for the limiting substrate, it can also restore the amino acid stores,
leading to a transitory inhibition of the biosynthesis of exoprotease. Such a blocking mechanism permits the cell to maintain the level of protease necessary to supply the cell with
a limiting substrate. In the absence of protein substrates,
as shown for Asp. candidus and Str. spheroides, biosynthesis

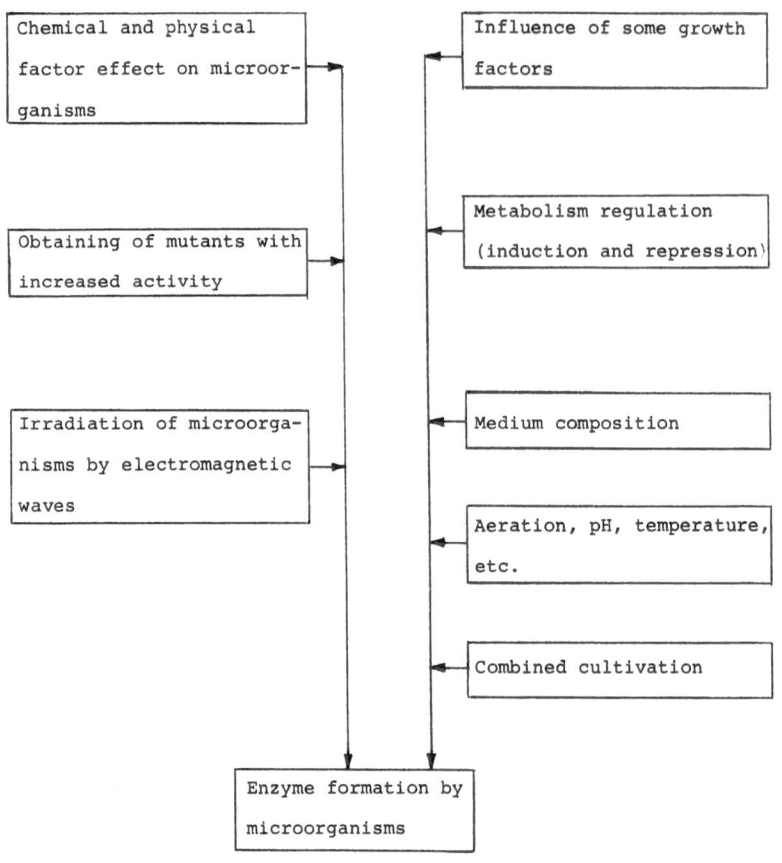

Fig. 8.2.   Some factors which can increase the formation of
            proteolytic enzymes by microorganisms.

of exoprotease is negatively controlled by the level of its
concentration in the medium [49, 50].

By studying the physiology of fibrinolytic enzyme producers and the primary controlling methods for their formation, some regularities have been established which permit
targetted transport and intensive synthesis of these enzymes
by microorganisms. In developing the biotechnology of enzyme production for medical purposes, based on microbic synthesis, it is necessary to consider the primary factors which
increase the formation of fibrinolytic enzyme. These factors
are depicted in Figure 8.2.

REFERENCES

1. B. A. Kudrjashov, G. V. Andreenko, N. S. Egorov, S. M. Strukova, and N. S. Landau, "Fibrinolytic agents isolated from cultures of some saprophytic fungi," Dokl. Akad. Nauk SSSR, 153, 939-941 (1963).

2. N. S. Egorov and N. S. Landau, "The influence of various glycerine concentrations and nitrogen sources on fibrinolytic substance biosynthesis by the culture Aspergillus oryzae, Moscow State University strain," Prikl. Biokhim. Mikrobiol., 1, 487-491 (1965).

3. N. S. Landau and N. S. Egorov, "Fibrinolytic agent Asp. oryzae formation (Moscow State University strain) upon fermentation on synthetic medium," Nauchn. Dokl. Vyssh. Shkol., Biol. Nauki, No. 3, 420-424 (1965).

4. N. S. Egorov and V. I. Ushakova, "On fibrinolytic and thrombolytic activity of some bacteria in vitro," Nauk. Dokl. Vyssh. Shkol., Biol. Nauki., No. 6, 96096 (1968).

5. N. S. Egorov, V. I. Ushakova, and L. M. Nikol'skii, "On the properties of certain microorganisms to form fibrinolytic substances," Dokl. Akad. Nauk SSSR, 165, 217-220 (1965).

6. N. S. Egorov and V. I. Ushakova, "On fibrinolytic and proteolytic activity of some Mycobacterium and Actinomyces," Prikl. Biokhim. Mikrobiol., 2, 595-599 (1966).

7. N. S. Egorov, S. N. Vybornykh, V. I. Ushakova, and N.S. Agre, "Formation of fibrinolytic enzymes by some thermophilic Actinomyces," Nauchn. Dokl. Vyssh. Shkol., Biol. Nauki, No. 9, 98-102 (1972).

8. V. U. Ushakova and N. S. Egorov, "Study of proteolytic enzymes formed by certain Actinomyces in connection with their fibrinolytic and thrombolytic activity," Nauchn. Dokl. Vyssh. Shkol., Biol. Nauki, No. 7, 84-88 (1969).

9. N. S. Egorov and N. S. Landau, "On fibrinolytic activity of proactinomyces," Nauchn. Dokl. Vyssh. Shkol., No. 9, 109-112 (1969).

10. N. S. Landau and N. S. Egorov, "Study of protease formation possessing fibrinolytic action in Nocardia species (strain 1)," Mikrobiologiya, 40, 829-833 (1971).

11. N. S. Egorov, N. I. Ushakova, and I. P. Smirnova, "Capability of microbacterial saprophytic and pathogenic species to form fibrinolytic substances," Prikl. Biokhim. Mikrobiol., 4, 549-553 (1968).

12. N. A. Andreeva, V. I. Ushakova, and N. S. Egorov, "Study of proteolytic enzymes from various strains of Penicillium lilacinum Thom. in connection with fibrinolytic activity," Mikrobiologiya, 41, 417-422 (1972).

13. N. S. Egorov, V. I. Ushakova, T. G. Mirchink, and V. V. Klechkovskaya, "On fibrinolytic and thrombolytic activity of fungi from Aspergillus genus," Nauchn. Dokl. Vyssh. Shkol., Biol. Nauki, No. 9, 103-108 (1973).

14. N. S. Egorov, V. I. Ushakova, and B. Prudlov, "Study of proteolytic enzyme formation by defective fungi from the genera Cladosporium, Fusarium, and Alternaria in connection with fibrinolytic activity," Mikrobiologiya, 40, 4-14 (1971).

15. S. N. Vybornykh and N. S. Egorov, "Study of proteolytic enzyme preparation from thermophilic actinomyce Act. thermovulgaris (strain T-54)," Prikl. Biokhim. Mikrobiol., 11, 829-832 (1975).

16. S. N. Vybornykh and N. S. Egorov, "Targetted synthesis of proteolytic enzymes in Act. thermovulgaris," Mikrobiologiya, 14, 993-998 (1975).

17. A. A. Imshenetskii and S. V. Brotskaya, "Selection of microorganisms possessing thrombolytic activity," Mikrobiologiya, 38, 823-829 (1969).

18. I. R. Schimi and M. W. Kelada, "Isolation of a fibrinolytic activator from metabolite of Bacillus subtilis," Arch. Microbiol., 50, 326-329 (1965).

19. L. R. Christensen, "Methods for measuring activity of the streptococcal fibrinolytic system, and streptococcal deoxyribonuclease," J. Clin. Invest., 28, 163-172 (1949).

20. N. S. Egorov and V. I. Ushakova, "Study of fibrinolytic enzyme formation by some microorganisms," in: Enzymes of Microorganisms, Nauka, Moscow (1973), pp. 63-70.

21. N. S. Landau and N. S. Egorov, "Fibrinolytic enzymes of nonpathogenic microorganisms," in: Streptokinase and Other Thrombolytic Enzymes, Minsk (1979), pp. 26-32.

22. N. S. Egorov, V. I. Ushakova, and V. A. Arakelova, "Influence of various nitric and carbon sources on fibrinolytic enzyme formation in Act. odorifer and its mutants," Mikrobiologiya, 41, 139-145 (1972).

23. N. S. Egorov, N. S. Landau, N. A. Sycheva, M. B. Golant, G. M. Okhokhonina, and A. K. Bryukhova, "Method for obtaining protease," Author's Certif. No. 506617; Byull. Isobret., No. 10 (1976).

24. N. S. Egorov, M. B. Golant, T. M. Okhokhonina, N. A. Sycheva, and A. K. Bryukhova, "Influence of electromagnetic waves within the millimeter range on the formation of proteases with fibrinolytic action in Asp. oryzae (Ahlb) Cohn.," Mikrol. Fitopatol., 11, 45-49 (1977).

25. N. S. Egorov, M. A. Al'-Nuri, N. S. Landau, and L. I.
     Buyak, "Method for obtaining protease inhibitor,"
     Author's Certif. No. 605827; Byull. Izobret., No. 17
     (1978).
26. N. S. Egorov, Zh. K. Loriia, and N. S. Landau, "Proteo-
     lytic enzymes of microorganisms in connection with their
     fibrinolytic and coagulase activity," in: Nuclease and
     Protease Biosynthesis by Microorganisms, Nauka, Moscow
     (1979), pp. 146-197.
27. N. S. Egorov, "Some methods for increasing protease bio-
     synthesis by microorganisms," in: Enzyme Biosynthesis
     by Microorganisms, Moscow (1979),   pp. 40-61.
28. N. S. Egorov, M. P. Kolesnikov, L. I. Buyak, and N. S.
     Landau, "On the question of the origin and properties
     of the factor stimulating exoprotease biosynthesis by
     the fungus Asp. kanagawaensis," Dokl. Akad. Nauk SSSR,
     24, 1497-1501 (1979).
29. I. I. Milovanova, N. S. Landau, and N. S. Egorov, "On
     the causes of increased fibrinolytic activity in coli-
     form bacteria in mixed cultures," Prikl. Biokhim. Mikro-
     biol., No. 2, 202-206 (1982).
30. N. S. Landau, N. S. Egorov, and L. I. Buyak, "Role of
     the microbic factor stimulating exoprotease biosynthesis
     of Asp. kanagawaensis during its growth and development,"
     Mikrobiologiya, 49, 919-923 (1980).
31. B. J. Bromke and J. M. Hammel, "Regulation of extracel-
     lular protease formation by Serratia marcescens," Can.
     J. Microbiol., 25, 47-55 (1979).
32, F. G. Priest, "Extracellular enzyme synthesis in the
     genus Bacillus," Bacteriol. Rev., 41, 711-735 (1977).
33. N. S. Egorov and V. I. Ushakova, "Fibrinolytic enzyme
     formation by Act. species depending on inorganic and
     organic nitrogen forms," Nauchn. Dokl. Vyssh. Shkol.,
     Biol. Nauki, No. 3, 97-102 (1970).
34. Zh. K. Loriia, B. Bryukner, and N. S. Egorov, "Induc-
     tion of intracellular protease synthesis from Serratia
     marcescens," Nauchn. Dokl. Vyssh. Shkol., Biol. Nauki,
     No. 6, 108-113 (1977).
35. Zh. K. Loriia, B. Bryukner, and N. S. Egorov, "On the
     nature of the true inducer of intracellular protease
     from Serratia marcescens," Mikrobiologiya, 46, 440-446
     (1977).
36. M. A. Al'-Nuri, G. B. Sattarova, and N. S. Egorov, "Exo-
     protease biosynthesis in Actinomyces spheroides under
     medium limitation by carbon, nitrogen, and sulfur
     sources," Mikrobiologiya, 49, 408-411 (1980).

37. M. A. Al'-Nuri, V. A. Ivanitsa, and N. S. Egorov, "Bio-
    synthesis of intracellular proteases in Asp. candidus
    in the absence of carbon and sulfur sources," Mikro-
    biologiya, 50, 1019-1023 (1981).
38. N. S. Egorov, T. G. Yudina, Zh. K. Loriia, and R. N.
    Zeleneva, "Medium for exoprotease synthesis in Bac.
    thuringiensis," Prikl. Biokhim. Mikrobiol., 17, 676-681
    (1981).
39. N. S. Egorov and I. P. Smirnova, "Biosynthesis of fibrin-
    olytic substance in Mycobacterium sp. strain 69 during
    growth on media with various carbon sources," Mikro-
    biologiya, 38, 442-446 (1969).
40. Zh. K. Loriia, A.I. Marchenkova, and N. S. Egorov, "On
    sporulation linked to intracellular protease synthesis
    in Bac. mesentericus," Mikrobiologiya, 46, 1014-1019
    (1977).
41. N. S. Egorov and N. I. Landau, "Biosynthesis of proteases
    with fibrinolytic action by some microorganisms," in:
    Streptokinase and Other Thrombolytic Enzymes, Minsk
    (1979), pp. 52-58.
42. N. S. Egorov, Zh. K. Loriia, and T. G. Yudina, "Influ-
    ence of various carbon sources and purine nucleotides
    on exoprotease synthesis in Bac. thuringiensis," Mikro-
    biologiya, 51, 43-47 (1982).
43. Y. Opheim Dennis, "Effect of ammonium ions on activity
    of hydrolytic enzymes during sporulation of yeast," J.
    Bacteriol., 138, 1022-1025 (1979).
44. Zh. K. Loriia, B. Bryukner, and N. S. Egorov, "Influence
    of amino acids on intracellular protease synthesis in
    Serratia marcescens," Mikrobiologiya, 46, 41-47 (1977).
45. S. N. Vybornykh, Zh. K. Loriia, and N. S. Egorov, "The
    effect of glucose and ammonium ions on exoprotease syn-
    thesis in Act. thermovulgaris T-54," Nauchn. Dokl. Vyssh.
    Shkol., Biol. Nauki, No. 8, 105-110 (1976).
46. S. N. Vybornykh, Zh. K. Loriia, and N. S. Egorov, "Amino
    acid influence on intracellular protease synthesis in
    Act. thermovulgaris," Mikrobiologiya, 46, 227-231 (1977).
47. V. I. Ushakov and N. S. Egorov, "Study of proteolytic
    enzymes in some Actinomyces in connection with their fib-
    rinolytic and thrombolytic activity," Nauchn. Dokl.
    Vyssh. Shkol., Biol. Nauki, No. 7, 84-91 (1969).
48. V. A. Ivanitsa, M. A. Al-Nuri, and N. S. Egorov, "Dynam-
    ics of amino acid pool and intracellular protease synthe-
    sis in Aspergillus candidus," Mikrobiologiya, 50, 801-
    806 (1981).

49.   V. A. Ivanitsa, "Intracellular protease biosynthesis in Aspergillus candidus," Master's Thesis, Moscow (1978).
50.   G. B. Ermolina (Sattarova), "Intracellular protease formation in Actinomyces spheroides 35," Master's Thesis, Moscow (1981).

PROSPECTS OF USING ANTIVITAMINS K
IN THE TREATMENT OF
INTRAVASCULAR THROMBOSIS

F. I. Komarov, I. N. Bokarev, and E. V. Kabaeva

First Moscow Medical Institute
Ministry of Public Health of the USSR, Moscow

Issues of anticoagulant therapy stir the minds of every physician in our contemporary world regardless of his or her specialization. Intravascular blood coagulation is a severe complication of many diseases and often causes death among the peoples of industrially developed countries. Population data indicate that 80 to 90% of those who die from ischemic heart disease and cerebral vascular disease have arterial thrombi [1]. Venous thrombi and pulmonary embolism often complicate many diseases and surgical interventions and also cause death. Pulmonary embolism is found in 1 to 14.7% of all autopsies [2] and is a direct cause of postoperative mortality in 10% of the cases [3]. Venous thromboses have been found, by radioisotopic methods and angiographic technique, in 35% of the patients who are operated on by the age of 40 and in 65% of those between 40 to 70 years old [4]. Leg vein thromboses occur in 23 to 38% of the patients with myocardial infarction and in 60% of those with stroke [3].

Today, physicans have a series of antithrombotic drugs to treat these complications. These drugs are divided into three groups according to their effect on hemostasis: fibrinolytics, anticoagulants, and inhibitors of platelet function. Oral anticoagulants, or antivitamins K, which are re-

viewed here, have been used for more than 40 years for prophylaxis of thrombus or prevention of its spreading. Antivitamins K (indirect anticoagulants), unlike other drugs such
as heparin and defibrinators, which inhibit fibrin formation,
affect this process during the synthesis of coagulation factors II, VIII, IX, and X. Antivitamins K affect blood coagulation
in vivo through proteins of the prothrombin complex; their
antagonist is vitamin K.

The synthesis of vitamin K-dependent clotting factors
and the mechanism of action of antivitamins K are described
below. The plasma clotting factors are formed in hepatocytes
in the presence of a sufficient amount of vitamin K, which
provides for the carboxylation of the molecular precursors
of the coagulation factors. The molecule formed is subjected
to glycosylation and becomes an active clotting factor. During carboxylation, vitamin K loses the COOH group and is converted to the vitamin K-epoxide. Normally, the inversion
of the latter to vitamin K proceeds in parallel with the above
reaction. Antivitamins K, however, inhibit the inversion of
vitamin K-epoxide, leading to an accumulation of it and a
decrease in the vitamin K stored. The carboxylation rate of
the precursor coagulation factors decreases and leads to glycosylation of the percursor molecules that have not been car-
boxylated previously. Thus, the proteins called PIVKA (protein induced in vitamin K absence) are formed; they are incapable of binding Ca^{2+} ions and thus lack procoagulant activity. Finally, this results in a deficiency of coagulation
factors and then hypocoagulability [6]. In this way, the
mechanism of antivitamins K acts by their effect on vitamin
K-dependent carboxylation of the precursor coagulation factors, thus preventing vitamin K regeneration from its epoxide.

Despite the long-term application of antivitamins K in
clinical practice, the indications for use are still a problem. In order to properly evaluate the effect produced by
antivitamins K, the characteristics of thrombogenesis in both
the arteries and veins need to be considered.

Arterial thrombogenesis differs from venous thrombogenesis. Recent studies have demonstrated that arterial thrombi
formed in rapid blood flow consist mainly of platelets with
an insignificant incorporation of fibrin and erythrocytes.
Venous thrombi are formed in slow blood flow and are composed
mainly of fibrin which builds up the thrombus, a few erythrocytes, and platelets [7, 8]. Until recently, antivitamins K

were regarded as efficient in cases of venous thrombosis and
as having no desired effect in cases of myocardial infarction, arterial thromboses, or cardiac valve prosthetics. We
assert that antivitamins K need to be used in treating pathologies such as venous thromboses and pulmonary embolism,
and also in treating arterial thromboses. The correctness of
this assertion has been proved both theoretically and practically. First, in examining patients with ischemic heart
disease, activation of both the fibrin formation and platelet
parts of the coagulation system was found to result from an
increase in the level of fibrin-monomer soluble complexes,
fibrin degradation products, and the level of platelet factor
4 [9]. An increase in the level of fibrinopeptide A in patients with this pathology has also been elucidated by Nichols et al. [10]. These data indicate that intravascular fib-
rinogenesis proceeds in parallel with the activation of the
platelet component. Second, activation of this unit simultaneously leads to the activation of fibrin formation. Factor 3 is
released from the platelets and is involved in the activation
of factor X, which favors the generation of thrombin to convert fibrinogen to fibrin. Third, the use of antivitamins K
inhibits the formation of thrombin itself, which is a significant factor in blood platelet aggregation. In addition,
thrombi are found by radioisotopic assay in leg veins of almost 30% of the patients with acute myocardial infarction.
Autopsy data indicate frequent occurrence of intracardiac
thrombi as a source of embolism in transmural myocardial infarctions. Thus, antivitamins K are indicated in cases of
ischemic heart disease for (a) the prevention of venous
thrombi and associated pulmonary emboli, (b) the prevention
of intracardiac thrombi and systemic emboli, and (c) the
limitation of proximal growth of coronary thrombi [11].

Analysis of clinical studies on the efficacy of antivitamin K in myocardial infarction showed that these drugs
reduced mortality by 20% in the hospital phase of acute myocardial infarction [12]. In the long term (for more than 5
years), mortality is reduced twofold and the frequency of reinfarction is decreased threefold [13]. A positive experience
with the use of antivitamins K in cases of myocardial infarction has been demonstrated by a number of studies [14-17].
There is also no doubt, at present, that it is reasonable to
prescribe oral anticoagulants to patients with prosthetic
heart valves [18], although recent data show a more pronounced
antithrombotic effect with the combined use of both antivitamins K and antiaggregating agents [18, 19]. This effective-

ness should be quite clear given the interrelationship of the many components of the hemostasis sytem.

Antivitamins K have been used for the prevention of post-operative thromboses as well. Although small doses of heparin is the priority method of prophylaxis against thrombus formation, there are certain clinical situations (for example, hip surgery and other operations followed by massive tissue destruction and considerable thromboplastin extraction into the blood flow) when the use of antivitamins K is preferable. This has been demonstrated by many clinical trials on groups of patients in whom antivitamins K produced the best prophylactic effect when compared to small doses of heparin [20-22].

Another factor which affects the efficacy of treatment with antivitamins K is the correct dosage. The antithrombotic effect of oral anticoagulants has been shown to strictly depend on the level of hypocoagulation. Evaluation of the results of prothrombin time quotients in comparison with the frequency of myocardial infarction has shown that only when prothrombin activity was decreased to 15-25% did the frequency of recurrence not exceed 50%. When prothrombin activity amounted to 28%, there was a 100% recurrence [11]. Studies to determine the optimum laboratory data for preventing pulmonary thromboembolism in patients with venous thrombosis treated with antivitamins K showed that the frequency of thromboemboli with a level of prothrombin activity ranging from 30 to 49% was 2.5 times as high as with that ranging from 10 to 29% [23]. Analysis of the efficacy of antivitamins K used by patients with prosthetic heart valves indicated that the frequency of thromboembolic complications after many years of followup was three times lower in patients in whom the indices of prothrombin time were within the therapeutic range (not less than 75%) for a long time than in patients treated adequately for a shorter period of time [18].

The difficulty with antivitamin K therapy is the need for preparing an optimum dose which will prevent thrombus formation without causing hemorrhage. The optimum dose value of the anticoagulant depends, however, on the patient's sex, age, diet, accompanying diseases, other drugs taken at the same time, and many other factors. The reasons for these variations seem to be endless, and the optimum dose of antivitamins K varies over the course of long-term treatment even within one patient.

The importance of laboratory control does not require proof, given the necessity of maintaining optimum hypocoagulation in the patient's body.

During the entire period of clinical application of antivitamins K, the main method of control is to determine the prothrombin time and its derivatives (P and P test, Thrombotest) according to Quick [24]. This method is based on the activation of blood coagulation by a tissue extract such as thromboplastin. Factors in the "extrinsic" cascade of fibrin formation, the content of which is mainly decreased by antivitamins K, are involved in this reaction. Some 30 thromboplastins are now used through the world to determine prothrombin time to activate plasma or blood coagulation. They differ in their sensitivity to the blood coagulation defects caused by antivitamins K; they are not likely to be inhibited by PIVKA; and they react differently with the blood coagulation factors [25, 26]. This results in a considerable divergence among the indices of prothrombin time obtained in various laboratories and it hinders the development of a single optimum level of hypocoagulation.

Intensive attempts to standardize the method of determining prothrombin time have yielded positive results: Based on a World Health Organization proposal, most common thromboplastins have been calibrated [27], a single standard scale of hypocoagulation intensity (with reference to the British comparative ratios for thromboplastin) has been created; and both synthetic reference plasmas and those obtained from the blood of patients treated with antivitamins K have been used to check the level of hypocoagulation [29]. The calibration constants determined by thromboplastin standardization permit comparison of prothrombin times, regardless of the specific thromboplastin used.

A similar study was carried out at our institute [30]. Thromboplastins produced by the Kirov Institute of Blood Transfusion and the Kaunas Enterprise of Bacterial Preparations are widely used by laboratories in the Soviet Union. Cadaverous human brain is a source of thromboplastins. Their relative calibration constant is 1. In comparing these thromboplastins with Simplastin (USA), the calibration constant was estimated as 1.35. Accordingly, the real calibration constant of thromboplastins prepared in the USSR, compared to the British reference thromboplastin, was calculated by an "indirect" method to be 0.46.

Standardization of the therapeutic level of hypocoagulation for the thromboplastins under study, based on the results
of both their calibration and frequency plasmas (Verify I-II-
USA), corresponds to the lengthening of prothrombin time by
a factor of 2.2-2.3. In the Soviet Union, the prothrombin
index is used to denote the results of prothrombin time. It
is expressed by $\dfrac{\text{donor prothrombin time}}{\text{prothrombin time of patient}} \cdot 100\%$. Optimum
therapeutic values for this index range between 30 and 45%.

In recent years, new possibilities for controlling antivitamin K therapy have become available due to the introduction of chromogenic substrates to clinical practice. Because
antivitamins K are known to decrease the synthesis of coagulation factors II, VII, IX, and X, a certain decrease in the
concentration of these factors may serve as a criterion for
the efficacy of an anticoagulant therapy. Measurement of the
concentration of three of the four vitamin K-dependent clotting factors by means of chromogenic substrates has now become
possible. The results of a series of clinical studies to determine factor X, using chromogenic substrates S-2222 and S-
2237 (AB Kabi, Sweden) as a control method in patients taking
antivitamins K, showed a good correlation with those of prothrombin time and its derivatives [31-35].

In our studies, we used both S-2222 and S-2237 chromogenic substrates for factor X determination. Factor X was
measured in parallel with prothrombin time using thromboplastin from human brain in 120 patients receiving long-term treatment with antivitamins K (Pelentan, Phenylin).* The following pathologies served as indications for therapy: prosthetic heart valves, myocardial infarction, cardiac fibrillation,
and recurring thrombophlebitis. The mean age of the patients
was 49.3 ± 10.7 years. In order to control the therapeutic
efficacy and to prevent hemorrhagic complications, the patients were examined thoroughly in the clinic. During the
course of treatment, the hemoglobin level in the patients'
blood and the amount of erythrocytes in their urine were
monitored, as well as other hemostatic characteristics (prothrombin time, factor X, fibrin degradation products, antithrombin III, and activated partial thromboplastin time).

---

*Pelentan is a Soviet trade name for neodicoumarin; Phenyl-
in, for indandione.

Both clinical and laboratory examinations were conducted
every 3 weeks, and investigations were conducted more frequently when necessary.

We found a good correlation (r = 0.75) between the results of factor X and prothrombin time determination (the latter was expressed in terms of the prothrombin coefficient).
The data obtained (n = 96) were compared with the level of
hypocoagulation: The results coincided in 72% and were partially discordant in 28%. None of our observations revealed
completely discordant evidence [36]. Similar positive results were obtained by Axelson et al. [37] in determining
factor II by Chromozym TH. There are analogous data on the
determination of factor VII by chromogenic substrates [38].
The authors of this study, however, conclude that this procedure is impractical because it is time-consuming and complicated.

By and large, the methods of determining certain proco-
agulants by chromogenic substrates can be readily standar-
dized and adapted to automatization. The endpoint of fibrin
formation, affected by many factors, is eliminated. The
presence of heparin in the blood, which is very important
when a patient undergoes other types of antithrombotic therapy, as well as the amount of antithrombin III and the PIVKA
level essentially do not affect the results. In line with
Lammle et al. [32], our experience in using one blood coagulation factor, X, as an anticoagulant control suggests wider
application for chromogenic substrates in the nearest future.
The decrease in the therapeutic level of factor X, determined
by various researchers from a number of countries, is practically the same and ranges from 15-18% to 28-32% [32-36].
Slight variations in therapeutic values of factor X are not
essential and are likely a result of using modifications of
the Quick test (and various thromboplastins) as a standard
control method. This is the cause of the deviations in intensity of hypocoagulation estimated by the level of factor X.
A simpler and more precise standardization of the method
for determining factor X, which would allow for the comparison of findings and the development of a single optimum therapeutic level, would encourage the prospects for using chromogenic substrates as a control for antivitamin K therapy.

The laboratory control methods mentioned here reflect
the capabilities of blood plasma to form fibrin. However,
they offer no information on current fibrin formation. This

requires the study of the relationship between the antithrombotic
effect of antivitamins K and the indices of intravascular
blood coagulation intensity in the patients under examination.
Harenberg et al.'s study [39] is noteworthy in this respect.
In this study, he measured fibrinopeptide A as a criterion
for thrombin activity in the blood of patients who had been
treated for a long time with phenprocoumon. No decrease in
fibrinopeptide A was observed in some patients despite effective (according to other control test data) anticoagulant
therapy.

In our followups of patients treated with antivitamins
K for a long time, determination of fibrin degradation products (FDP), taken as the index of intensity of intravascular
thrombogenesis, did not always indicate the expected antithrombotic effect. In a group of patients followed and treated with antivitamins K, comparison of the mean levels of FDP
before and after treatment showed a statistically significant
decrease (p < 0.05) after treatment. The FDP values were
24.4 ± 7.2 µg/ml and 7.1 ± 1.6 µg/ml, respectively. In analysis of the dynamics of FDP during treatment with antivitamins K, we also revealed an increase to 60-120 µg/ml in
some patients. In a number of cases the increase in FDP level
coincided with new attacks of the disease that called for
antivitamin K treatment. These data demonstrate that increased
thrombin activity is not eliminated and the intensity of
intravascular fibrin formation is not reduced in every patient by means of antivitamins K. In order to recognize the
patients at increased risk of thrombogenesis, we must study
the indices of intensity. These data once again testify
to the need for a complex approach to the heterogeneous nature
of thrombogenesis.

Questions of the clinical significance of a decrease in
antithrombin III and the treatment under study in recent years
provide one more indication for the use of antivitamins K.
With antivitamin K therapy, the level of antithrombin III has
been observed to increase. This phenomenon has been described for the patients with both hereditary and acquired antithrombin-III deficiency [40]. These observations promote the
prescription of antivitamins K for prophylactic and therapeutic purposes to patients with acute nephrotic syndrome and
a decreased level of antithrombin III [41].

The present authors have also had a positive experience
in treating a patient with antithrombin-III deficiency. A

woman aged 44 years who was suffering from recurrent thrombophlebitis of the deep leg veins was observed. The disease
resulted from the intake of oral contraceptive pills prescribed to her for a therapeutic purpose. Laboratory examination
showed that she had a deficiency of antithrombin III, with a
level, measured by chromogenic substrate S-2238 (AB Kabi, Sweden),
of 46%. Although she was not allowed to use estrogen-con-
taining preparations, the thrombophlebitis progressed. Antivitamins K (Sincumar*) were prescribed. Two months after beginning of oral anticoagulant therapy, her level of antithrombin III increased to 82%. The remission was observed
in the clinic. After half a year of antivitamin K treatment,
the therapeutic effect was preserved, but the level of antithrombin III was 86%. This result supports the prescription
of antivitamins K in cases of a decrease in anthithrombin III,
both to prevent intravascular thrombogenesis and to potenti-
ate the antithrombotic effect of heparin.

The indications for antivitamin K prescription can be
expected to grow based on reports of success in studying tissue factor. Antivitamins K were found to affect not only the
production of prothrombin complex factors by the liver, but
also to inhibit tissue factor formation [42]. Identification
of tissue factor as a specific glycoprotein, its localization
in cell membranes, and many painless determinations in vivo
have led to the observation of a significant increase in this
substance in a series of tumors, inflammations, and other
diseases. It is hoped that, if the decrease in tissue factor due to antivitamins K in such diseases is shown to coincide often with an improvement in the patient's state, then
new possibilities for using antivitamin K in the clinic will
be opened up in the coming years.

In closing, we emphasize once again that it is vital to
pursue the questions raised on the use of antivitamins K in
clinical practice. The range of indications for their application will undoubtedly expand and additional methods for
controlling the efficacy of these drugs will be devised.

REFERENCES

1. Yearbook of World Sanitary Statistics, Moscow (1978).

---

*Sincumar is Soviet trade name for a drug belonging to
the coumarin family.

2. N. G. Gataullin, F. F. Muhametrahimov, and I. I. Semenov, "Experience in the treatment of pulmonary embolism," in: Diagnostics and Treatment of Pulmonary Embolism, Moscow (1980), pp. 33-35.

3. D. A. Natradze, "Our experience in diagnostics and treatment of pulmonary embolism," in: Advances in Surgery, Moscow (1976), pp. 65-66.

4. H. Vinazzer, "Neue methoden zur freiherkennung thromboemolischer erkrankungen," Folia Haematol. (Leipzig), 102, 130-139 (1975).

5. A. Gallus and J. Y. Hirsh, "^{125}I-Fibrinogen leg scanning," in: Prophylactic Therapy of Deep Vein Thrombosis and Pulmonary Embolism, J. Frantoni and S. Wessler (eds.), Washington (1975), pp. 77-82.

6. J. W. Suttie, "How coumarin anticoagulants work," Drug Ther., 9, 63-71 (1979).

7. S. Sevitt, "Venous thrombosis in injured patients (with some observations on pathogenesis)," in: Thrombosis and Thrombolysis, S. Sherry (ed.), Washington (1969), pp. 29-33.

8. O. Having, "Pathogenesis of peripheral venous thrombosis. An autopsy study," in: Proceedings of Fourth International Congress on Thrombosis and Haemostasis, Vienna (1973), p. 275.

9. I. N. Bokarev, Abstract, Doctoral Dissertation, First Moscow Medical Institute, Moscow (1980), p. 27.

10. A. Nichols, J. Owen, and K. Kaplan, "Plasma β-thromboglobulin, platelet factor 4, and fibrinopeptide A in patients with coronary artery disease," Thromb. Haemostas., 46, Abstr. No. 0845 (1981).

11. Ch. J. Bjerkelund, "Anticoagulant therapy in myocardial infarction," Heart Lung, 4, 61-68 (1975).

12. T. C. Chalmers, R. J. Matta, H. Smith, Jr., and A. M. Kunzler, "Evidence favoring the use of anticoagulants in the hospital phase of acute myocardial infarction," New Engl. J. Med., 297, 1091-1096 (1977).

13. E. A. Loeliger, "Oral anticoagulation after myocardial infarction," Scand. J. Haematol., 27 (Suppl. 38), 87-95 (1981).

14. "Report of the Sixty Plus Reinfarction Study Research Group, a double-blind trial to assess long-term oral anticoagulant therapy in elderly patients after myocardial infarction," Lancet, 11, 989-993 (1980).

15. E. I. Chazov, A. V. Vinogradov, A. M. Vikhert, et al., Myocardial Infarction, Meditsina Press, Moscow (1971).

16. A. Kher and M. Samama, "La prevention secondaire de l'in-
    fartus du myocarde par les anticoagulants oraux: etude
    critique des essais therapeutiques," Rev. Med., 19, 821-
    830 (1978).

17. M. C. Rozenberg, H. Kronenberg, and B. G. Firkin,
    "'Thrombotest' and prothrombin time. A controlled clin-
    ical trial," Aust. Ann. Med., 14, 3-11 (1965).

18. J. Dale and E. Myhre, "Incidence and prevention of ar-
    terial thromboembolic complications in patients with
    prosthetic heart valves," Scand. J. Haematol., 27
    (Suppl. 38), 121-129 (1981).

19. P. Didisheim and V. Fuster, "Actions and clinical status
    of platelet-suppressive agents," Sem. Haematol., 15, 55-
    72 (1978).

20. D. J. Pinto, "Controlled trial of anticoagulant (war-
    farin sodium) in the prevention of venous thrombosis fol-
    lowing hip surgery," Br. J. Surg., 57, 349 (1970).

21. A. Bronge, S. Dahlgren, and B. Lindqvist, "Prophylaxis
    against thrombosis in femoral neck fractures — a com-
    parison between dextran 70 and dicoumarol," Acta Chir.
    Scand., 137, 29 (1971).

22. D. Bergqvist and S. Dahlgren, "Leg vein thrombosis diag-
    nosed by ^{125}I-fibrinogen test in patients with fracture
    of the hip: a study of the effect of early prophylaxis
    with dicoumarol or dextran 70," Vasa, 2, 121-126 (1973).

23. W. W. Coon, P. W. Willis, and M. J. Symons, "Assessment
    of anticoagulant treatment of venous thromboembolism,"
    Ann. Surg., 170, 559-568 (1969).

24. A. J. Quick, "The prothrombin in hemophilia and in ob-
    structive jaundice," J. Biol. Chem., 109, 73 (1935).

25. H. C. Hemker, J. J. Veltkamp, and E. A. Loeliger, "Ki-
    netic aspects of the interaction of blood clotting en-
    zymes. III. Demonstration of an inhibitor of prothrom-
    bin conversion in vitamin K deficiency," Thromb. Diath.
    Haemorr., 19, 346-363 (1968).

26. E. A. Loeliger, "Progress in the control of oral antico-
    agulation," Thromb. Diath. Haemorrh., 28, 109-119 (1972).

27. ICTH/ICSH, "Prothrombin time standardization: report
    of the expert panel on oral anticoagulant control,"
    Thromb. Haemostas., 42, 1073-1114 (1979).

28. J. B. Miale and J. W. Kent, "Standardization of the ther-
    apeutic range of oral anticoagulants based on standard
    reference plasmas," Am. J. Clin. Patho.., 57, 80-88 (1972).

29. E. A. Loeliger and L. P. van Halem-Visser, "Biological
    properties of the thromboplastins and plasmas included
    in the ISCH/ICTH international cooperative prothrombin

234   F. I. KOMAROV ET AL.

time standardization study," Thromb. Haemostas., 42,
1128 (1979).
30. E. V. Kabaeva, Zh. N. Torik, I. N. Bokarev, and I. M.
Dynkina, "Experience of thromboplastin standardization
and determination of hypocoagulation optimum level in
treatment with antivitamins K," in: Second National
Conference on Antithrombotic Therapy in Clinical Practice. Advances in Theory, Diagnostics, and Therapy,
Moscow (1982), p. 100.
31. L. L. Aurell, P. Friberger, G. Karlsson, and G. Claeson,
"A new sensitive and highly specific chromogenic peptide
substrate for factor Xa," Thromb. Res., 11, 595-609
(1977).
32. B. Lammle, R. Eichlisberger, L. Haenni, et al., "Kon-
trolle der oralen anticoagulation. Vergleich zwischen
quik und kolorimetrischer factor X-bestimmung bei 109
patienten," Schweiz Med. Worchenschr., 109, 1115-1119
(1979).
33. J. G. Erskine, I. D. Walker, and J. F. Davidson, "Maintenance control of oral anticoagulant therapy by a chromogenic substrate assay for factor X," J. Clin. Pathol.,
33, 445-448 (1980).
34. A. A. Famodu, G. I. C. Ingram, and S. C. Darby, "Anticoagulant control with chromogenic measurements of factors X and VII (abstract)," Thromb. Haemostas., 42, 292
(1979).
35. J. Conard, A. Gradelet, and M. Samama, "Dosage du fac-
teur stuart a l'aide d'un substrat synthetique au cours
des traitements anticoagulants oraux. Resultants pre-
liminaires," Ann. Biol. Clin., 39, 81-83 (1981).
36. I. N. Bokarev, E. Berggren, V. S. Smolenskii, et al.,
"Measurement of coagulation factor X by means of chromogenic substrates as a control method by antivitamins K,"
in: Second National Conference on Antithrombotic Therapy in Clinical Practice. Advances in Theory, Diagnostics, and Therapy, Moscow (1982), pp. 97-98.
37. G. Axelson, K. Korsan-Gengsten, and J. Waldenström,
"Prothrombin determination by means of a chromogenic
peptide substrate," Thromb. Haemostas., 36, 517 (1976).
38. L. Poller and Z. S. Latallo, "An assessment of an amidolytic assay for factor VII in the laboratory control of
oral anticoagulants," Br. J. Haematol., 50, 688 (1982).
39. J. Harenberg, R. Haas, and R. Zimmermann, "Measurement
of fibrinopeptide A in patients treated with phenprocoumon," Thromb. Haemostas., 45, 282-284 (1981).

40. E. Marciniak, C. H. Farley, and P. A. De Simone, "Famili-
     al thrombosis due to antithrombin-III deficiency," Blood,
     43, 219-231 (1974).
41. E. Thaler, E. Balzar, H. Kopsa, and W. F. Pinggera,
     "Acquired antithrombin-III deficiency in patients with
     glomerular proteinuria," Haemostasis, 7, 257-272 (1978).
42. L. R. Zacharski, R. Rosenstein, and F. G. Phillips,
     "Tissue factor: a vitamin K-dependent clotting factor,"
     Ann. N. Y. Acad. Sci., 370, 311-324 (1981).

APHERESIS TECHNIQUES (GRAVITATION SURGERY) IN THE REGULATION OF THE HYPERCOAGULABLE (BLOOD AGGREGATE) STATE

O. K. Gavrilov

Central Research Institute of Hematology and Blood Transfusion
Ministry of Public Health of the USSR, Moscow

A. O. Gavrilov

Research Institute of Transplantation and Artificial Organs
Ministry of Public Health of the USSR, Moscow

ABSTRACT

A fundamentally new approach to the control of intravascular
thrombogenesis has been developed. This approach uses blood
fractionation and the removal of excessive amounts of various elements to retard the clotting processes, reduce blood
viscosity, and normalize the microcirculation of blood flow.
The elements removed include plasmatic and cellular factors
in hemostasis, lipids, plasma protein and fatty structures,
and macroglobulins which cause high viscosity and low fluidity of blood. This new approach involves the use of gravitation forces for separating blood fractions and components and
isolating the factors of cellular and plasmatic hemostasis
from the blood in continuous flow.

REGULATION OF THE HYPERCOAGULABLE STATE

The appearance of a specific, internal liquid medium in
the evolutionary development of animals raised the possibility of moving from water to land and of developing more perfect metabolic processes [1]. In the evolutionary process
in the animal world it was necessary to develop physiological

237

mechanisms that could maintain this internal medium in an optimal aggregate state, even during loss of the medium in case
of a local disturbance in blood vessel integrity. This de-
manded the development of a functional system to regulate the
hypercoagulable (or blood aggregate) state. We term this system the "blood aggregate state regulation" (BASR) system.
When the animal has normal vital activity, the BASR system
provides for blood fluidity, restores capillary and vessel
walls that are damaged in the normal functioning of organs
and systems, and preserves the necessary level of blood clotting factors. In extreme conditions, the BASR system forms
thrombi in damaged vessels, prevents blood effusion, and promotes connective tissue components in the healing of organ
injuries [2].

This system for regulating the hypercoagulable state has
a complicated morphological structure which includes (a) central organs (bone marrow, liver, spleen), (b) peripheral
formations (mast cells, endothelium and other layers of the
blood vessel wall, capillary endothelium in the microcirculatory bed, mechanisms of water-electrolyte exchange, and blood
cells); (c) local regulatory systems (vessel reflexogenic
zones with chemoreceptors, and the heart, lungs, kidneys,
uterus, prostate, and digestive organs) and (d) central regulators such as the internal secretion glands (the adrenals,
hypophysis, thyroid gland, etc.) and the autonomic nervous
system and subcortical and cortical structures of the brain.
Thus, the BASR system, as a functional system, contains a
variety of components that differ from each other in structure, tissue property, and chemical specificity.

What regulates the variety of components in this system?
And what are the criteria for performing this regulation?
The main criteria in the BASR system are determined by the
final result of its activity, i.e., the potential for hemostasis, providing for the preservation of blood fluidity or
coagulation depending on the demands of the organism.

The hemostatic potential is defined as the integral
property of blood to maintain fluidity under normal conditions and to coagulate (stop blood flow) under certain, and
usually extreme, situations. This property of the blood involves many elements. Some provide for the aggregate state
of the blood — its fluidity, optimal level of viscosity, and
the ability to penetrate the smallest arterioles, capillaries,
and venules and to realize metabolic processes there. Other

elements create the necessary conditions for the formation
of stable fibrin and thrombi. These elements cause the cessation of blood flow through the vessels and blood effusion
from damaged arteries and veins.

Under any conditions, the functioning BASR system decides what should be done and then how and when to do it.
This system selects the hemostatic potential, chooses the
time and the mechanisms that should participate in achieving
the potential, and then decides which body system should confirm that the result is sufficient.

If the result is insufficient, activating mechanisms are
stimulated, new components are selected, the individual freedom of the acting factors is changed, and, finally, after
trial and error, a completely sufficient adaptation is
achieved.

None of the adaptations of the BASR system can be obtained at the expense of any of the anatomical morphologic
structures. All of the components adhere to a principle of
mutual aid (even those factors that may be contrary to the
character of the action needed). In the modern systemic approach, the functional BASR system is studied not by focusing
attention on some anatomic sign of a participating component,
but by understanding the principles of organization that are
shared by many components from various anatomic systems in
order to obtain the results desired from this extensive heterogeneous system.

The functional BASR system is composed of dynamically
mobilized structures within the entire organism, and an ex-
ceptional effect of a certain anatomic component within the
structure does not influence the activities and final result
of the system. The components of any of the anatomic structures are mobilized and involved in the BASR system only to
the extent that they can assist in achieving the programmed
hemostatic potential. As part of the functioning system,
they lose their excessive degrees of freedom. Only those remain that can assist in achieving the positive result, since
the activities (or function) in general are truly continuous
with the results.

Like any functional system, the BASR system possesses
an ability to instantaneously mobilize the structural elements of the organism in accordance with the continuous de-

mands which the function requires from the structure. This
ability to mobilize implies the possibility of rapidly co-
ordinating fractional combinations to provide for a useful
adaptation of the system. If the structures of the BASR system had no such potential for instantly mobilizing in any arrangement, then instantaneous reorganization of the system
would not be possible and the adaptive result would be imper-
fect [3].

The BASR system is a hierarchy of subsystems which relate
to each other at the level of action. The supersystem, as
a whole, acts with the necessary adaptive response, its excitation occupies efferent pathways of the subsystems, and it
establishes a sufficient level of hemostatic potential. Each
subsystem has to resist a hemostatic potential resulting from
the supersystem's activities that is either too high or crit-
ically low.

Reverse afferentation in this adaptive effect (result)
for life can come by nerve communications in the central organs of the regulatory system or by a complex chain of molecular processes in the cells, where the system stabilizes
this or other exchange processes.

Thus, the BASR system is a complex organization of selectively involved components whose interaction and relationships, regardless of their contrary nature, are mutually help-
ful in obtaining a focused and useful result — maintaining
the hemostatic potential by providing for blood fluidity and
coagulation. These final results are the decisive points in
the BASR system — the major tool for creating the regulated
interaction between all of the components.

The hemostatic potential in all parts of the blood flow
can be positive, negative, or neutral. Quantitative indices
of the factors in the BASR system can be very diverse. For
instance, a neutral hemostatic potential can be established
at both low and maximally high indices of the BASR system.

A neutral hemostatic potential (such as the pH of the medium)
occurs when the coagulation and anticoagulation potentials
are mutually destroyed. Neutrality can develop only at the
same level of the coagulation and anticoagulation potentials.

With shifts of the hemostatic potentials to a positive
(+) or negative (−) side, i.e., to the relative predominance

of the coagulation over the anticoagulation potential, the
same difference in these potentials can be established at the
most varied initial levels of the coagulation and anticoagulation potentials. A similar difference in the coagulation and
anticoagulation potentials can occur at high, medium, or low
initial levels. However, the effective hemostatic potential
at a similar difference in the coagulation and anticoagulation potentials will be the same functionally.

   Fluctuations of the hemostatic potential within certain
values are considered normal. In normal fluctuations, the
BASR system corrects the final response in a shift to a positive potential by increasing the concentration of anticoagulation factors or by decreasing the concentration of coagulation factors. In a negative shift of the hemostatic potential, correction of the final response occurs due to an increase in the concentration of coagulation factors and a decrease in the concentration of anticoagulation factors.

   Thermodynamically, the BASR system is an open system.
An open system is in a stationary state and cannot change into a balanced state. Transition into a balanced state is
characteristic of closed systems. In this case, entropy accumulates to the maximal level, i.e., the system loses the
ability to function.

   The main internal contradiction in the development of
regulatory processes of the hypercoagulable state consists
in the shift of the blood aggregate state from the necessary
optimal level, which activates the regulating devices which,
through a number of processes, provide for its decrease (negative feedback). These internal shifts from the final adaptive effect stimulate a switching on of the mechanisms of compensation, restoration of which returns the system to the set
level.

   The set level is programmed by the action acceptor based
on afferent synthesis.

   Lack of coordination in the system can be recorded at
the level of any of the major mechanisms composing it. In
any case, the result will be distorted, i.e., the hemostatic
potential will no longer correspond to a programmed potential
by the action acceptor and to conditions of the external and
internal medium.

242   O. K. GAVRILOV ET AL.

The large number of data accumulated in clinical studies
of hemostasiology testify that the regulation of the hypercoagulable state is tessellated, i.e., the hemostatic potential
is not the same in different areas of blood flow and in different organs [4-6]. This is the natural, normal state of
the BASR system. This important biological regularity in the
regulation of fluidity and coagulation has been observed in
our laboratories. The BASR system is discrete, and its various parts, differently and at different levels, determine the
necessary hemostatic potential in the circulating blood. In
every circulatory region, the blood participates in metabolism and is subject to different effects. Its hemostatic
potential therefore changes, reflecting the variety of functions and processes in the organism and the metabolism of its
organs, systems, and separate regions [7-9]. Our studies
show that this regularity is preserved in pathologic conditions. This regularity is very important for the organism.
A critical situation arises when the BASR system, because of
the lack of coordination among its subsystems, does not provide for discrete hemostatic potentials in different parts
of the blood flow.

Elimination of the tessellated nature of hemostatic potentials is usually accompanied by a general increase in the
blood's ability to coagulate and form thrombi, first in the
microcirculatory chains of the circulatory system. A similar
situation occurs during and after major operations which involve devices for extracorporeal circulation, in certain severe diseases of the heart and lungs, and in obstetric pathology [9-11].

APHERESIS TECHNIQUES (GRAVITATION SURGERY)
AND THE BASR SYSTEM

Thromboses are currently prevented and treated by the
suppression of blood clotting activity through the use of
anticoagulant activation of fibrinolytic properties of the
blood by natural and artificial activators, and inhibition
of platelet aggregation [12-15]. All of these methods are
associated with the use of drugs that change the qualitative
indices of the BASR system, but which have a weak effect on
its quantitative composition — the relationships between separate components of the system [13, 16, 17].

A radical effect in the regulation of the hypercoagulable state is possible by fundamentally new methods which we
call "blood gravitation surgery" (apheresis).

Different tissues of the human organism react differently to the forces of gravitation. This property has been used
in developing apheresis techniques.

Various techniques for tissue and organ removal have dif-
fered sharply from each other throughout the history of surgery. At first, a simple surgical scalpel was used, while
later methods included electrocoagulation, thermocoagulation,
ultrasound, cryotherapy, and the use of lasers. Unfortunately, none of these methods can be used for the removal of malignant cells and pathologically changed plasma substrates
from peripheral blood and bone marrow. Only by the forces
of gravitation is it possible to separate blood into components of different weight. In this way, pathologically changed
components can be removed and released from the peripheral
blood.

Devices that create gravitation forces for separation
of blood into components were first developed in the 1960s.
Today, there are a variety of blood fractionators based on
gravitation. The complete process is performed in a sterile,
closed system. Continuous and discrete whole blood is taken
from a person, divided into components. Tumor cells, plasmatic proteins, and cholesterol are concentrated and isolated,
and the remaining components are returned to the patient.
The differences between blood fractionation systems depend
on the rotor mechanism.

The field of application of blood fractionators continues to widen. These fractionators were first used to obtain leukocyte concentrates from donors. Later, platelet
concentrates and plasma were obtained. Today, blood fractionators based on gravitation are widely applied in different diseases for the removal of pathologic cells or plasma
containing immune and toxic agents.

We include this method in the complex of procedures for
correcting thrombotic states of the BASR in ischemic heart
disease. The theory of systemic regulation of the hypercoagulable state establishes a close interdependence between the
hemostatic potentials of blood and the metabolic processes
developing in the organism. Hypoxia caused by cardiac muscle

diseases and atherosclerotic injuries to vessels is necesar-
ily accompanied by an enhancement of the blood's hemostatic
properties, especially in those organs which are most sensitive to hypoxia. As a result of hypoxia, the content of cholesterol, fibrinogen, triglycerides, and thromboplastinoactive
proteins increases in the blood of patients with ischemic
heart disease. The more severe the ischemic heart disease,
the more marked the changes in the BASR system.

By including apheresis techniques ("gravitation surgery") for the correction of the hypercoagulable state in the
treatment of patients with ischemic heart disease, we sought
to reduce the amount of protein, cholesterol, fibrinogen,
lipids, triglycerides, and platelets circulating in the blood;
to replace them with plasma substitutes; and to improve the
rheologic properties of blood for normalization of microcirculatory blood flow.

Apheresis was used in patients with different manifestations of ischemic heart disease and marked disseminated intravascular coagulation (DIC). Therapy was initiated as early
as possible after the onset of disease. Our experience shows
that direct indications for operation are as follows: (a)
myocardial infarction complicated by cardiogenic shock, (b)
preinfarct angina and drug-resistant angina, (c) repeat myocardial infarction, (d) myocardial infarction with severe
disturbances in cardiac rhythm, and (e) myocardial infarction
complicated by thrombosis of the small branches of the pulmonary artery.

In analyzing the data obtained, we concluded that this
method is promising in patients with hypertension, obliterat-
ing endarteritis, and other occlusive vessel lesions when
other surgical methods cannot be used.

APHERESIS TECHNIQUE FOR ISCHEMIC HEART DISEASE

Once the indications for correction of hypercoagulabil-
ity are determined, a patient is placed in an operating room
that is especially equipped for the treatment of patients
with complicated forms of myocardial infarction and which includes all the necessary devices (cardiostimulators, defib-
rillator, pulmonary ventilators, cardiomonitor, and drugs).
The patient undergoes bilateral catheterization of the subclavian veins using the standard technique with administra-

tion of serial homemade catheters 1.4 mm in diameter. A patient is connected to a monitoring system by precardial
needle electrodes; blood pressure is controlled by the Korot-
kov method.

Correction of the hypercoagulable state was performed
using the PF-0.5 apparatus for plasmapheresis. This apparatus works at a continuous, periodic rate. Under the action
of a pump, the patient's blood travels along the magistral
system through the subclavian catheter, and is mixed with an
anticoagulant solution delivered by a second pump in the magistral system in direct proximity to the catheter. The blood
then enters the rotor where it is separated into two fractions: plasma and erythrocyte mass. The erythrocyte mass
is delivered in a return system from the rotor. Rheologic
solution is added to the same system, and the erythrocytes
are diluted within the parameter required.

Plasma flows from the rotor into a special cup and is
removed from the vascular bed. By checking the rate of blood
delivery, the rate of erythrocyte selection, and the rotation
of the rotor, it is possible to obtain the concentration of
plasma required and to enrich it with platelets, lymphocytes,
macroglobulins, and other components.

Our express laboratory enables us to maintain constant
control over the patient's state (acid—alkali state, hemoglobin, and concentrations of plasma components and centrifugate
elements). Removed plasma is replaced by the rheological preparations in a volume ratio of 1:1 or 1:2, depending on the
severity of thrombosis.

CLINICAL RESULTS WITH APHERESIS

Myocardial Infarction
(Without Cardiogenic Shock)

Apheresis of the hypercoagulable state was performed in
20 patients who had acute myocardial infarction not accompanied by cardiogenic shock. Changes in the biochemical indices as a result of apheresis are presented in Table 1.

The table shows that, before the correction, all of the
patients had hypercholesterolemia, elevated amounts of β-
lipoproteins and fibrinogen, and total protein that was close

Table 10.1.   Apheresis Correction of Biochemical Indices of
              the Blood in Patients with Myocardial Infarction
              (without cardiogenic shock)

Index	Normal value	Before correction	After correction
Fibrinogen, mg%	200-400	723 ± 70	232 ± 10
Cholesterol, mg%	160-200	256.4 ± 13	163.3 ± 7
β-Lipoproteins, mg%	350-600	706.9 ± 14	378.7 ± 10
Total protein, g%	6.5-8.5	8.39 ± 0.67	6.23 ± 0.32

to the upper limit of normal. By removing part of these substances from the circulating blood and by replacing them with
rheologic preparations, it was possible to decrease and normalize their concentrations and to achieve a stable improvement in the patients' clinical state.

Myocardial Infarction
with Cardiogenic Shock

     We also used apheresis to treat four patients with acute
myocardial infarction complicated by cardiogenic shock and
complete atrioventricular heart block. The combined therapy
for cardiogenic shock included apheresis of the hypercoagulable state, temporary transvenous endocardial heart stimulation, and the use of cardiac glycosides, direct-acting anticoagulants (heparin), narcotic analgesics, hormones, and
plasma substitute and antishock solutions.

     Apheresis consisted of the separation of blood into eryth-
rocytes and plasma. Blood plasma enriched with platelets,
leukocytes, lymphoid cells, fibrinogen, and cholesterol was
removed from the circulating blood and substituted with disaggregating solutions (rheopolyglucin, physiological solution
with heparin) and plasma substitutes (polyglucin, gelatinol,
albumin). As a result, blood viscosity and coagulational
activity significantly decreased, dense erythrocyte blasts
were lysed, and blood rheologic properties improved.

     Before the correction, the state of the patients was ex-
tremely severe: Cyanosis and paleness of skin was observed,
the patients were covered with a sticky cold sweat, and they

were inert and adynamic. Two patients complained of pain and burning in the chest. Their pulse rates were lower than 40 beats per minute. Systolic pressure was below 90 mm Hg. In the first patient, the ECG showed acute and extensive transmural myocardial infarction in the posterior wall of the left ventricle and complete atrioventricular heart block. In the second patient, we observed recurrent, extensive transmural myocardial infarction in the anterior wall of the left ventricle with macrofocal scars in the posterior wall of the left ventricle and complete atrioventricular heart block.

In spite of the antishock therapy using hormones, narcotic analgesics, glycosides, and electrical stimulation of the heart by an external cardiostimulator with electrodes placed in the right ventricular cavity, the hemodynamic indices did not stabilize and a gradual aggravation of the shock was observed. The cardiogenic shock was eliminated completely, however, after a 4-hour apheresis of the blood. This result was manifested by a disappearance of pain and a significant decrease in cyanosis. The skin surfaces became warmer and had normal color, and arterial pressure increased to 110/70 and 115/70 mm Hg. A sharply positive diuresis of 1400 and 1600 ml of urine was observed. In the first patient, sinus rhythm was restored by the end of the correction. In the second patient, atrioventricular block decreased to the second degree.

The following biochemical changes in venous blood indices were observed: total protein decreased by 20%, urea by 7.35%, fibrinogen by 51.7%, and cholesterol by 30.12%. No marked changes in enzymatic activity (creatinine phosphokinase, aspartate transaminase, alanine transaminase) were observed. During the correction, 2000 ml of blood were treated; about 50-60 g of plasma protein structures were removed: $1.8 \cdot 10^6$ granulocytes, $21 \cdot 10^8$ platelets, $8 \cdot 10^5$ lymphocytes, and $4.2 \cdot 10^5$ monocytes. Blood coagulation time increased by 15- and 10-fold, respectively.

Control of the rates of delivery for the solutions infused into the magistrals and return of the erythrocyte mass to the patients enabled us to regulate the hypovolemia developing during cardiogenic shock and to maintain the blood pressure for adequate perfusion of the kidneys. In cases of low central venous pressure, the PF-0.5 apparatus can provide for rapid filling of the vascular bed with low-molecular-weight dextran to the necessary level (under blood pressure control);

simultaneous removal of cholesterol, fibrinogen, thrombocytes, and other components of hemostasis; and significant improvement in microcirculatory blood flow. In the treatment of cardiogenic shock, we achieved a stable positive effect in two patients who were discharged from the hospital. Hemodynamics were also stabilized temporarily in another two patients.

Acute Myocardial Infarction with Thrombosis of Pulmonary Artery Branches

Many patients with acute myocardial infarction develop symptoms of thrombosis in the small branches of the pulmonary artery in the early days after onset of the disease [2]. Clinicians term this state hypostatic pneumonia or thromboembolism of the small branches of the pulmonary artery. We frequently expect to see pathological changes in the lungs similar to those with pulmonary edema in patients with repeat myocardial infarction, and in patients with marked respiratory insufficiency, cyanosis, hypoxia, and hypoxemia. The incidence of lesions of the lung increases in patients with extensive injuries of the heart and usually depends on myocardial pump function.

Apheresis of the hypercoagulable state was included in the treatment program for patients with acute myocardial infarction with thrombosis of the small branches of the pulmonary artery. One example is patient I. Yu. P., aged 55 years (record No. 14172), who was admitted on June 4, 1981, and discharged the following month, on July 24. This patient was treated in the department of emergency cardiosurgery with a diagnosis of ischemic heart disease (IHD) and acute repeat myocardial infarction (on June 1, 1981) in the posterior wall of the left ventricle. This was complicated by thrombosis of the small branches of the pulmonary artery of the right lung, right-sided and infarcted.pneumonia, and complete atrioventricular heart block. Atherosclerosis of the aorta and coronary arteries was also observed, with atherosclerotic and macrofocal postinfarction cardiosclerosis, and disseminated intravascular thrombogenesis in transition to a thrombohemorrhagic syndrome.

The patient had been ill for 5 years and had had two macrofocal myocardial infarctions. After the second one he was treated in a sanatorium, where the third myocardial infarction occurred and he was again taken to the hospital. Routine anti-infarction therapy was ineffective and the patient was admitted to the department of emergency cardiosurgery.

This patient complained of chest pain, dyspnea, hemoptysis, pain in the right chest on breathing, and absence of urine
for about 20 h. His general condition was severe; dyspnea
was 30 respirations per minute, and cyanosis and acrocyanosis
were observed. A high content of fibrinogen, cholesterol,
lipids, and total protein was found in his blood. The presence of acute myocardial infarction was confirmed on the ECG
and enzymatically. An x-ray showed total right-sided pneumonia.

In order to achieve clinical remission, eliminate thrombosis in the microcirculatory system, and break through the
pathophysiologic mechanisms of the thrombohemorrhagic syndrome, we included apheresis of the hypercoagulable state in
his treatment. The patient was taken to an operating room
especially equipped for the treatment of complicated forms
of myocardial infarction. Double-sided puncture of the subclavian veins was performed by the standard technique. The
patient was then connected to the PF-0.5 apparatus and apheresis was initiated. Plasma with a high concentration of factors
causing sharp disturbances in the blood rheologic properties
(fibrinogen, cholesterol, lipoproteins, macroglobulins) was
removed with simultaneous replacement of the plasma by rheologic preparations. We removed 1450 ml of plasma, and 2500
ml of rheologically active preparations were injected.

By correcting the hypercoagulable state, clinical remission was achieved, manifested by the disappearance of dyspnea,
hemoptysis, and pains behind and in the right part of the
chest. The clinical effect followed the normalization of
microcirculatory blood flow by decreasing the blood coagulation factors and the plasmatic components and increasing blood
viscosity (Table 10.2). The improvement and positive clinical effect was manifested not only subjectively (disappearance
of dyspnea, coughing, hemoptysis, and pain) but also in the
x-ray picture. The total right-sided pneumonia caused by
microcirculatory thrombosis disappeared. After the operation,
the patient released 1600 ml of urine, thus confirming the
normalization of microcirculation.

The patient's later condition was also satisfactory with
no pains behind the chest, complete disappearance of the signs
of thrombosis in the small branches of the pulmonary artery,
and no elevation of fibrinogen, cholesterol, β-lipoproteins,
or total protein. These indices, which together reflect the
microcirculatory processes, were within normal levels within
30 days after the correction.

Table 10.2.   Changes in Biochemical Indices of Patient I. Yu.
             P., Aged 55 Years, as a Result of Apheresis Cor-
             rection of the Hypercoagulable State

Index	Normal value	Before correction	After correction	1 month later
Fibrinogen, mg%	200-400	620	320	300
Cholesterol, mg%	160-200	232	171	192
β-Lipoproteins, mg%	350-600	974	373	331
Total protein, g%	6.5-8.5	8.17	6.42	7.0

Repeat Myocardial Infarction
Complicated by Pulmonary Edema

Patient A. N. S., aged 73 years (record No. 22869), was
admitted on September 12, 1981, and discharged the following
month, on October 14.   This patient was treated in the department of emergency cardiosurgery with a diagnosis of IHD
and acute repeat myocardial infarction (on September 12,
1981) complicated by lung edema, angina at rest and on effort,
atherosclerosis of the aorta and coronary arteries, atherosclerotic and postinfarction cardiosclerosis, and disseminated intravascular coagulation (DIC).

The patient had had IHD for a long time and had sustained
macrofocal myocardial infarction.   On September 12, he had a
hard pain attack behind the chest, which was complicated by
lung edema.   The patient was admitted to a cardiosurgery department where his acute symptoms were eliminated after intensive therapy.   But, the patient complained of pains in the
chest, had to take up to 60 nitroglycerin pills, and, from
time to time, used narcotics and analgesics.   He showed no
marked clinical improvement and had a high content of total
protein, fibrinogen cholesterol, and β-lipoproteins in his
blood.   Special studies confirmed the presence of marked intravascular thrombogenesis (DIC syndrome).

Due to the absence of an effect with the treatment provided, the patient was transferred to the department of emergency cardiosurgery for apheresis of his hypercoagulable state.

Table 10.3.   Changes in Biochemical Indices of Patient A.N.S.,
             Aged 73 Years, as a Result of Apheresis Correc-
             tion of the Hypercoagulable State

Index	Normal value	Before correction	After correction
Fibrinogen, mg%	200-400	725	250
Cholesterol, mg%	160-200	274	163
β-Lipoproteins, mg%	350-600	815	489
Total protein, g%	6.5-8.5	8.17	6.12

The patient received apheresis twice to remove plasma enriched
with the factors causing an increase in coagulation and
viscosity. The plasma was replaced by rheologic preparations.
We removed 2000 ml of plasma and injected 2000 ml of rheologic preparations.

    As a result of the correction, the content of fibrinogen,
cholesterol, β-lipoproteins, and total protein was reduced
(Table 10.3). Normalization of the concentration of factors
which caused disturbances in blood coagulation effected improvement in the blood supply of his heart, manifested by a
stable remission (disappearance of pain both at rest and with
physical exercise). After correction, it was no longer necessary to inject narcotics or analgesics into the patient or to
use nitroglycerin.

Acute Disturbances of Coronary Blood Circulation
and Cardiac Rhythm

    Patient K. S. M., aged 49 years (record No. 19654), was
admitted on August 5, 1981, and discharged the next month,
on September 3. The patient was treated in the department of
emergency cardiosurgery with a diagnosis of IHD, acute disturbance of the coronary circulation in the anterior wall of
the left ventricle, extrasystole, atherosclerosis of the aorta
and coronary arteries, atherosclerotic and postinfarction
cardiosclerosis, arterial hypertension, and chronic nephritis.

    The patient had had IHD for more than 7 years and had
sustained an acute transmural myocardial infarction in the
region of the posterior wall of the left ventricle. He was

Table 10.4. Changes in Biochemical Indices of Patient K.S.M.,
Aged 49 Years, as a Result of Apheresis Correction of the Hypercoagulable State

Index	Normal value	Before correction	After correction
Fibrinogen, mg%	200-400	520	355
Cholesterol, mg%	160-200	259	156
β-Lipoproteins, mg%	350-600	640	373
Total protein, g%	6.5-8.5	8.17	6.72

admitted to a cardiology department with intensive pains behind the breast, which could not be eliminated by narcotics
or analgesics. The ECG showed extrasystole and signs of acute
disturbance of the coronary circulation in the anterior wall
of the left ventricle.

Due to the absence of a clinical effect with routine
therapy, the patient was transferred to the department of
emergency cardiosurgery. A high content of fibrinogen, cholesterol, β-lipoproteins, and total protein was found in his
blood when he was admitted. The ECG showed signs of acute
disturbance of the coronary circulation in the anterior wall
of the left ventricle (ST-segment elevation in leads V_{1-3},
frequent ventricular extrasystole, and scarring in the region
of the posterior wall of the left ventricle). Apheresis was
performed to remove plasma enriched with the factors causing
an increase in blood viscosity and coagulation activity. The
plasma was replaced by rheologic preparations. We removed
1600 ml of plasma and injected 1600 ml of rheologic preparations.

As a result of the correction, the content of fibrinogen,
cholesterol, β-lipoproteins, and total protein was decreased
(Table 10.4). The disturbance of the blood microcirculation
in the cardiac muscle was manifested by different rhythm and
conduction disorders, pains behind the breast, and ECG signs
of a disturbance in coronary circulation.

Due to the normalization of the factors responsible for
microcirculatory processes, the blood supply of the heart was

improved, manifested by a stable remission (disappearance of pains behind the breast and extrasystole) and normalization of ECG signs (decrease of ST-segment elevation in the right thoracic leads, disappearance of extrasystole). After correction, it was no longer necessary to inject narcotics and analgesics into the patient and the use of glycerin was reduced significantly.

The positive effect obtained with apheresis of the hypercoagulable state in complex treatment of ischemic heart disease shows promise, provides cardiologists with a new and powerful means for regulating hemostasis, and extends the prospects for a therapeutic effect.

REFERENCES

1. B. A. Kudrjashov, Biological Problems in the Regulation of Blood Fluidity and Its Coagulation, Meditsina, Moscow (1975), p. 488.
2. V. P. Baluda, "Hemostasis as one of the factors in hemostasis preservation," in: P. D. Gorizontov (ed.), Homeostasis, Meditsina, Moscow (1976), pp. 314-340.
3. P. K. Anokhin, Articles on the Physiology of Functional Systems, Meditsina, Moscow (1975), p. 446.
4. Z. S. Barkagan, Clinical Study of Hemostasis System, Medical Institute, Barnaul (1975), p. 186.
5. O. K. Gavrilov, Problems and Hypotheses in Studies of Blood Coagulability, Meditsina, Moscow (1981).
6. A. O. Gavrilov and A. M. Shilov, Problems of Hematology and Blood Transfusion, Meditsina, Moscow (1979), pp. 16-18.
7. V. A. German (ed.), Hemostasis System in Normal and Pathological Conditions, Medical Institute, Kuibyshev (1977).
8. B. I. Kuznik and V. P. Skypetrov, Formed Elements of the Blood and Vascular Wall, Hemostasis and Thrombosis, Meditsina, Moscow (1974), p. 306.
9. M. S. Machabeli, Coagulopathic Syndromes, Meditsina, Moscow (1970), p. 410.
10. A. Y. Grytsuk and W. Y. Shchygelsky, "Prethrombotic state and its diagnosis in major cardiovascular diseases," Ter. Arkh., $\underline{1}$, 26-31 (1979).
11. E. I. Chazov and K. M. Lakin, Anticoagulants and Fibrinolytic Drugs, Meditsina, Moscow (1977), p. 308.
12. J. Gradson, "Problems of the myocardial microcirculation," Acta Cardiol. [Suppl.] (Brux), $\underline{19}$, 109-118 (1974).

13. J. C. Giddings, "The investigation of hereditary coagu-
    lation disorders," in: Blood Coagulation and Haemo-
    stasis (A Practical Guide), J. M. Thompson (ed.), Chur-
    chill Livingstone, New York (1980), pp. 48-116.
14. J. W. Haerem, "Sudden, unexpected coronary death. The
    occurrence of platelet aggregation in the epicardial and
    myocardial vessels of man," Acta Pathol. Microbiol. Im-
    munol. Scand. [A], 265, 321-329 (1978).
15. L. A. Harker, R. Ross, and S. Glomset, "Role of the
    platelet in atherogenesis," Ann. NY Acad. Sci., 275,
    321-329 (1976).
16. J. Maruch, "Novsie patofyziologicke poznatky o pluchej
    tromboembolii a ich klinicky vyznane," Bratisl. Lek.
    Listy, 69, 216-226 (1978).
17. I. S. Wright, C. Merple, and D. F. Beck, Myocardial In-
    farction; Its Clinical Manifestations and Treatment with
    Anticoagulants, Grune & Stratton, New York (1954).

PHARMACODYNAMIC CHARACTERISTICS
OF UROKININ

Z. D. Fedorova, G. V. Samsonov,
S. V. Kol'tsova, and N. A. Bynyaeva

Research Institute of Hematology and Blood Transfusion
Leningrad

Institute of High-Molecular-Weight Compounds
Academy of Sciences of the USSR, Leningrad

ABSTRACT

Urokinase preparations have been increasingly applied in clinical practice in recent years. Their physicochemical properties and thrombolytic effect are being studied and various methods of active enzyme isolation have been suggested. Experimental study of urokinase preparations (urokinin) has demonstrated that they have a high specific activity. A therapeutic effect has been obtained in patients with both newly formed thrombi and with thrombi as old as 3 weeks. Studies that are noteworthy have revealed the possibility of producing complex preparations having prolonged action that are based on urokinase and that have both a fibrinolytic and an anticoagulant effect, the latter resulting from the inhibition of thrombin action.

INTRODUCTION

In 1947, Macfarlane and Pilling showed that human urine contained a substance that possessed kinase activity and the ability to change the proenzyme plasminogen into the enzyme plasmin. This latter activity is responsible for fibrinolysis. The substance that was extracted from urine was later termed urokinase [1-3]. It was shown [4, 5] that the enzyme is synthesized in the endothelial cells of renal vessels.

This finding was substantiated by the isolation of urokinase from the culture fluid of renal cells.

Urokinase is an endopeptidase with specific action. Purified by gel filtration on a Sephadex G-100 column, urokinase shows one peak; in polyacrylamide gel electrophoresis, two subpeaks with a molecular weight of 30,000-33,000 and 50,000-55,000 daltons, respectively, are seen [7-9].

Plasminogen activation by urokinase is an enzymatic reaction caused by the splitting of the arginine—valine bond, followed by the formation of molecules with two polypeptide chains linked by disulfide bridges. Streptokinase, urokinase, and other substances in tissue and plasma are known to activate the plasminogen zymogen, having a single polypeptide chain, into plasmin composed of heavy and light chains [8-10]. The heavy chain is a derivative of the NH_2-terminal, and the light chain is the derivative of the COOH-terminal of plasminogen.

Purified plasminogen contains lysine in the NH_2-terminal and asparagine in the COOH-terminal. Plasmin contains lysine and valine in the NH_2-terminals and arginine and asparaginic acids in the COOH-terminals. The molecular weight of plasminogen has been shown by ultracentrifugation to be 83,200-92,000, while that of plasmin is 75,400-85,000. Three different forms of human plasminogen containing lysine, valine, and glutamic acid residues in the NH_2-terminal have been identified by isoelectric focusing.

It has also been demonstrated that at least two peptide bonds are broken in plasminogen during its activation into plasmin by urokinase. Proteolysis of the bond in Glu-plasminogen changes it mainly into intermediate Lys-plasminogen, releasing the preactivation NH_2-terminal peptide. Urokinase breaks down the arginine—valine bond in the inactive intermediate to form the plasmin molecule composed of two polypeptide chains covalently bound by a disulfide linkage. The molecular weights of these chains are 55,000-60,000 and 24,000-26,000 daltons, respectively. Plasmin itself is the agent responsible for the splitting of the first bond in the molecule of plasminogen. The release of the plasminogen NH_2-terminal is not necessary to break down the second bond by urokinase.

The resulting plasmin affects proteolytically both fibrin and fibrinogen as well as other coagulation factors,
growth hormone, somatotropin, angiotensin, γ-globulin, complement factors, and mucoproteins. However, it shows a distinct affinity for fibrin in vivo, while it affects both fibrinogen and fibrin in purified systems at an approximately
equal rate. Plasmin converts insoluble fibrin into soluble
products by breaking some of the peptide bonds.

Urokinase substantially affects the activation mechanism
of fibrinolysis.

FIBRINOLYTIC POTENTIAL OF UROKINASE

Fibrinolysis is a complicated process during which fibrin, an insoluble substance formed at the end of coagulation,
is decomposed into soluble products by plasmin. In various
stress situations and pathologic states, the fibrinolytic potential of blood may either be sharply activated or blocked,
leading to thromboses or acute hemorrhages.

The complex mechanism of thrombolysis with urokinase has
been described by many authors [4, 9, 11, 12] who have put
forward a number of hypotheses. Some authors [4, 9] suggest
that plasminogen and fibrinogen are present in the blood in
almost equal concentrations and that plasminogen is incorporated into the fibrin at the moment of clot formation. Exposed
to the activator, the formed plasmin lyses fibrin. In
this case, antiplasmins, which are always present in the blood
(or are part and parcel of the blood), are not able to inhibit the plasmin which is inside the clot where it manifests a
lysing effect. Other researchers [11, 12], however, propose
that plasminogen may be activated directly in the vascular
bed. The plasmin formed in such a way binds to the antiplasmin and starts circulating in the blood as a reversible complex. Due to its high affinity for fibrin, plasmin separates
from the complex and binds to fibrin while antiplasmin remains
unbound in the blood flow. From this hypothesis, the important role of antiplasmins in preventing the lysis of other
blood proteins, and coagulation factors in particular, has
been established.

A number of other authors [13, 14] assume that the thrombus is lysed not by plasmin but by the diffusion of plasminogen and its activator from the blood into the thrombus. The

administration of urokinase into the blood flow changes plasminogen to plasmin, which is immediately inactivated by antiplasmins. Residual plasminogen and urokinase diffuse into
the clot and the plasmin formed in it displays a lysing effect
inside the thrombus.

Study of the metabolism of urokinase has shown [14] that
the half-life of the enzyme labeled by ^{125}I is 2 min. Fifteen
minutes after an injection of the labeled enzyme, 50% of the
radioactive dose was found in the liver and kidneys, and by
the fourth hour amounted to less than 5%. Later studies [15-
17] followed ^{131}I-urokinase in the animal organism. After
7.5 min, 37% of the radioactive dose was found in the blood
and, after 15 min, 43% was detected in the liver and kidneys.
Less than 1% of the radioactivity was found in the heart,
lungs, and spleen. Only 2.5% of urokinase was excreted with
the bile. Urokinase excreted mainly in the urine amounted to
63% in rats and 43% in dogs. Also, the radioactivity of the
enzyme was almost completely bound to a biologically inert
molecule of the low-molecular-weight enzyme.

In the organism, the urokinase administered is either
unbound or bound to proteins. The percentage of radioactivity in the high-molecular-weight fraction of blood decreases with
time, while that in the low-molecular-weight fractions increases.
This observation confirms that urokinase disintegrates into
smaller fragments in the blood.

Study of the metabolism of human urokinase by determining fibrinolytic activity in the blood [18-20] shows that its
half-life is 9-16 min. It should be noted that the half-life
of urokinase increased when drop and continuous infusion was
used. Within 24 h, 300 to 900 units of urokinase activity
were excreted with urine in 78% of virtually healthy people,
the former amount in those with high NaCl diets and the latter
amount in those with low NaCl diets. In patients with malig-
nancies, cardiac insufficiency, deep burns, and trauma, excretion of urokinase is sharply decreased [20].

Urokinase preparations excreted from human urine are used
in clinical practice to treat venous and arterial thromboses
and thromboemboli. Many researchers hold that urokinase preparations are the most promising thrombolytic treatment because
of the availability of sources, the absence of antigenicity,
and its high efficacy. Because isolation and purification of
the enzyme are time consuming, however, study and clinical

application of urokinase are hindered. Urokinase preparations are currently produced in the USA, Great Britain, Denmark, France, Italy, Japan, and the USSR.

Extensive literature is devoted to the experimental study
of urokinase. Following the lysis of clots formed from human
plasma as a result of the incorporation of urokinase, a number
of authors [10, 21, 22] have concluded that the degree of
lysis of artificially induced thrombi depends on the concentrations of fibrin and plasmin in the plasma. Thrombolysis
in plasma increases with an increase in urokinase, although
high enzyme concentrations also produce an inhibiting effect
on thrombolysis. It is supposed that partial thrombolysis
is associated with an insufficient formation of plasmin in
the thrombus, but with an excess of plasminogen in the plasma
it might diffuse into the thrombus, increasing thrombolysis
up to complete lysis of the thrombus.

Comparative study of tissue plasminogen activator and
urokinase in purified systems, unseparated human plasma, and
in a system with fibrinogen labeled by ^{125}I has shown that
these are two immunologically different activators. The tissue activator possesses a higher thrombolytic effect, a low
activity threshold, and an activity that increases with time.

A number of authors [10, 23] have demonstrated that urokinase has pronounced activator and fibrinolytic properties
in intravenous injection. These data have been confirmed in
experiments with cerebral and femoral artery emboli with both
a single-dose or drop-by-drop administration of the preparation.

A sufficient amount of experimental material and clinical observations have been accumulated thus far on the therapeutic efficacy of urokinase preparations [10, 23, 25].

UROKINASE ISOLATION ON HIGHLY PERMEABLE BIOSORBENTS

The urokinase preparation urokinin has been extracted
from human urine by ion-exchange sorption on highly permeable
ionosorbents at the Scientific Research Institute of Hematology and Blood Transfusion in Leningrad in cooperation with
the Institute of High-Molecular-Weight Compounds, USSR Academy of
Sciences (author's certificates Nos. 363917, October 5, 1972,
and No. 584034, August 22, 1977).

To isolate urokinase, sorption procedures are presently used in addition to traditional methods of protein precipitation by salting out and decreasing solubility at the isoelectric point, or the use of these methods together with ultrafiltration or gel chromatography. As with the isolation of other proteins, the following sorbents are applied for reversible sorption: mineral sorbent — kaolin [17], mineral gel sorbent with a rather stable structure — silica gel [26], and sorbents capable of interacting with protein macromolecules at the fiber periphery — acrylonitrile fibers [27] and phosphate cellulose [28]. With the latter, one can not only perform reversible sorption of urokinase, but also achieve a sufficiently high rate of interphase mass exchange as well.

New possibilities for the isolation of proteins and enzymes appeared when the entire range of problems on selectivity, reversibility, and kinetics of protein sorption at high sorption capacity had been solved. Certain principles were established which enabled the production of a new type of highly permeable biosorbents and which set the conditions for their use in protein isolation by sorption chromatography [29-32]. The high absorption capacity of biosorbents depends on the porosity of their reticular structure formed by copolymer synthesis under conditions of precipitative copolymerization. The process leads to the formation of compact reticular structures as well as channels and cavities available for protein macromolecules. Biosorbents capable of absorbing 5-10 g of protein per 1 g of dry sorbent were obtained on this basis.

Selective sorption was achieved by polyfunctional interaction between proteins and the porous copolymer matrix having a high content of the cross-agent. The high cross-linking of structure attained under specific conditions of biosorbent synthesis does not lead to loss of porosity and permeability of their reticular structure.

The polyfunctional nature of the interaction lies not only in the capability of the biosorbent structure to form electrovalent bonds with proteins but also in weak interactions of the van der Waals and hydrophobic types. Such specificity allows one to isolate proteins with high sorption capacity from salt solutions of high ionic strength. This is of great interest for the development of a preparative

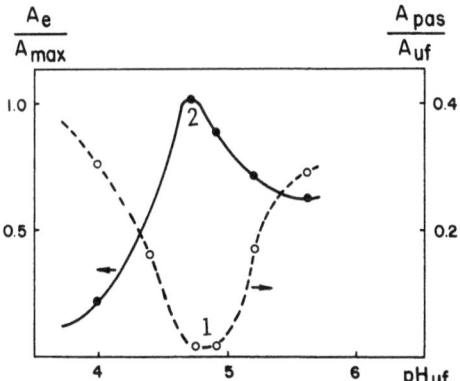

Fig. 11.1.   Dependence of urokinase sorption on cat-
             ionite CMT on the pH of urine filtrate.
             v/v_{sorp} = 150.

technique for isolating protein under conditions of competition between protein and other ions in the solution. Protein
sorption by the new type of biosorbents leads to an increase
in the system entropy. This means that, with a high specificity of protein sorption, the value of the bond's energy
is not so high that it favorably affects the reversibility
of protein sorption. To acquire this effect, the new type
of biosorbents are given a structural stability comprised of
a slight variability in ionite volume swelling with a marked
change in the pH and ionic strength of the solution. The
complex of all of these properties is characteristic of highly permeable carboxyl biosorbents produced under the name of
CMT and Biocarbs T and D (etc.).*

    Isolation of urokinase on highly specific biosorbents
involves the main stage — preparative chromatography on carboxyl biosorbents CMT or Biocarb T or D — and a preliminary
frontal purification on polycondensational anionite FAF, permitting the isolation of pigments and other impurities before

------

*CMT is a commercial carboxylic cationite manufactured
in the USSR, the product of precipitative copolymerization of
methacrylic acid and hexahydro-1,3,5-acryloyltriazine. Biocarbs are commercial heteroporous carboxylic cationites. FAF
is a commercial anionite produced in the USSR, the product of
copolymerization of trimethylphenoxyethylammonia and phenol
with formaldehyde.

Z. D. FEDOROVA ET AL.

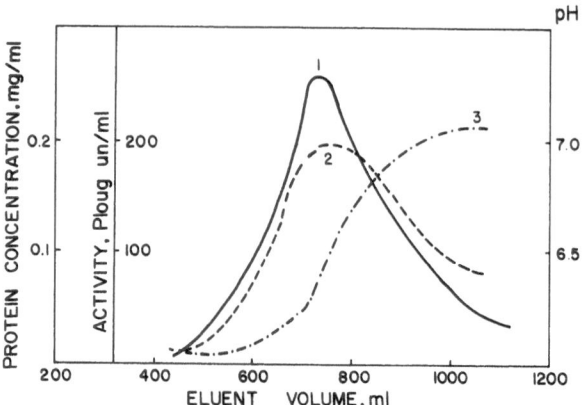

Fig. 11.2. Urokinase desorption from cationite CMT by 0.1 M phosphate buffer solution at pH 7.2. v/v_{sorp} = 300. 1) Protein concentration; 2) activity; 3) pH.

the onset of the main stage of the process. The isolation of highly active urokinase on highly permeable biosorbents is a four-stage process. Isolation of urokinase from human urine is as follows. Ten liters of initial material are alkalized to pH 8.2-8.5; the precipitate is then separated and the filtrate acidified to pH 4.75-4.80. The solution obtained passes through an anionic FAF column (5 × 20 cm) and then through a column with the highly permeable carboxylic cationite CMT (3 × 10 cm) at the rate of 100 ml/h per 1 cm². When sorption is finished, the column with cationite is washed with 0.1 M phosphate buffer solution, pH 6.0-6.1, at the rate of 30-40 ml/h per 1 cm². Urokinase is desorbed by 0.1 M phosphate buffer solution, pH 7.1-7.3, at the rate of 20 ml/h per 1 cm². and 50-ml fractions are selected. For each fraction we determined pH, optical density at 280 nm (E_{280}), protein concentration [35] using as standard trypsin from "Spofa" (Czechoslovakia) purified by gel chromatography on Sephadex G-50, and specific activity of urokinase [36] using either urokinase from "Koch-Light Lab" (England), or streptase from "Bechring Werke" (FRG), as standards. Active fractions are pooled, dialyzed against 0.1 M NaCl solution, and lyophilized. All column processes as well as dialysis are carried out at 5-7°C Activity of the final preparation is 100,000-200,000 IU/mg of protein.

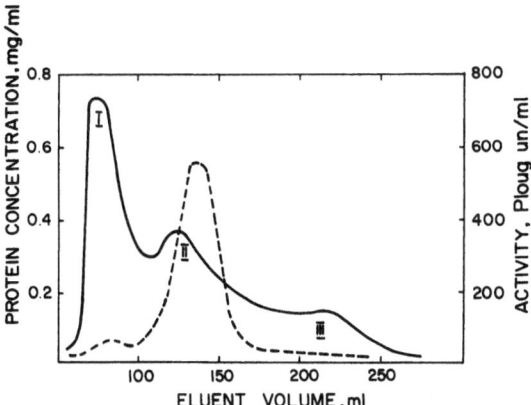

Fig. 11.3. Urokinase preparation by gel chromatography on Sephadex G-100, pH 6.0, ionic strength 0.5, temperature 5°C. 1) Protein concentration; 2) activity.

Figure 11.1 shows the dependence of the relative sorption capacity of urokinase on cationic CMT on the pH of urine filtrate (UF). These data are presented, first, as the ratio of activity in the passing (A_{pas}) to the activity in the urine filtrate (A_{uf}) (curve 1) and, second, by the relation between the activity in the eluate after the sorption of urokinase at the given pH (A_e) and maximum eluate activity (A_{max}) resulting from the sorption of urokinase from UF at pH 4.75-4.80 (curve 2). Both relationships are obtained using a series of columns with a constant volume of sorbent (v_{sorb}) and of passing urine filtrate (v). They also demonstrate a high selectivity of urokinase sorption on a highly permeable cationic CMT which is manifested at a narrow pH optimum of enzyme sorption (4.7-4.8). In this case, the active component of UF is fully adsorbed on cationic CMT, since independent experiments showed that the A_{pas}/A_{uf} ratio was 1 at the stage of UF depigmentation of the FAF anionite at a pH interval from 4.5 to 6.0. Besides high selectivity, porous cationic CMT also provides for the highly specific sorption capacity of urokinase from UF at an optimum pH of sorption. In the system studied, the v/v_{sorb} ratio may be very high (up to 1000) without a decrease in the specific yield of the end product. Desorption of urokinase from cationic CMT by 0.1 M phosphate buffer solution with a stepwise increase in pH from 6.0 (washing stage) to 7.2 showed that the active component is desorbed as a highly purified product with a peak concentra-

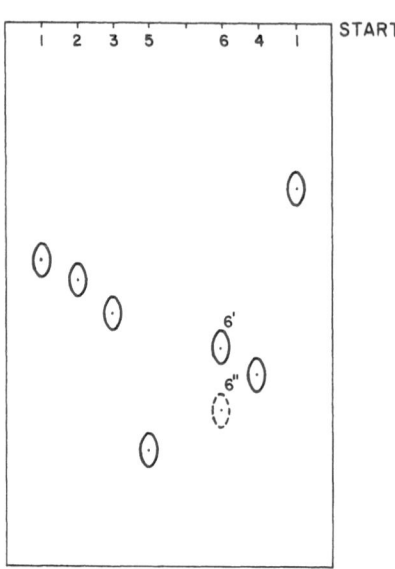

Fig. 11.4. Study of purified urokinase by thin-layer gel
chromatography on Sephadex G-100 (Superfine), pH
6.0, ionic strength 0.3. Reference proteins: 1)
cytochrome C; 2) myoglobin; 3) chymotrypsinogen;
4) egg albumin; 5) human serum albumin; 6) purified urokinase.

tion at pH 6.5 (see Fig. 11.2 where the protein peak practically
coincides with that of specific activity).

Gel chromatography of the urokinase preparation obtained
by ion-exchange sorption on porous cationic CMT showed three
protein fractions (Fig. 11.3), two of which (I and II) offer specific activity.

It should be noted that the purity of the above urokinase preparation is close to that obtained by sorption on
acrylonitrile fibers. In fraction II, more than 90% of the
activity of the initial urokinase preparation is eluted.
This highly active fraction (5000-10,000 Ploug units/mg of
protein), termed "a purified urokinase" after rechromatography,
was used for further investigations of its homogeneity and
for determination of certain characteristics, in particular
molecular weight, pI, and $E \frac{1\%}{1 \text{ cm}}$.

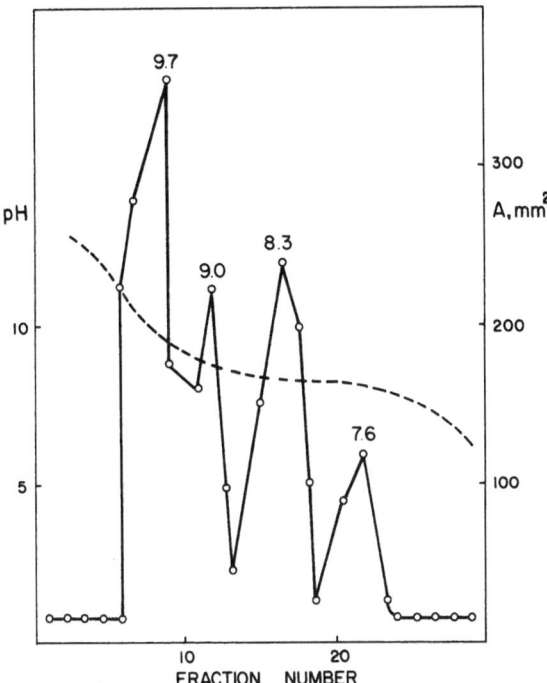

Fig. 11.5.   Isoelectric focusing of purified urokinase within a pH of 3-10.   Column, 110 ml; time, 30 h;
temperature, 5°C.

       Study of purified urokinase in a native state by thin-
layer gel chromatography (TLGC) on Sephadex G-100 (Superfine),
pH 6.0-8 and ionic strength 0.3, pointed to the existence of
two protein components (the main and the minor one) with molecular weights of 33,000 and 53,900 daltons, respectively
(designated in Fig. 11.4 as 6' and 6"). Both values for the
molecular weight agree with data in the literature regarding
"light" and "heavy" forms of urokinase [26, 27, 37, 38].
These results were confirmed by disc electrophoresis (DE) of
purified urokinase in polyacrylamide gel at pH 8.3 in the
presence of sodium dodecyl sulfate, ethylenediamine tetra-
acetic acid (EDTA), and dithiothritol according to [39]. The
preparation showed two zones: the main zone with a molecular
weight of 33,100 daltons and the minor zone with a molecular
weight of 20,900 daltons. The "heavy" form of urokinase is
known to be the product of association between the "light"
urokinase and an inert protein with a molecular weight of
18,000-20,000 daltons [38, 40, 41]. Under conditions of DE

[39], high-molecular-weight urokinase should dissociate into
two polypeptide chains and, thus, the minor zone is likely
to be a low-molecular-weight component of the "heavy" urokinase. A similar result was obtained by using TLGC in the
analysis of purified urokinase after denaturation in 6 M of
guanidine-chloride solution with further reduction by β-mer-
captoethanol and conversion to the CM-peptide [42].

In isoelectric focusing (IEF) of purified urokinase [43],
four components with a pH of 7.6, 8.3, 9.0, and 9.7, respectively, and possessing specific activity, were obtained (Fig.
5). This result corresponds to the data of others who have
studied multiple molecular forms of urokinase [43]. Molecular forms of urokinase were previously assumed to be characterized by the presence of different amounts of amino saccharides of sialic acids as determined for plasminogen and plasmin
[27, 40]. Recently [44], however, the pI of isoforms has
been shown to remain unchanged in the treatment of urokinase
with sialidase, and sialic acids were not found in direct
analysis. The low concentration of amino saccharides could
not shift the pI to an alkaline domain insofar as the number
of acidic amino acids for both forms of urokinase is 1.7- to
1.9-fold greater than the number of basic amino acids. Multiple forms of urokinase with different pI are imagined [38]
to be dissimilar molecules in amide forms of acidic amino
acids. The same study shows that only the "light" urokinase
has several isoforms (with a pH of 6.9-9.3), while the "heavy"
urokinase in IEF produces a component with a pI of 9.7. These
data suggest that the peak, with a pH of 9.7, resulting from
IEF of purified urokinase, as shown in Figure 11.5, is attributed to the "heavy" urokinase present in the sample, although the "light" urokinase has an isoform with a pI of 9.6.
In any case, the IEF data confirm the previous conclusion
that the purified urokinase under investigation has a low-
molecular-weight form as the main component.

The UV spectra of purified urokinase taken at 200-320 nm
have two absorption bands characteristic of proteins: at
280 nm in 0.1 M phosphate buffer solution, pH 7.8, E $\frac{1\%}{1\ cm}$ is
12.5, within the known values (11.5-13.2) for "light" urokinase [37, 38].

The preparative-column method for separating highly purified urokinase with a molecular weight of 33,000 daltons and
activity 30,000 Ploug units/mg protein was developed based

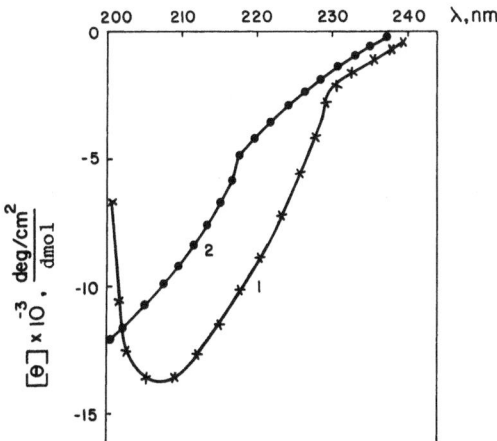

Fig. 11.6. CD spectra of urokinase with a molecular weight
of 33,000 daltons in 0.1 M phosphate buffer solution. 1) pH 3.0-11.6; 2) pH 1.7.

on the TLGC data given in Figure 11.4, which demonstrated the
possibility of separating two forms of urokinase in the native
state on Sephadex (Superfine). The development of this method has led to a new stage in the study of urokinase, i.e.,
to the investigation of its structure and conformational stability under experimental conditions.

The secondary structure of "light" urokinase within a
wide range of pH and temperature was studied using the circular dichroism (CD) and ORD methods. Using CD, the calculation of secondary structure parameters was carried out using
the least-squares method by minimization of the differences
between the calculated and experimental values of the CD
spectra for certain protein structures in the wavelength
range of 206-240 nm. Calculations were performed for three
structures (α-helical, β structures, and irregular segments)
[45], for four structures including β bends [46], and for the
five β parallel and β antiparallel structures isolated in the
β structure [47].

The ORD data obtained at 350-600 nm were processed using
the Drude monomial equation. The percentage of α helices
in the enzyme [48] was determined according to λ_c. Figure
11.6 shows the CD spectra of urokinase at various pH levels.
Within pH 3.0-11.6, the CD spectra practically do not change,
pointing to the conformational stability of the enzyme in

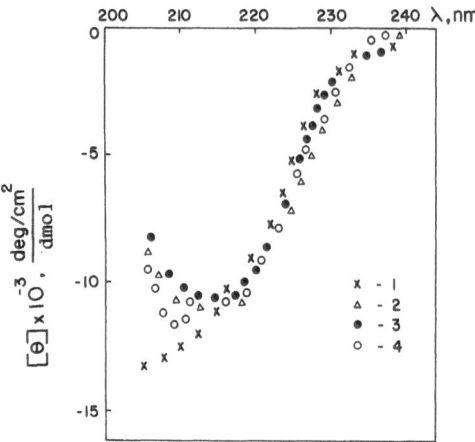

Fig. 11.7. Experimental and calculated CD spectra of urokinase in 0.1 M phosphate buffer solution at pH 6.1.
1) Experimental spectrum; 2-4) calculated spectra
by reper spectra for three, four, and five structures, respectively.

this pH range of the medium. At pH < 3, urokinase undergoes
a conformational transition involving destruction of the secondary structure of the protein. The same figure presents
the CD spectrum at pH 1.7 when conformational transition is
completed.

The calculated versus experimental CD spectra of urokinase at pH 6.1 are shown in Figure 11.7. It is evident
that, under the conditions of the study, the convergence of
the calculated and experimental spectra could not be considered fair for urokinase; the calculated spectra for five structures are somewhat better than those for three or four structures. Divergence of the spectra at 206-240 nm might be associated with a somewhat anomalous appearance of a CD spectrum because of extremely low molecular ellipticity $[\theta]$ at
206 nm.

We compared the amino acid composition of urokinase with
that of standard proteins, used for calculation of the reper
spectra, and of a number of globular proteins. In the latter
case, the parameters of the secondary structure from the CD
spectra correlate well with those from x-ray data. Urokinase
did not appear to differ from these proteins in concentration of aromatic amino acids and it had a higher content

of proline than the other proteins. Urokinase contains 6.37%
of proline, which accounts for 17 proline residues of 276
amino acids in the molecule. The proline content in the other
proteins studied does not exceed 4.3%, and it is much lower
in many proteins. Additional interactions resulting from the
presence of proline in the urokinase polypeptide chain and not
considered in the reference CD spectra are likely to occur in
the enzyme molecule. Parameters of the urokinase secondary
structure are calculated with approximate accuracy, as shown
in Figure 11.7. Urokinase with a molecular weight of 33,000
daltons comprises about 30% of the amino acid residues included in α helical regions, and nearly the same amount incorporated in the β structure (mainly the parallel one). Most of
the amino acids are in β bends. The value of α helices in
the urokinase molecule at pH 6.1 is borne out by the ORD data.
In passing to pH 1.7, the secondary structure of urokinase
alters and the content of α helices and β structure decreases
while the content of irregular segments increases.

Study of the conformational stability of urokinase kept
in solution under a wide range of experimental conditions
showed that urokinase retained the native state for a long
time at 10-30°C, preserving the wide range of pH. Study of
the dependence on temperature of specific optical rotation
$[\alpha]_{450}$ of urokinase in 0.1 M phosphate buffer solution at pH
6.6 did not reveal any conformational changes in the enzyme
within 20-80°C.

The technique used for isolation of urokinase enables
one to obtain a preparation with an activity no lower than
100,000 IU or 2000 Ploug units/mg protein. The amount of protein in the ampul is 3.0-4.0 mg. By its properties and composition, urokinin is exactly like the preparations of "Uron-
ase" (in Japan) and "Urokinase" (in England).

EXPERIMENTAL STUDY OF "UROKININ"

Specific Activity of Urokinin

The specific activity of the preparation was studied by
Astrup's method using human fibrinogen. The activating action
of the preparation was manifested in its ability to lyse the
unheated fibrinous plates. The absence of direct proteolytic
activity was demonstrated by the absence of lysis on the heated plates with inactivated plasminogen. The high lysing ca-

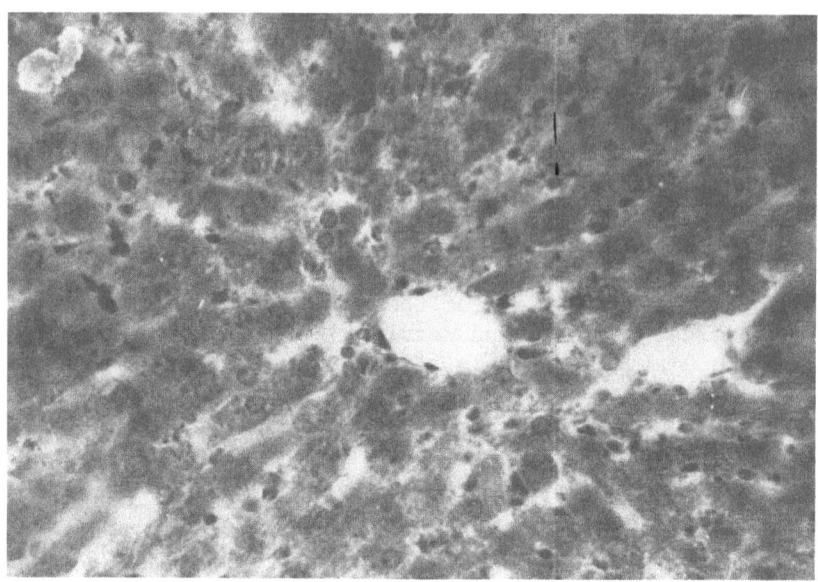

Fig. 11.8. Liver of a rat killed after termination of a
            course of urokinin administration. Hematoxylin-
            eosin stain. Magnification × 200.

pacity of the preparation was also substantiated by acceleration of the lysis of a clot obtained from the plasma euglobulin fraction. However, the rate of lysis depended directly
on the amount of preparation administered. After 10, 30, and
60 minutes of incubation of the clot in fibrinogen solution,
the thrombin time was significantly lengthened. It was observed that the longer the incubation time, the greater the
degree of fibrinogen destruction. This finding could be explained not only by the activity of the plasmin formed, but
also by the occurrence of fibrinogen "fragments" (antithrombin VI) offering anticoagulant properties. The pronounced
lytic effect of urokinin on the fibrin clot could be demonstrated visually on the thromboelastograph, which recorded
the coagulation process. A typical spindle-shaped curve showing complete convergence of the branches of the thromboelastogram was obtained.

    Acute Toxicity. The preparation was injected intrave-
nously into three animal species in a dose of 25 ml/kg of body
weight. Although urokinin essentially has no toxic properties, a slight decrease in arterial pressure was observed in

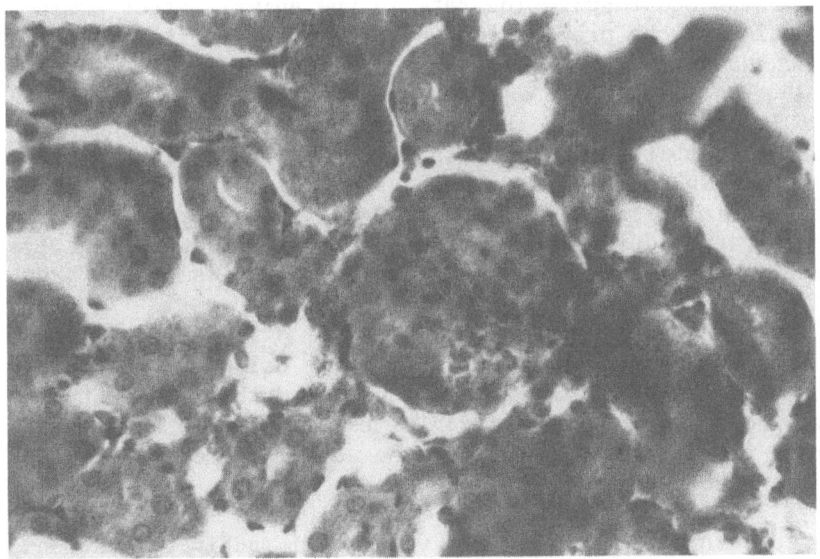

Fig. 11.9. Kidney of a rat killed after administration of
            urokinin was terminated.   Hematoxylin-eosin stain.
            Magnification × 200.

some animals. This points to activation of the kallikrein—
kinin system during the administration of fibrinolytic preparations. We failed to develop the LD_{50} because the dose of
the preparation used, exceeding therapeutic dose by 4000
times, did not cause the death of the experimental animals.

Neither chronic toxicity nor pyrogenicity of the preparation were found. Upon repeated administration of the preparation, no morphologic changes in the formed elements of the
blood were observed. Histomorphologic study of hepatic (Figure 11.8), renal, splenic, pulmonary, and adrenal tissues and
cardiac muscle revealed neither inflammatory infiltrates nor
destructive changes, and no local irritation from urokinin
was observed at the sites of administration.

In studying the renal preparations, differences between
the cortical and medullar parts of the kidney were found.
Glomeruli of the uroexcretive tubules showed no peculiarities.
Filling of the vessels in the interstitial tissue between
straight tubules was observed (Figure 11.9).

Pulse and respiration rate in the animals after administration of the preparation was within the limits of physiologic changes.

In order to study the antigenic properties of the preparation, rabbits were immunized according to a certain scheme.
The urokinin solution was taken as antigen, and blood serum
from the immunized animals served as a source of antibodies.
Precipitation during the entire research period was negative,
pointing to the absence of antigenic properties of the preparation.

Urokonin Thromboplastic Activity. Using the generally
accepted time-of-plasma-recalcification method, this study
revealed a marked prolongation of thromboplastic activity
versus the control, indicating the active formation of plasmin and the decay of fibrinogen.

Urokinin Proteolytic Activity. This activity has not
been demonstrated as evident by the lack of uropepsin in the
preparation.

Lysis of Experimentally Induced Thrombi in Animals

The success of thrombolytic therapy is known to depend
primarily on the amount of plasminogen adsorbed on the clot
and the amount of activator capable of converting this proenzyme into active plasmin. Plasmin circulating in plasma
participates to a lesser extent in thrombolysis insofar as
it is usually bound to antiplasmin. However, the administration of excessive doses of activator is likely to result in
a depletion of antiplasmin in the plasma, thus inducing hemorrhage. Accordingly, we maintained the same dosage of the
preparation and followed the same method of administration.
Before and after administration of the preparation, a strict
control for coagulation was also performed, with special regard to the concentration of fibrinogen.

Experimental thrombi were induced in an isolated segment
of the dog femoral vein by the administration of 1 ml of
thrombin suspension (100 units). Therapy was begun 1, 3, and
24 h after the thrombus had been formed. Urokinin was injected intravenously in a dose of 1000 IU/kg of animal body
weight. Urokinin was shown to have a pronounced thrombolytic

effect. Patency of the vessel was marked in almost all animals throughout the observations. Activation of fibrinolysis and a decrease in the concentration of fibrinogen were revealed.

In addition, purified urokinase, oligomeric forms synthesized on its basis, and soluble conjugates with dextran [49] which had 30-50% of the specific activity of nonmodified urokinase and a molecular weight of 60,000-200,000 daltons, were used for treating experimentally induced hyphemias in rabbits. Hyphemias were obtained by administering 0.3 ml of autologous blood from the auricular vein of the animal into the anterior chamber of the eye. The preparations were administered subconjunctively with a novocaine solution at dosages that were equivalent in specific activity. Injections of urokinase and its polymeric forms caused no local irritation on the tissues.

All of the preparations investigated demonstrated a high therapeutic efficacy in that the time of hyphemia resorption was sharply decreased as compared to the control. In addition, the advantages of immobilized forms of urokinase were revealed, consisting of a decrease in the number of injections by three to six times versus the native enzyme in order to obtain the same therapeutic effect. This result is indicative of the prolonged action of the high-molecular-weight forms of urokinase.

Despite a marked thrombolytic effect, urokinase still fails to induce the desired hypocoagulation in the organism. This requires the administration of heparin at the very onset of thrombolytic therapy (up to 20,000 units within 24 h). Heparin, a unique direct-acting anticoagulant affecting the blood coagulation factors, is also an inhibitor of thrombin, the pathogenetic factor in thrombogenesis.

Considering the marked capability of urokinase and heparin in complex, experiments were performed to produce a complex preparation with both fibrinolytic and anticoagulant action. The molecular weight of the complex was 48,000 ± 2000 daltons. In data from coagulation recorders, the maximum amplitude index with administration of the complex decreased more than fourfold while the reaction time increased threefold as compared to the control.

Fibrinolytic activity of the complex proved to be four times higher than that of urokinin. Therapeutic tests on ex-

perimentally induced thrombosis in the femoral vein of dogs
in which the complex was administered intravenously for 24 h
after the onset of thrombogenesis (1000-2000 IU/kg of body
weight) showed a considerable decrease in the activity of prothrombin complex factors and in fibrinogen concentration,
fibrinolysis activation, and prolongation of thrombin time.
These data clearly demonstrate that the production of combined preparations with both a fibrinolytic and an anticoagulant effect is a valid endeavor and deserves attention.

Clinical Application of
Urokinase Preparations

A large number of publications have recently appeared
on the use of urokinase preparations in myocardial infarctions, thrombosis of the retinal vessels, thrombophlebitis,
and thromboembolic states, and in the prophylaxis and treatment of shunt thromboses in dialysis, disseminated intravascular coagulation, and other pathologic states. It should
be noted that the data from some clinics are based on the
followup of a small number of patients and, thus, lead to
some contradictions. These discrepancies also result from
the use of different therapeutic schemes and recommended dosages, as well as from variances in the indications for use
of the preparations. Thus, for instance, one group has suggested that doses which increase the fibrinolytic activity
of a patient's plasma 300-fold within 8 h should be used.
Conservative therapy is applied more widely and results in
the decrease of fibrinogen up to 1 g/liter and the prolongation of thrombin time two- to fourfold compared to the norm.

Most authors are also inclined to believe that urokinase
preparations are only capable of lysing newly formed thrombi.
However, in the treatment with urokinase of 250 patients with
iliac and femoral vein thrombosis as old as 3-4 weeks, quite
satisfactory results were obtained [10, 50, 51]. One must
therefore assume that some contradictions might be accounted
for by the age of the thrombus as well as by the size and
location of the vessel in which it was formed.

The degree of purification of the preparation has also
been shown to be important. Urokinase with a higher molecular weight had the best therapeutic effect because of its
complexing with tissue thromboplastin, kallikrein, and other
impurities, the nature of which is presently under study [52,
53].

Best results have been achieved with urokinase combined with heparin and arvin [16, 51]. In these cases, administration of the anticoagulant preceded urokinase. Such an approach permits fibrinogen to be decreased to 35% of the initial level, and plasminogen and α_2-antiplasmin by 60%.

In most clinics the dose of urokinase used varies depending on the area of injury. Thus, in myocardial infarction, the total dose amounts to 260,000 IU, in cerebral thrombosis to 130,000 IU, and in thrombosis of the retinal vessels to 80,000 IU.

Side effects such as sickness, chill, urticaria, headaches, and a decrease in arterial pressure are seldom observed and are accounted for by insufficient purity of the urokinase preparations during their production. In some cases, hemorrhages resulted from overdoses of the preparation, lack of regard for (or underestimation of) individual sensitivity, and the amount of plasminogen in these patients. In such cases, an overdosage of heparin during fibrinolytic therapy is not excluded and, sometimes, the dose amounted to 120,000 units per 24 h.

A urokinin preparation has been used in patients with acute venous thromboses [54]. Six of the sixteen patients who were followed up had acute thrombophlebitis of the superficial veins of the lower extremities, five patients had acute thrombophlebitis of the deep veins, four had thrombosis of the axillary and subclavian veins, and one patient had postoperative thrombosis of the arterial shunt.

The lyophilized urokinin preparation containing 20,000-50,000 IU per ampul was used to treat acute thromboses. Before use, the content of the ampul was dissolved in 100-150 ml of NaCl physiological solution. Heparin (2500 units per 500 ml of the physiological solution) was added to the system. The solution was injected intravenously by drops at a rate of 15-25 drops/min over 1-2 h. The physiological solution with heparin was then administered over 3-4 h. During the following 5-6 days, heparin (2500-5000 units) was administered intramuscularly every 4-6 h. The rate of administration of the solution varied so that the coagulation time would exceed the initial level by two- or threefold and would not be longer than 30 min.

The thrombolytic effect due to the administration of urokinin was evaluated by coagulative and fibrinolytic indices and by roentgenologic examination. Urokinin caused a considerable prolongation in coagulation time, plasma recalcification, and thrombin time, and a decrease in the Thrombotest and in the activity of certain coagulation factors. Fibrinogen and the activity of factors V, VII, and XII decreased most significantly. All coagulation factors retained their activated state during the entire period of administration and returned to the initial level within 24 h. Fibrinolytic activity in the urine and plasma increased considerably (the time of lysis of the euglobulin fraction decreased to 79 min), especially within the first 1.5 h (by a factor of 28) from the beginning of administration. The formation of fibrin degradation products was detected by immunoprecipitation. A decrease in antiplasmin was observed, and there were no considerable changes in the content of antithrombin III. The administration of urokinin did not lead to the formation of antibodies in the patients.

With regard to side effects, 10 patients had an increase in temperature by 1-2 degrees 30-40 min after injection, but their temperature returned to normal within 4 h. It should be noted that the preliminary use of desensitizing preparations (such as urbason, dimedrol, pipolphen) removed any side reactions.*

No edemas were observed in any patients on the second to fourth days after the administration of urokinin, and indurations along the thrombosed veins disappeared by day 7. Pain during palpation was not experienced within the first 2 days. Three to four hours after administration, the patients felt considerable relief: the strain on tissue vanished, skin color changed, and pain abated. Good and satisfactory results were obtained in all patients.

The best result was achieved in patient K., aged 24 years, who had thrombosis of the right axillary and subclavian veins (Paget—Schroetter syndrome). He was hospitalized on the third day after thrombosis with the following symptoms: edema pain, heaviness, and feeling as if something

---

*Urbason is an analog of prednisone produced in Yugoslavia. Dimedrol is an antihistamine preparation produced in the USSR. Pipolphen is a derivative of phenothiazine produced in Hungary.

were about to burst in his right upper extremity. He associated the onset of his disease with heaving. We observed cyanosis of the right upper extremity and marked strain and swelling of the veins spreading over the breast. The difference
in circumference between the left and right arms was 5 cm and
that of the forearm 3 cm. Therapy was performed according to
the scheme described above, with a subsequent shift to the
use of indirect-acting anticoagulants. On the following day
the edema abated and neither pain nor strain were felt in
either the tissues or veins. Within 11 days, the control
roentgenogram showed complete recovery of the patency of the
brachial and subclavian veins, preserving the valvular apparatus. Rheovasographic study demonstrated that venous outflow
in the affected extremity was normal after the treatment with
urokinin and direct- and indirect-acting anticoagulants.

Of interest are the studies [52] that reveal the clinical efficacy of urokinin and heparin in combination to experimentally treat the Sanarelli—Schwartzman phenomenon. One to two
hours after beginning administration of the preparation,
hemorrhages stopped in all animals, activity appeared, and the
survival rate was 100%. These experimental findings have
served as the basis for using these preparations in children
with hemolytic-uremic syndrome. By the end of the first day,
the state of the children improved significantly, urination
increased, nasal bleeding stopped, and appetite appeared.
Active treatment was carried on for 3-4 days, permitting the
physician to cope with acute manifestations of the disease
and to achieve good clinical results. Further followup showed
no recurrences in the children.

The above data indicate that urokinin, an active physiologic activator of fibrinolysis, has a pronounced specific
property — it increases fibrinolytic activity in the blood,
causes a decrease in the activity of blood coagulation factors, and provides for fast lysis of newly formed thrombi. Further observations have shown that urokinin has an expressed
clinical effect in the treatment of patients with thrombosis
of up to 3 weeks' duration.

Efficacious treatment with urokinin and determination
of the dosage is achieved in the laboratory by checking the
antifibrinogenic sera, which enables one to establish the degree of fibrinogen and fibrin degradation and the quantitation of their degradation fragments (FDF and fragment D-D).

Urokinase preparations are thus being increasingly used in clinical practice. The success of therapy depends on the initial material used in making the preparation (human urine, renal tissue culture), the degree of purification, and the time elapsed from the onset of thrombus formation. Studies directed at the use of complex preparations of urokinase—heparin—arvin, together with specific drugs, should be considered most promising. Studies by Japanese scientists [20], who have proposed combined therapy with antitumor preparations (mitomycin with urokinase, in particular) (2000 units), are worth noting. Tumor regression in their experiment was 52.5% compared to 27.3% in the control, thus permitting the clinical effect to be highly appreciated.

The urokinase preparation is most promising if we consider its biological basis and affinity for the human organism, the application of moderate doses to obtain a clinical effect, and the possibility of reuse in weeks or months should the thrombosis relapse [54-56]. The high ability of urokinase to complex with many proteins and drugs is also noteworthy. This ability considerably extends the possibilities for using these preparations to delay and prevent thrombogenesis. The production of urokinase preparations, if properly organized, is considerably more economical and effective than the production of streptokinase or other fibrinolytic preparations which involve either special cultures or expensive initial material (human plasma, trypsin, etc.).

REFERENCES

1. G. V. Andreenko. Fibrinolysis, Moscow (1979).
2. B. Wiman and P. Wallen, "Activation of human plasminogen by an insoluble derivative of urokinase. Structural changes of plasminogen in the course of activation to plasmin and demonstration of a possible intermediate compound," Eur. J. Biochem., 36, 25-29 (1973).
3. P. Fanti, "Plasminogen activity of plasma and serum," Science, 135, 767-791 (1962).
4. A. P. Kaplan, F. J. Castellino, D. Collen, B. Wiman, and F. B. Taylor, "Molecular mechanisms of fibrinolysis in man," Thromb. Haemostas., 39, 263-279 (1978).
5. H. Saito, O. D. Ratnoff, R. Waldmann, and T. P. Abraham, "Fitzgerald trait. Deficiency of a hitherto unrecognized agent, Fitzgerald factor, participating in surface-mediated reactions of clotting, fibrinolysis, generation

of kinins and the property of diluted plasma enhancing vascular permeability," J. Clin. Invest., 55, 1082-1090 (1974).

6. A. P. Fletcher, "Fibrinogenolytic dysfibrinogenemias," Fed. Proc., 24, 822-826 (1965).

7. N. Alkjaerrig, A. P. Fletcher, and S. Sherry, "The mechanism of clot dissolution by plasmin," J. Clin. Invest., 38. 1086-1090 (1959).

8. G. Murano and D. L. Aronson, "High- and low-molecular-weight urokinase," Thromb. Haemostas., 42, 1066-1068 (1979).

9. M. Samama, B. Cazenave, and A. M. Otero, "Urokinase I and II activity," Thromb. Haemostas., 40, 578-580 (1978).

10. O. Matsuo, T. Kosugi, and H. Mihara, "Is the hypotensive action of urokinase beneficial for thrombolytic therapy?" Thromb. Haemostas., 39, 526-527 (1978).

11. B. Wiman, "Primary structure of peptides released during activation of human plasminogen by urokinase," Eur. J. Biochem., 39, 1-4 (1973).

12. P. J. Walthar, H. N. Steinman, R. L. Hill, et al., "Activation of human plasminogen by urokinase. Partial characterization of a pre-activation peptide," J. Biol. Chem., 249, 1173-1181 (1974).

13. P. Vanhav, M. V. Donati, H. Clalys, R. Verhaegh, and G. Vermylen, "Action of brinase on human fibrinogen and plasminogen," Thromb. Haemostas., 42, 571-581 (1979).

14. B. A. Janik and S. E. Papaiannocc, "A comparative ex vivo study of plasminogen activators and proteases for fibrinolysis activity and side effect in rabbits," Thromb. Haemostas., 37, 154-160 (1977).

15. A. Ya. Ivleva and S. N. Zolotukhin, Farmakol. Toksikol., 6, 749-752 (1971).

16. E. G. Vairel, G. Gorelay, and G. Chroay, "Urokinase thrombolytic heparinate," US Patent No. 2,246,969, April (1973).

17. D. M. Zubairov and M. G. Assadulin, "Isolation and some properties of human urokinase," Biochemistry, 39, 378-387 (1974).

18. G. W. Sobel, S. R. Mohler, N. W. Jones, A. B. Dowdy, and M. U. Anesi. "Urokinase, an activator of plasma profibrinolisin extracted from urine," Am. J. Physiol., 171, 768-772 (1952).

19. D. R. Celander and M. M. Guest, "Urokinase. Biochemistry and physiology," Arch. Biochem. Biophys., 55, 286-290 (1955).

20. T. J. Lasshi, "Isolation of urokinase from human urine by affinity chromatography," Japanese Patent, 20(3/4), 280 (1973).

21. E. I. Chazov and K. M. Lakin, Anticoagulants and Fibrin-
    olytic Preparations [in Russian], Meditsina, Moscow
    (1977).
22. S. Niewiarowski, A. Z. Budsynski, and B. Lipinski, "Sig-
    nificance of the intact polypeptide chains of human fib-
    rinogen in ADP-induced platelet aggregation," Blood, 49,
    635-644 (1977).
23. A. Larcan, H. Lambert, and P. Alexandro, "Les coagulo-
    pathies aigues de consommation 120 observation," La
    Nuov. Presse Med., 41, 2771-2778 (1976).
24. Z. D. Fedorova, A. A. Klement, I. K. Slobozhankina, and
    S. N. Osipov, "The use of urokinase for treatment of
    acute venous thromboses," Vestn. Khir., No. 6, 64-67
    (1973).
25. S. N. Martinov and N. S. Stasiuk, The Blood Coagulation
    System and Fibinolysis [in Russian], Saratov (1975).
26. N. O. Kjeldgaard and J. Ploug, "Recovery of urokinase
    from urine," British Patent, 802, 326-338 (1958).
27. N. Ogawa, H. Yamamoto, T. Katamine, and H. Tajima, "Pur-
    ification and some properties of urokinase," Thromb.
    Diath. Haemorrh., 34, 194-209 (1975).
28. Pharmexbio. — Netherland's Application, Preparation of
    Urokinase, Patent 6, 350-354 (1966).
29. G. V. Samsonov, "Sorption and protein chromatography on
    ion exchangers and gel-filtration technique," in:
    Physicochemical Methods of Study, Analysis, and Frac-
    tionation of Biopolymers, Nauka, Moscow—Leningrad (1966),
    pp. 187-205.
30. L. K. Shartaeva, N. N. Kuznetsova, and G. E. El'kin,
    Carboxyl Cationites in Biology, Nauka, Leningrad (1979).
31. G. V. Samsonov, E. B. Trostyanskaya, and G. E. El'kin,
    Ion Exchange. Sorption of Organic Substances, Nauka,
    Leningrad (1969).
32. L. K. Shataeva and G. V. Samsonov, "The use of porous
    carboxyl cationites for protein isolation," Khim. Farm.
    Zh., No. 4, 78-90 (1977).
33. G. V. Samsonov, S. V. Kol'tsova, L. K. Shataeva, N. N.
    Kuznetsova, K. M. Rozhetskaya, I. K. Paradeeva, and
    Z. D. L. Fedorova, "Method for isolation of thrombolytic
    enzyme," Author's Certificate 584034, No. 46 (1977).
34. S. V. Kol'tosova, L. K. Shataeva, T. F. Sukhareva, N. A.
    Bynyaeva, Z. D. Fedorova, and G. V. Samsonov, "Isolation
    and purification of urokinase," Vopr. Med. Khim., No. 5,
    623-626 (1981).
35. O. H. Lowry, N. J. Rosenbroudh, A. L. Farr, and R. J.
    Randall, "Protein measurement with the folin phenol re-
    agent," J. Biol. Chem., 193, 265-275 (1951).

36. T. Astrup and S. Mullertz, "The fibrin plate method for estimating fibrinolytic activity," Arch. Biochem. Biophys., 40, 346-350 (1952).

37. W. F. White, G. H. Barlow, and M. M. Mozen, "The isolation and characterization of plasminogen activators (urokinase) from human urine," Biochemistry, 5, 2160-2169 (1966).

38. M. Nobuhara, M. Sakamaki, H. Ohnihi, and V. Suzuki, "A comparative study of high-molecular-weight urokinase and low-molecular weight urokinase," J. Biochem. (Tokyo), 90, 225-232 (1981).

39. K. Weber and M. Osborn, "The reliability of molecular weight determination by dodecyl sulfate–polyacrylamide gel electrophoresis," J. Biol. Chem., 244, 4406-4412 (1969).

40. M. E. Soberano, E. G. Ong, A. J. Johnson, M. Levy, and C. Schoellmann, "Purification and characterization of two forms of urokinase," Biochim. Biophys. Acta, 445, 763-773 (1976).

41. L. Holmberg, B. Bladh, and B. Astedt, "Purification of urokinase by affinity chromatography," Biochim. Biophys. Acta, 445, 215-222 (1976).

42. B. G. Belen'kii, E. S. Gankina, M. A. Aniskova, T. B. Anikina, and A. N. Krasovskii, "Ultramicro method for determination of molecular weights of polypeptide chains and protein subunits based on thin-layer gel chromatography," Bioorg. Khim., 1, 396-401 (1975).

43. O. Vesterberg and H. Svensson, "Isoelectric fractionation, analysis, and characterization of ampholytes in natural pH gradients," Acta Chem. Scand., 20, 820-834 (1966).

44. L. Summaria, L. Arzadon, P. Bernabe, and K. C. Robbins, "Studies on the isolation of the multiple molecular forms of human plasminogen and plasmin by isoelectric focusing methods," J. Biol. Chem., 247, 4691-4702 (1972).

45. I. A. Bolotina, V. O. Chekhov, V. Yu. Lugauskas, A. V. Finkel'shtein, and O. B. Ptitsyn, "Determination of the secondary structure of proteins from the circular dichroism spectra. I. Protein reference spectra for α- and β-, and irregular structures," Mol. Biol. (Mosk.), 14, 891-901 (1980).

46. I. A. Bolotina, V. O. Chekhov, V. Yu. Lugauskas, and O. B. Ptitsyn, "Determination of the secondary structure of proteins from the circular dichroism spectra. II. Consideration of the contribution of β bends," Mol. Biol. (Mosk.), 14, 902-909 (1980).

47. I. A. Bolotina, V. O. Chekhov, V. Yu. Lugauskas, and
    O. B. Ptitsyn, "Determination of the secondary struc-
    ture of proteins from circular dichroism spectra. III.
    Protein-derived reference spectra for antiparallel and
    parallel β-structures," Mol. Biol. (Mosk.), 15, 167-175
    (1981).
48. J. T. Yang, "Optical rotatory dispersion of polypeptides
    and proteins," Tetrahedron, 13, 143-165 (1961).
49. G. V. Samsonov, S. V. Kol'tsova, G. I. Shelykh, and V.F.
    Danilichev, "Preparation and biological study of urokinase
    and its immobilized form," in: Thrombolysis and Strepto-
    kinase Enzymology, Minsk (1982), pp. 107-112.
50. R. K. Zimmerman and T. Lubins, "Die ensymatische Kopp-
    lung von fibrinogenese und fibrinolyse," Klin. Wochenschr.,
    56, 781-788 (1978).
51. A. D. Schreiber and K. F. Austen, "Interrelationships
    of the fibrinolytic, coagulation, kinin generation, and
    complement systems," Ser. Haematol., 6, 593-600 (1973).
52. A. D. Schreiber, A. P. Kaplan, and K. F. Austen, "In-
    hibition of ClINH of Hageman factor fragment activation
    of coagulation, fibrinolysis, and kinin generation,"
    Clin. Invest., 521, 1402-1409 (1973).
53. W. T. Woodard, Jr., E. D. Day, and S. Silver, "The fate
    of infused urokinase," Surgery, 68, 692-694 (1970).
54. Z. D. Fedorova, I. K. Slobozhankina, A. A. Klement, and
    S. N. Osipov, "Treatment of acute venous thromboses with
    preparation of fibrinolytic action, urokinin," in: Sec-
    ond Meeting of the European and African Division of the
    International Society of Haematology. Abstracts, Vol.
    2, Prague, August 27-29, 1973, p. 203.
55. Z. D. Fedorova, A. V. Papayan, L. P. Papayan, T. A.
    Odesskaya, E. M. Petrova, E. J. Valkovitch, and I. K.
    Slobozhankina, "Treatment of generalized intravascular
    coagulation with heparin and fibrinolysis activator uro-
    kinin," in: Second Meeting of the European and African
    Division of the International Society of Haematology.
    Abstracts, Vol. 2, Prague, August 27-29, 1973, p. 504.
56. L. S. Latallo, "Perspectives of thrombolytic therapy,"
    in: Second Meeting of the European and African Division
    of the International Society of Haematology. Abstracts,
    Vol. 2, Prague, August 27-29, 1973, p. 502.

PLASMAKINASE: A NEW THROMBOLYTIC PREPARATION FROM POSTMORTEM PLASMA

V. B. Khvatov

N. V. Sklifosovsky Scientific-Research Institute of Emergency Aid, Moscow

ABSTRACT

The blood of sudden death victims is a specific material for obtaining fibrinolytic drugs, and an additional source of albumin and gamma globulin. A technique for producing a thrombolytic agent called plasmakinase has been developed under serial production conditions. The technique is based on the fractionation of postmortem plasma by ethanol at a given temperature and pH. Plasmakinase is an enriched (five- to ninefold) β-globulin fraction of cadaver plasma which contains fibrinolytic enzymes. Plasminogen activator accounts for 90-95% of plasmakinase activity, and plasmin for 5-10%. The specific antagonists of the preparation are ε-aminocaproic acid, Contrykal,* and human α_2-macroglobulin. Administration of plasmakinase to animals leads to an increase in fibrinolytic activity and to a decrease in coagulation and antiproteinase potential of the blood. The preparation has a prominent thrombolytic effect when infused into animals with venous and arterial thromboses. Intracoronary treatment of dogs with acute coronary artery thrombosis yields the same results, corroborated by visual, angiographic, biochemical, and morphologic examinations. The lytic effect of plasmakinase is due both to the activator and to direct proteolytic action on the thrombus — which constitutes its advantage over other fibrinolytic drugs.

---

*An antiproteinase drug produced in East Germany.

INTRODUCTION

Attempts to improve the effectiveness of thrombolytic therapy in modern medicine focus on three areas: (a) combined application of existing anticoagulating, fibrinolytic, and antiaggregating agents; (b) development of endovascular techniques for administering thrombolytic drugs; and (c) elaboration and adoption of new, specific antithrombotic drugs. In the last respect, the blood of sudden death victims appears to be a promising source of natural fibrinolytic enzymes.

Fibrinolytically active plasma (FAP) of cadaver blood is used as a highly efficacious thrombolytic drug for treating acute myocardial infarction [1-4], acute venous thrombosis [5-7], burns [8], and other cardiovascular [9-11] and children's diseases [12]. However, a number of factors limit wide clinical application of FAP: the relative deficiency of plasma with very high fibrinolytic activity, the necessity of storing and transporting it in a frozen state, and the possibility of a substantial protein load conditioned by the administration of a large volume of plasma to the patient [6].

Isolation of the components responsible for the fibrinolytic properties of FAP, therefore, seems to be a reasonable solution. Several "fibrinolytically active substances" have already been obtained from the plasma of sudden death victims: "fibrinogenase" [13-15], "cryoprecipitate of fibrinolytically active plasma" [16], "angiokinase" [171], "fibrinolysin" [10], and "cadaver factor" [18, 19]. The high efficacy of these agents has been shown experimentally, but the real outcome of these studies is more theoretical than practical. The physicochemical characteristics of the substances are not yet known, and production techniques are far from economical; consequently, these agents cannot be recommended for serial production.

In the laboratory of tissue conservation and transfusion (N. V. Sklifosovsky Scientific Research Institute of Emergency Aid), the characteristics of cadaver plasma proteins have been studied for many years. Criteria have been developed for selecting samples for fractionation under industrial conditions in order to obtain preparations with directed action [20-27]. This chapter characterizes the plasma of sudden death victims and the thrombolytic drug obtained from it.

CADAVER PLASMA CHARACTERISTICS

Blood was taken from men within 6 h after sudden death (from acute cardiac insufficiency, mechanical asphyxia, and alcohol poisoning [39]). Whole blood was immediately drained from cadaver vessels into 250-ml sterile flasks with 50 ml of 7[b] conservant (developed at the Central Institute of Hematology and Blood Transfusion, Moscow). Washed blood was obtained by infusing a special solution containing sodium citrate into the cadaver arterial bed.

After centrifugation (at 1500 rpm, 4-6°C), we obtained whole and washed plasma of cadaver blood. Sterile samples, which passed serological and bacteriological control, were used for clinical application. We rejected plasma that had very low fibrinolytic activity (less than 0.5 FU/ml) or a high fibrinogen content (more than 2 g/liter), as well as that which was hemolyzed (more than 31 moles/liter of free hemoglobin) or had a high concentration of total lipids (more than 1 g/liter) since both of these spoil the quality of preparations obtained from plasma. We averaged two to four flasks (400-800 ml) of whole plasma and three to five flasks (600-1000 ml) of washed plasma from one cadaver.

The average fibrinolytic potential of cadaver plasma, studied by various methods and expressed in different units, exceeded that of donor plasma by 173 to 187 times (Table 12.1). It should be noted that fibrinolytic activity of postmortem blood exceeds that of donor blood by two to eight times after the infusion of 2.5-3.0 million IU (international units) of streptase, pointing to the evident activation of the fibrinolytic system after sudden death.

Caseinolytic and BAPNA*-esterase activity of cadaver plasma exceeds that of donor plasma by 9.9 and 3.8 times, respectively. As shown in the table, the increase in proteolytic activity is significantly lower than that in fibrinolytic activity, suggesting that the activator mechanism plays a role in the fibrinolytic action of postmortem blood. On the other hand, there is a marked hypocoagulation of cadaver plasma conditioned by the decrease and deficiency of fibrinogen, and the factors responsible for the formation of thromboplastin and thrombin.

---

*BAPNA = N-benzoyl-D,L-arginine-p-nitroanilide.

Table 12.1. Proteolytic and Antiproteinase Potentials of Whole Postmortem Plasma

Type of activity	Units*	Donor plasma (n = 40)	Whole postmortem plasma			
			n	M ± m	difference	p
Fibrinolytic	FU/ml	0.015 ± 0.003	1372	2.80 ± 0.04	+187	0.01
	Ploug/ml	0.08 ± 0.02	100	14.5 ± 1.3	+181	0.01
	IU/ml	8.1 ± 2.3	100	1400 ± 95	+173	0.01
Caseinolytic	mCU/ml	37 ± 2	240	365 ± 10	+9.9	0.01
BAPNA-esterase	mIU/ml	2.2 ± 0.3	267	8.39 ± 0.17	+3.8	0.01
Total antiproteinase	IU/ml	87 ± 2	200	38 ± 1	-2.3	0.01
Fibrinolytic System Components						
Plasminogen activator	AU/ml**	0.60 ± 0.10	240	51.7 ± 0.9	+86	0.01
Plasmin	PU/ml	0.10 ± 0.02	240	2.62 ± 0.06	+26	0.01
Plasminogen	mCU/ml	4000 ± 300	240	1050 ± 20	-3.8	0.01
α_1-Antitrypsin	mg%	210 ± 9	220	137 ± 2	-35	0.01
α_2-Macroglobulin	mIU/ml	80 ± 4	158	64 ± 1	-20	0.01

*FU = fibrinolytic units; IU = international units; CU = caseinolytic units; AU = activator units; PU = plasmin units.
**In this experiment, 1 activator unit is equivalent to 26-29 IU of streptase or 0.26-0.32 Ploug units of urokinase.

Analysis of the components of the fibrinolytic system showed that the activity of plasma and plasminogen activator is sharply increased in the cadaver whole plasma, while the level of fibrinolysis inhibitors and plasminogen activity are decreased (Table 12.1). It was found that the main component in determining enhanced fibrinolytic activity in cadaver plasma is plasminogen activator, which is responsible for more than 90% of the lytic potential.

Electrophoretic and immunochemical characteristics of the protein in whole cadaver plasma are given in Table 12.2. The albumin and γ-globulin content in whole cadaver plasma is identical with that in donor plasma, while the concentration of β-globulins is increased by 37.1% and the amount of α_1- and α_2-globulins is decreased by 31.9 and 26.4%, respectively. Identification of fibrinolytic activity in electrophoretic fractions showed that it is mainly linked to the β-globulins. One can assume that components of the activated fibrinolytic system of cadaver blood, as well as those of donor blood, are of a globulin nature. This assumption determines the methodological approaches to the production of their "concentrate."

The concentrations of α_1-antitrypsin and antithrombin III in cadaver plasma are decreased by 30 and 23.8%, respectively, which correlates with the reduction in antiproteinase potential. The content of fibrinogen degradation products is considerably increased: the D-fragment to 2.5 g/liter and the E-fragment to 1.0 g/liter, on the average. The concentration of plasminogen activator averaged 2.0 g/liter. Studies of proteins in cadaver plasma by gel chromatography and immunophoresis showed that they are eluted in the native form according to their molecular weight, and that the activator of plasminogen and plasmin is detected in two forms: "free" and in combination with macroglobulins (α_2-macroglobulin, immunoglobulins G and M, and fibrinogen degradation products). Studies of washed cadaver plasma yield the same results.

These data indicate that the plasma of sudden death victims is a unique, specific material for obtaining fibrinolytic drugs and fibrinogen degradation products, as well as another source (in addition to donor plasma) of albumin and gamma globulin. More than 90% of the samples of cadaver plasma, which passed an appropriate control, can be used for making preparations that have directed action.

Table 12.2. Comparative Electrophoretic and Immunochemical Characteristics of Donor
and Whole Postmortem Plasma

Proteins analyzed	Units of activity	Donor plasma (n = 25) M ± m	Whole postmortem plasma M ± m	%	p
Total protein	g/liter	69.4 ± 0.8	67.0 ± 1.2	−3.4	0.1
Electrophoretical Analysis					
Albumins	%	57.9 ± 0.6	58.2 ± 0.8	0.5	0.1
α_1-Globulins	%	4.7 ± 0.1	3.2 ± 0.1	−31.9	0.01
α_2-Globulins	%	8.7 ± 0.3	6.4 ± 0.2	−26.4	0.01
β-Globulins	%	10.5 ± 0.2	14.4 ± 0.4	37.1	0.01
γ-Globulins	%	18.2 ± 1.0	17.8 ± 0.9	−2.2	0.1
Immunochemical Analysis					
α_1-Acidic glycoprotein	g/liter	0.98 ± 0.10	0.90 ± 0.15	−8.2	0.1
α_1-Antitrypsin	g/liter	2.50 ± 0.20	1.75 ± 0.23	−30.0	0.02
Antithrombin III	g/liter	0.17 ± 0.02	0.11 ± 0.02	−23.8	0.05

Ceruloplasmin	g/liter	0.40 ± 0.03	0.41 ± 0.04	2.5	0.1
Gc-globulin	g/liter	0.35 ± 0.04	0.32 ± 0.04	-8.6	0.1
α_2-HS-globulin	g/liter	0.61 ± 0.07	0.63 ± 0.05	3.2	0.1
α_2-Macroglobulin	g/liter	3.70 ± 0.26	3.26 ± 0.18	-12.0	0.1
Plasminogen + plasmin antigen	g/liter	0.20 ± 0.03	0.11 ± 0.01	-45.0	0.01
Transferrin	g/liter	3.10 ± 0.25	3.37 ± 0.27	8.7	0.1
Immunoglobulin A	g/liter	1.90 ± 0.15	1.93 ± 0.17	1.6	0.1
Immunoglobulin G	g/liter	11.0 ± 1.0	9.9 ± 1.1	-10.0	0.1
Immunoglobulin M	g/liter	0.80 ± 0.20	0.84 ± 0.19	-5.0	0.1
D-fragment of fibrinogen*	g/liter	0	2.51 ± 0.12		0.01
E-fragment of fibrinogen*	g/liter	0	1.10 ± 0.10		0.01
Plasminogen activator**	g/liter	traces	2.00 ± 0.13		0.01

*Donor blood serum and defibrinogen cadaver plasma were analyzed.
**Analysis was carried out using a monospecific antiserum to plasmic activator of human plasminogen.

Table 12.3. Comparative Electrophoretic and Immunochemical Characteristics of Whole Postmortem Plasma and Plasmakinase Obtained From It

Proteins analyzed	Units of activity	Postmortem plasma (n = 30) M ± m	Plasmakinase (n = 25) M ± m	%	p
Total protein	g/liter	67.0 ± 1.2	25.0 ± 1.0		
Electrophoretic Analysis					
Albumins	%	58.2 ± 0.8	2.8 ± 0.2	−95.2	0.01
α_1-Globulins	%	3.2 ± 0.1	0.2 ± 0.02	−93.7	0.01
α_2-Globulins	%	6.4 ± 0.2	1.0 ± 0.05	−84.4	0.01
β-Globulins	%	14.4 ± 0.4	86.0 ± 0.8	+497	0.01
γ-Globulins	%	17.8 ± 0.9	10.0 ± 0.4	−43.8	0.01

		Immunochemical Analysis			
α_1-Acidic glycoprotein	%	1.3 ± 0.2	0	−100	0.01
α_1-Antitrypsin	%	2.5 ± 0.4	0	−100	0.01
Antithrombin III	%	0.3 ± 0.1	traces	−99.9	0.01
Ceruloplasmin	%	0.6 ± 0.1	0	−100	0.01
α_2-Macroglobulin	%	4.8 ± 0.3	traces	−99.9	0.01
α_2-HS-glycoprotein	%	0.9 ± 0.2		−99.9	0.01
Plasminogen + plasmin antigen	%	0.2 ± 0.03	2.5 ± 0.5	+1150	0.01
Transferrin	%	5.0 ± 0.4	0	−100	0.01
Immunoglobulin A	%	2.9 ± 0.3	0.3 ± 0.1	−89.6	0.01
Immunoglobulin M	%	1.2 ± 0.2	0.7 ± 0.2	−41.7	0.05
Immunoglobulin G	%	14.8 ± 1.0	9.9 ± 0.9	−33.1	0.01
D-fragment of fibrinogen	%	3.8 ± 0.6	8.8 ± 1.0	+132	0.01
E-fragment of fibrinogen	%	1.6 ± 0.4	2.7 ± 0.6	+67.0	0.1
Plasminogen activator	%	59.3 ± 6.7	+1877	0.01	0.01

Additional studies were aimed at isolating, purifying, and describing the physicochemical characteristics of plasminogen activator [23, 28], and at obtaining various fibrinolytically active fractions which were then evaluated for their biological effects [21, 29]. Based on this investigation, we elaborated a technique for producing a "concentrate" of fibrinolytic enzymes; this technique is based on the differentiated fractionation of whole and washed cadaver plasma at a given temperature and pH [24, 30]. The technique has been approved under the conditions of conventional production. We call the new drug plasmakinase. From this biological material we also obtained albumin and gamma globulin whose physicochemical and biological characteristics correspond to the preparations isolated from donor plasma. Because of the factors mentioned above, the production of plasmakinase involves an integrated technological cycle, and so the drug is much cheaper to produce than either streptase or urokinase.

PHYSICOCHEMICAL CHARACTERISTICS
OF PLASMAKINASE

Plasmakinase is an enriched (five- to ninefold) β-globulin fraction of cadaver plasma with an admixture of albumin and gamma globulins (Table 12.3). Some 90-95% of its fibrinolytic activity is conditioned by plasminogen activator and some 5-10% of plasmin. The specific activity of plasmakinase is 0.7-1.3 FU/mg, which exceeds that of the initial plasma by 25 to 38 times.

Plasmakinase is a white, lyophilized powder (residual moisture less than 2%) that is soluble in 0.85% NaCl buffer solutions. It is a sterile, nonpyrogenic preparation produced experimentally in flasks containing 1000-3000 FU, which is equivalent to 250,000-750,000 IU of streptokinase. The lyophilized preparation remains active for 3 years at 4-6°C.

β-Globulins, whose content varies from 68 to 95% (averaging 85%) are the main electrophoretic fraction of the preparation. There is also an admixture of albumin (2.8%) and gamma globulins (10%). During isolation of plasmakinase it is possible to remove as much as 84-94% of the α-globulins, which are known to contain the inhibitors of proteinases. The antiproteinase activity of the preparation is, therefore, not more than 1%.

Various techniques of plasma fractionation make it possible to obtain plasmakinase that is completely devoid of albumin and gamma globulins. However, this leads to the loss of stability of the preparation and fibrinolytic activity, and the preparation yield, judging by total activity, does not exceed 30%. The decrease to 2-3% in content of immunoglobulin G (IgG) in plasmakinase preparations does not result in a substantial increase in specific activity. IgG was found to be the main "stabilizer" of plasmin activator activity, making it possible to retain enzymatic activity in the solution at room temperature for 18-25 h. We therefore decided that an additional purification of plasmakinase from albumin and gamma globulins was not necessary.

The protein content of plasmakinase was evaluated by immunochemical analysis (Table 12.3). Plasminogen activator was the main protein involved (comprising 47-90% of the total protein content). Plasminogen activator of whole plasma averaged 59.3%, and that for washed plasma averaged 76.7%. Plasminogen and plasmin antigens accounted for 1.6 and 2.5%, respectively, of the protein content, correlating with the activity of plasmin in plasmakinase preparations.

Fibrinogen degradation products are found in plasmakinase(2-14%). Gamma globulins are mostly represented by IgG (9.9%), the content of which is 33.1% less than in the initial plasma.

Comparative analysis of the plasma and the plasmakinase isolated from it showed that fibrinolytic activity (mostly activator) is revealed in identical electrophoretic fractions and corresponds to the plasminogen activator [23].

BIOLOGICAL CHARACTERISTICS OF PLASMAKINASE

Thrombolytic activity of plasmakinase (Table 12.4) was studied using the technique [22] which models the mural thrombolysis of circulating blood in vitro. The study showed that the preparation possesses a pronounced thrombolytic action that is conditioned by dosage.

Comparative study of plasmakinase preparations, fibrinolytically active plasma (FAP), streptase, and urokinase demonstrated that certain doses of these drugs have the same thrombolytic action. Thus, 1 FU of plasmakinase is equivalent (in

in vitro thrombolytic activity) to 200-300 IU of streptase
and to 50-60 Ploug units of urokinase.

The activator action of fibrinolytic agents on plasminogen was studied by analyzing the activity of the proenzyme
in plasma at 80-95% thrombolysis. It was found that, within
4 h of incubation following the effect of streptase, the
level of plasminogen decreased by 10- to 20-fold compared to
the initial level (3.8 CU/ml). The preparations of urokinase
and plasmakinase, in equivalent doses, caused a decrease in
the proenzyme by 7-25% only. One can assume that, unlike
streptase, plasmakinase mostly "attacks" the plasminogen of
a thrombus, while the proenzyme in plasma is more "resistant"
to its action.

Study of the thrombolytic effect of plasmakinase on
blood clots in various animals (rats, rabbits, dogs) and man
revealed a relative specificity in its action. The dose of
the preparation, which caused 85-92% lysis of human blood
clots, had a lesser effect on clots in dogs and rats (45-55%),
and a still lesser effect on clots in rabbits (10-22%). These
data were considered when the effect of the preparation was
studied in animal experiments.

Plasmakinase displays antithrombin action, as indicated
by the presence of fibrinogen degradation products, while
antithrombin III is absent.

The effect of inhibitors on fibrinolytic and thrombolytic activity of plasmakinase was studied. It was found
that 1 FU of plasmakinase is completely inhibited by 0.5-1.0
mg ε-aminocaproic acid, 50-100 antithrombin units, or 0.3-0.7
units of human α_2-macroglobulin. This inhibition is manifested by the absence of the lysis of standard blood clots, and
the decomposition of the thrombus in a modeled thrombolysis
system within 24 h.

Based on these studies and the clinical data with FAP,
we have determined tentative doses of the preparation for
animal experiments: 50-200 FU/kg of body weight, which is
equivalent to 10,000-50,000 IU/kg of streptokinase. Analysis
of the biological characteristics of plasmakinase was performed according to the recommendations of the Pharmacological Committee of the USSR Ministry of Health. The preparation was found to be nontoxic and nonpyrogenic, and its per-
missible dose exceeds the therapeutic dose by 20- to 35-fold.

Table 12.4. Thrombolytic Effect of Fibrinolytic Preparations on the Blood of Healthy Subjects

Fibrinolytic preparations	No. of studies	Concentration of preparations (in ml of blood)	Thrombolytic activity (% lysis of standard thrombus in 4 h of incubation)	
			Limits	M ± m
Streptase (FRG) Op. No. 366 B,	8	500 IU	78-95	86 ± 3
D380/E 5556	10	1000 IU	81-100	92 ± 3
Urokinase (USA) No. 672123	6	20 Ploug U	25-40	36 ± 3
Lot. 400738	5	40 Ploug U	54-64	59 ± 2
Fibrinolysin (USSR)	9	40 Ploug U	15-38	27 ± 2
No. 1730874/c. No. 2082	7	200 Ploug U	50-80	68 ± 5
Fibrinolytically active plasma (Sklifosovksy SRI of Emergency Aid)	6	0.6 FU	36-43	40 ± 1
	10	1.2 FU	76-86	82 ± 1
Plasmakinase (series Nos. 5, 15, 12, 8)	15	1.0 FU	50-68	62 ± 1
	12	2.0 FU	80-96	91 ± 2
	13	4.0 FU	96-100	99 ± 1

Intravenous administration of plasmakinase to intact
animals (rats, dogs) leads to an increase in fibrinolytic
activity in the blood and to a decrease in the coagulating
and antiproteinase potential; these changes in hemostasis
are normalized within 2.5-3.0 h.

Inactivation and elimination of plasmakinase (50 and
200 FU/kg) from rat blood were studied enzymatically and immunochemically (Figure 12.1). Depending on the dose, the
time for elimination of half of the preparation averaged 75.6
and 226.3 min, respectively. For urokinase, this time does
not exceed 11-16 min. According to the calculations, the
rate of elimination of plasmakinase from the blood varies
from 19.6 to 25.5 mg/kg/h. The rate of inactivation exceeds
this by 1.7- to 2.4-fold and equals 32.8-62.3 FU/kg/h, which
accords with equivalent data on streptokinase [31]. Immunochemical analysis of the distribution of the drug in rat
organs and tissues 5 h following administration showed that
plasminogen activator is accumulated mainly in the liver (377
and 1600 μg/g of raw tissue, depending on the dose), spleen
(249 and 670 μg/g), lungs (229 and 906 μg/g), and heart (185
and 500 μg/g). Plasminogen activator is practically absent
in the brain. Plasmakinase was detected in urine (190-463
μg/ml), probably associated with the removal of plasminogen
activator fragments containing immunochemical determinants.

THROMBOLYTIC EFFECT OF PLASMAKINASE
IN EXPERIMENTAL THROMBOSIS

Experimental thrombosis of the jugular vein was induced
in rats by an injection of thrombin (thrombus age, 2-3 h).
Circulation was restored 30-60 min following intravenous injection of plasmakinase against the background of a three-
to tenfold increase in fibrinolytic activity of the blood.
Thrombolysis was corroborated by morphologic analyses. The
results of these experiments have been published elsewhere
[32].

Thrombolytic action of plasmakinase during experimental
arterial thrombosis has also been studied in dogs. Thrombosis
of the femoral artery or of its branches was caused by ligat-
ing the vessel and injecting thrombin solution (15 U/ml) into
the isolated vessel segment. Three hours after thrombi formation, 100 FU/kg of plasmakinase (experiment) or 0.85% NaCl
(control) were administered once intra-arterially. The lytic
effect was evaluated angiographically and biochemically.

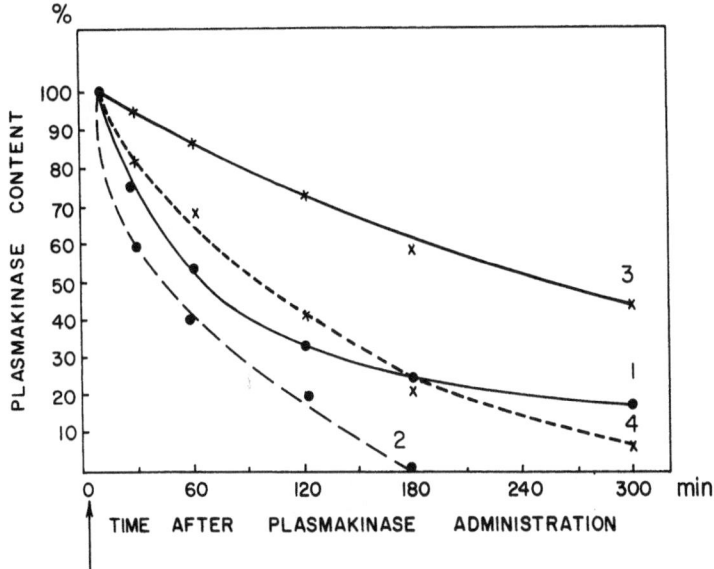

Fig. 12.1.   Enzymatic and immunochemical analyses of the rate
              of inactivation and elimination of plasmakinase
              from rat blood. Dashed lines denote enzymatic
              analysis; continuous lines denote immunochemical
              analysis. 1, 2) Following injection of 50 FU/kg
              of plasmakinase; 3, 4) following injection of 200
              FU/kg of plasmakinase. The arrow points to the
              moment of intravenous administration of plasma-
              kinase.

    No statistically significant changes in the coagulating
or fibrinolytic characteristics were detected in the venous
circulation within 3 h following thrombosis or in the subsequent 3 h after administration of 0.85% NaCl. Recalcification time, prothrombin index, thromboelastogram indices (R,
K, mA), fibrinogen and plasminogen levels, and fibrinolytic
activity were 84-105% of the values prior to injection of
the physiologic solution and compared to the initial level.
Injection of the physiologic solution did not restore blood
flow in the clotted artery for 6-8 h of observation. Intra-
arterial injection of plasmakinase, however, led to the restoration of blood flow 10-15 min after the injection (as recorded angiographically (Figure 12.2).

    Biochemical studies showed that thrombolysis occurs
against the background of hypocoagulation as manifested by

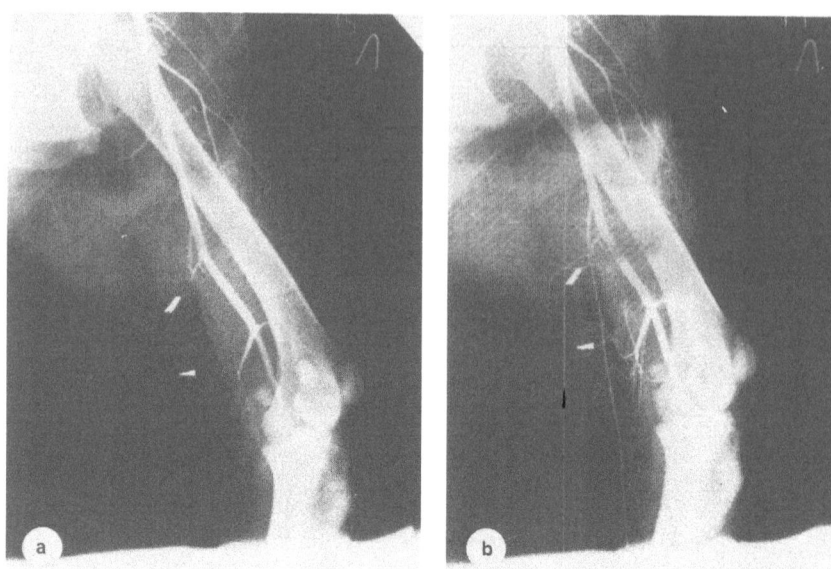

Fig. 12.2.   Angiogram of a dog prior to and following injection of plasmakinase. a) Angiogram of dog with
experimental thrombosis of the left femoral artery
branch: vessel "amputation" symptom; b) angiogram of dog 10 min after plasmakinase administration (100 FU/kg): complete restoration of patency in the clotted femoral artery branch.

the prolongation of recalcification time and the R and K indices on the thromboelastogram (by 32-89%, 45-83%, and 74-
380%, respectively); and by the decrease in prothrombin index, mA on the thromboelastogram, and fibrinogen concentration (by 9-29%, 17-37%, and 8-54%, respectively), as compared
to the same indices prior to injection and to those in the
control animals. One can assume that a five- to ninefold increase in fibrinolytic potential of the blood against a background of a decrease in antiproteinase activity (by 21-30%)
conditions the restoration of vessel patency.

A 17-31% reduction in the level of plasma plasminogen
was detected enzymatically and immunochemically within 2 h
after the injection of plasmakinase, substantiating the role
of the activator mechanism in thrombolytic drug action. During this period, fibrinogen/fibrin degradation products were
found in the blood serum and urine (60-200 µg/ml). Infusion

of plasmakinase led to a reduction in antiproteinase potential in dog blood due to the decrease in content and activity
of α_1-antitrypsin and α_2-macroglobulin by 16-29% and 9-16%,
respectively, as compared to the control. These data demonstrate the significance of major proteinase inhibitors in
the regulation of fibrinolytic potential.

In order to study the thrombolytic effect of plasmakinase during experimental thrombosis of the coronary artery,
25 mongrel dogs weighing 18-25 kg were examined. The animals
were separated into three groups: group 1 received no injections, group 2 received 0.85% NaCl + heparin, and group 3 received plasmakinase + heparin. Multiple anesthesia were used:
Ketalar (ketamine hydrochloride), droperidol, sodium hydroxy-
butyrate, neuroleptanalgesia, and artificial lung ventilation
with nitrogen oxide mixture. Thirty to 60 minutes after gain-
ing access to the left anterior descending coronary artery
via transternal thoracotomy, experimental thrombosis was induced in the middle third of the artery by injecting pituitrin
(1-2 U) and thrombin (50 U/ml) following the application of
a tourniquet. In 1-1.5 h, the ligatures were removed. Six
hours following thrombi formation, plasmakinase (25-35 FU/kg)
with heparin (25 U/kg) or 0.85% NaCl (1.5 ml/kg) with heparin
(25 U/kg) were administered using superselective catheterization of the coronary artery. The optimum rate of plasmakinase administration was 1 ml/min (20-30 FU). When this rate
was increased (5-7 ml/min), bradycardia and extrasystoles developed. We regarded this as a cardiodepressive effect that
is also observed with other thrombolytic agents [33].

Pulse, arterial pressure, and a three-lead ECG were recorded continuously during the experiment. Blood for analysis was drawn with a catheter from the venous cardiac sinus
and femoral artery; and was stabilized with a 3.8% sodium
citrate solution in the proportion 4:1. Recalcification time,
prothrombin index, free heparin, fibrinolytic activity, and
fibrinogen were studied, and thromboelastographic analysis
was performed [34]. Using the caseinolytic technique [35],
we studied the activity of plasmin, plasminogen, and proteolysis inhibitors; the composition of fibrinogen/fibrin degradation products was analyzed immunochemically [36].

Throughout the observation period, statistically significant changes in the coagulating, antiproteinase, and fibrinolytic characteristics were revealed in the blood of the femoral artery from group 1 animals. Only a small decrease in

fibrinogen concentration and the mA index on the thromboelastogram was revealed, which is likely due to some loss of blood
during thoracotomy. Hypercoagulation and a decrease in fibrinolytic potential were detected in the blood of the venous
sinus 2-3 h following thrombosis. These phenomena progressed
by 4-6 h, as manifested by the shortening of the temporal R
and K indices on the thromboelastogram by 69 and 51%, respectively, a four- to fivefold decrease in free heparin, a 73%
reduction in fibrinolytic activity, and a 20-30% increase in
antiproteinase blood potential as compared to the initial
level. These data are highly informative and indicate that
the study of hemostasis in blood flowing from the cardiac
venous sinus appears to be important. Against the background
of increasing hypercoagulation, we observed by 6-7 h ascending thrombosis of the coronary artery, disturbance in cardiac
function, a fall in blood pressure, and, finally, death of
the animals. In order to prevent these severe complications
in the animals of group 2, we injected smll doses of heparin
1 and 4 h after production of the thrombosis; this made it
possible to preserve the coagulating, fibrinolytic, and antiproteinase potential of the venous blood at a level near
physiologic level, and to block the development of ascending
coronary thrombosis.

After production of thrombosis in the left anterior descending coronary artery, exsanguination of the adjoining
myocardial region was detected in all animals. The number of
heart contractions within 1 h was 200-250 per min, and tachycardia reached 360 ± 40 beats per min by 6 min following thrombosis. Arterial pressure (AP) showed a tendency to decrease
and, by 6 h, systolic and diastolic AP were 11.5 ± 1.3 kPa and
7.5 ± 0.5 kPa, respectively. Cardiac output and stroke volume
averaged a 50% decrease after 6 h of thrombosis, as compared to
initial levels.

The ECG of all animals showed ischemic signs of myocardial damage 10 min after coronary thrombosis: Lead 1 showed a
rise in the ST segment above the isoline, and leads 2-3 showed
a decrease in the ST segment. Within 1 h, the ECG showed pronounced sinus tachycardia, single ventricular extrasystoles,
and a pronounced shift in the ST segment from the isoline.
Six hours later, myocardial ischemia had worsened: Group
ventricular extrasystoles evolved against a background of the
remaining and even increasing tachycardia (lead 1, ST-segment
elevation was 1 mm; leads 2 and 3, ST segment was decreased
2-3 mm below the isoline). During this period, creatine
phosphokinase activity in the blood increased two- to three-

fold compared to the initial level. Coronary angiography revealed a symptom of vessel "amputation" in the middle third
of the artery in all animals, proving the presence of coronary thrombosis.

In the control animals, a dark-red thrombus was detected
macroscopically in the lumen of the coronary artery 1.5-2.0
cm away from the opening and stretching to the distal parts
of the artery. The thrombus was lightly adherent to the
arterial intima. Light optic and electron microscopy studies
showed that this thrombus was a dense network of fibrin
threads with formed elements between them. These elements
were represented by numerous platelets in a "viscous metamorphosis" stage, altered erythrocytes, and single polymorphonuclear leukocytes (Figure 12.3). Morphologic study of the
myocardial region adjoining the clotted artery revealed the
presence of disseminated coagulation from the arterioles to
the large veins while other vessels remained plethoric. Multiple interstitial hemorrhages were detected in this part of
the myocardium. Most of the muscle fibers revealed dystroph-
ic changes, and some even showed early necrotic changes.

Against the background and immediately after administration of plasmakinase, coloration in this myocardial area
normalized. Within 1 h, we recorded a 23-24% fall in the
pulse rate (p < 0.05), an increase in systolic and diastolic
pressure (by 18-24%, p < 0.05, and 32-44%, p < 0.01, respectively), and an increase in stroke volume and cardiac output
by 141-197% and 92-125%, respectively (p < 0.01) as compared
to these parameters prior to the injection of plasmakinase
and to those of the animals in group 2 after the injection
of 0.85% NaCl. The work of the heart was further normalized
over the next 2-3 h: Systolic and diastolic pressures were
15.6 ± 0.8 kPa and 8.6 ± 0.6 kPa, respectively, and stroke
volume and cardiac output were 1.04 ± 0.09 ml/kg and 208 ±
18 ml/min/kg, respectively, i.e., they were within 88-92% of
the values prior to clotting of the coronary artery. In the
dogs in group 2, disturbances in heart function increased progressively over this period.

Ten minutes after the administration of plasmakinase,
the ECG showed positive dynamics: Against the background of
the remaining tachycardia, ventricular extrasystoles disappeared and the ST segment showed a tendency to decrease.
After another 60 min we observed a slower heart rhythm and a
shift in the ST segment in lead 1 to the isoline and in leads 2

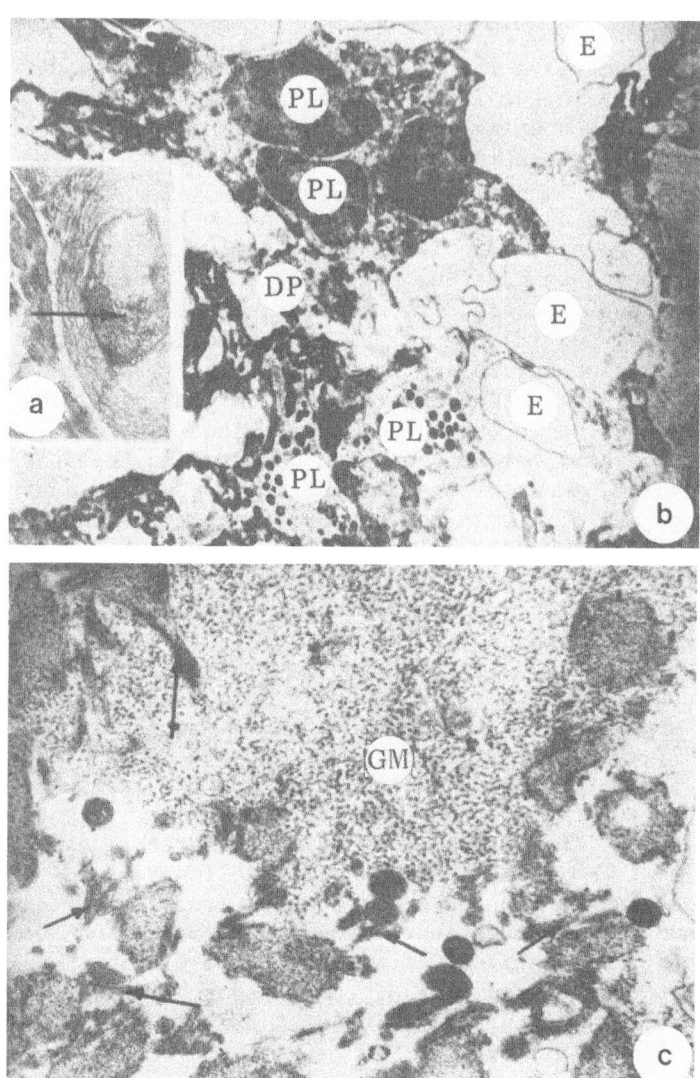

and 3 approaching the isoline. Within 5-6 h, the ECG became
almost similar to the initial reading, i.e., an injection of
the drug removed the ECG signs of ischemic damage to the myocardium.

Coronary angiography revealed the restoration of blood
flow in the clotted vessel in all animals 10-30 min after the
administration of plasmakinase.

Ten minutes after intracoronary administration of the
preparations, marked hypocoagulation was revealed in the venous blood of animals in groups 2 and 3. This was manifested
by a considerable increase in recalcification time conditioned by both heparin and plasmakinase.

In order to elucidate the role of plasmakinase in the
decrease in coagulation, we studied the blood by thromboelastography after neutralization of the free heparin with
protamine sulfate. We found that plasmakinase had already
caused a prolongation of the R and K indices on the thromboelastogram as compared to the controls by 77 and 373%, respectively (Table 12.5). Ten minutes after administration of
plasmakinase, a 6.3-fold increase in fibrinolytic activity
in the blood was revealed, as compared to the control. This
activity remained at a high level for 2-3 h. An increased
fibrinolytic potential of the blood against the background
of a decrease in antiproteinase activity conditioned the
thrombolytic action (Table 12.5), as corroborated by the following changes detected 60 min after the administration of
plasmakinase. Plasminogen and fibrinogen levels dropped (by

Fig. 12.3.   Pathomorphology of coronary artery thrombus in
              dogs prior to and following the administration
              of plasmakinase. a) Coronary artery lumen oc-
              cluded by the thrombus; fibrin staining accord-
              ing to Mallory, magnification ×100. b) Ultra-
              structure of the thrombus without plasmakinase
              application; compact fibrin bundles are seen be-
              tween the formed elements. PL) Polymorphonuclear
              leukocytes; E) erythrocytes devoid of hemoglobin;
              DP) degranulated platelets, magnification ×12,000.
              c) Ultrastructure of the thrombus after plasma-
              kinase treatment; vast fields of globular materi-
              al (GM) with fibrin fibrils (arrows) are seen on
              the periphery, magnification ×18,000.

304

Table 12.5. Hemostasis Indices in the Venous Circulation of
kinase Administration

Time of examination	Groups	Recalci- fication	Thromboelastogram	
			R (min)	K (min)
30-60 min after thoracotomy	C	135 ± 14	1.1 ± 0.1	2.3 ± 0.2
	Ex	139 ± 16	1.2 ± 0.2	2.2 ± 0.2
1 h after thrombosis	C	157 ± 14	1.5 ± 0.2	2.8 ± 0.3
	Ex	161 ± 12	1.3 ± 0.1	2.8 ± 0.3
6 h after thrombosis	C	138 ± 12	1.7 ± 0.1	3.1 ± 0.5
	Ex	153 ± 16	2.1 ± 0.2	4.1 ± 0.5
Time after administration of preparations:				
10 min**	C	2000	4.9 ± 0.4	4.8 ± 0.5
	Ex	2000	8.7 ± 0.8*	22.7 ± 3.3*
60 min	C	317 ± 27	10.9 ± 3.1	4.9 ± 0.5
	Ex	636 ± 56*	27.3 ± 6.3	undeterm.
120 min	C	209 ± 21	3.7 ± 0.3	5.3 ± 0.6
	Ex	535 ± 60*	10.6 ± 1.9*	11.1 ± 1.3*
240 min	C	149 ± 18	2.7 ± 0.3	4.0 ± 0.4
	Ex	207 ± 19*	4.5 ± 0.6*	8.5 ± 1.6*

Note: C) control group (0.85% NaCl + heparin); Ex) experi-
*p < 0.05.
**R, K, mA, and fibrinogen are calculated after neutraliza-

40 and 36%, respectively), and the mA on the thromboelastogram decreased by 43% as compared to the same indices prior
to drug administration and to the animals in group 2 after
injection of 0.85% NaCl. Fibrinogen degradation products appeared in the blood serum, amounting to 30-120 µg/ml (the
control was 0-5 µg/ml).

Only slight changes in hypocoagulation were revealed in
the femoral artery 1-2 h after the injection of plasmakinase.
No statistically significant changes were detected in the
levels of plasminogen, fibrinogen, plasmin, and fibrinolytic
and antiproteinase activity, thus making it possible to assume that intracoronary administration of small doses of the

Dogs with Experimental Coronary Thrombosis Following Plasma-

	Fibrinogen	Fibrinolytic activity	Plasmino-gen	Total antipro-teinase activity
mA (mm)	mg%	(mm²)	(CU/ml)	(IU/ml)
48 ± 2	267 ± 25	13.9 ± 1.4	1.3 ± 0.1	32 ± 1
47 ± 2	283 ± 30	14.7 ± 1.6	1.4 ± 0.1	32 ± 2
40 ± 2	229 ± 23	12.4 ± 1.6	1.2 ± 0.1	31 ± 2
42 ± 2	238 ± 21	13.2 ± 1.5	1.3 ± 0.1	30 ± 2
32 ± 2	206 ± 17	10.6 ± 1.5	0.9 ± 0.1	30 ± 2
37 ± 3	218 ± 17	11.3 ± 1.7	1.0 ± 0.1	35 ± 2
31 ± 2	217 ± 17	11.1 ± 2.0	0.9 ± 0.1	31 ± 2
36 ± 3	207 ± 23	70.3 ± 9.8*	0.9 ± 0.1	34 ± 2
28 ± 2	213 ± 19	13.1 ± 1.4	1.0 ± 0.1	32 ± 2
16 ± 4*	137 ± 18*	39.4 ± 5.7*	0.6 ± 0.1*	23 ± 2*
31 ± 2	222 ± 21	14.3 ± 2.1	1.1 ± 0.1	29 ± 2
23 ± 3*	162 ± 15*	29.1 ± 4.3*	0.6 ± 0.2*	23 ± 2*
38 ± 3	247 ± 18	12.7 ± 1.6	1.2 ± 0.1	30 ± 2
28.3*	175 ± 22*	19.3 ± 2.2*	0.8 ± 0.1*	24 ± 1*

mental group (plasminogen + heparin).

tion of plasma heparin with protamine sulfate.

fibrinolytic drug prevents sharp alterations in total hemostasis, in addition to the thrombolytic effect.

Macroscopic analysis of the heart 8-10 h after administration of plasmakinase showed that the lumen of the coronary
artery was clear and that the intima was smooth and glossy.
A fibrin film slightly adherent to the intima was found only
in the paracentetic trauma zone of the left anterior descending artery. The main differences in the structure of the
thrombus in this group, as compared to that of the animals
receiving no thrombolytic injection, were a fall in the number
of erythrocytes and an increase in the polymorphonuclear
leukocyte content; fibrin bundles became looser and vast fields

of globular material emerged (Figure 12.3).

Thus, physiologic, angiographic, electrocardiographic, biochemical, and morphologic data show that plasmakinase possesses a prominent thrombolytic action. The efficacy of experimental treatment with plasmakinase accords with clinical data on the treatment of thrombosis with fibrinolytically active plasma [2-5, 7], and with results of intracoronary administration of fibrinolysin with heparin and streptokinase [33, 37].

This study also demonstrated the importance of intracoronary drug administration, as noted previously by other researchers [38]. This method allows one to decrease the concentration of plasmakinase by three- to sixfold, as compared to administration into the general circulation, when treating experimental venous thrombosis in rats [32]. And, yet, plasmakinase is brought directly to the thrombus without much disturbance in total hemostasis.

The presence of natural (for humans) fibrinolytic enzymes (plasminogen activator, plasmin, plasminogen) in the drug and the way in which it affects the hemostasis of experimental animals make it possible to suggest several mechanisms for its lytic action. With regard to the activator mechanism, the preparation is able to convert thrombus plasminogen into plasmin, thus performing mostly endogenic fibrinolysis. The existence of such a mechanism is corroborated by an increase in plasminogen activator and plasmin levels in the blood of experimental animals, as determined enzymatically and immunochemically. The increase in plasmin level after injection of plasmakinase points to a direct proteolytic action conditioned by the human plasmin contained in the drug. The presence of human plasminogen in the preparation makes it possible to supplement the specific substrate for the plasminogen activator, thus facilitating the maintenance of a high level of fibrinolytic activity for a longer period of time. Consequently, one can assume that the thrombolytic action of plasmakinase belongs to the well-known "urokinase + fibrinolysin" type, which gives this an advantage.

Thus, this study demonstrates that plasmakinase, a fibrinolytic agent made from postmortem plasma, is an effective preparation having an activator thrombolytic action, which opens up wide prospects for its clinical application in treating venous and arterial thromboses.

REFERENCES

1. K. I. Zykova and L. I. Petrova, "Fibrinolytic proper-
    ties of postmortem blood plasma and its clinical ap-
    plication during myocardial infarction," Abstract, in:
    Twelfth International Congress of Blood Transfusion,
    A. E. Kiselev (ed.), Meditsina, Moscow, p. 74.
2. G. A. Pafomov, "Biological characteristics of the blood
    of sudden death victims and its application in surgical
    practice," Thesis, Sklifosovsky Scientific Research In-
    stitute of Emergency Aid, Moscow (1971).
3. A. S. Smetnev, L. I. Petrova, G. A. Pafomov, K. I.
    Zykova, A. G. Ivashchenko, and S. A. Potemkina, "Ex-
    perience with treatment of myocardial infarction by fi-
    brinolytically active plasma from postmortem blood,"
    Ter. Arkh., 45, No. 8, 14-19 (1973).
4. A. P. Golikov and S. A. Koroleva, "Hemostasis rehabilita-
    tion in acute myocardial infarction complicated by
    cardiogenic shock," in: Topical Problems of Hemostasiol-
    ogy, B. V. Petrovskii, E. I. Chazov, and S. V. Andreev
    (eds.), Nauka, Moscow (1981), pp. 362-371.
5. E. P. Dumpe, G. A. Pafomov, V. A. Shestakov, E. G. Jablo-
    kev, and V. N. Ilyin, "Treatment of acute thromboses of
    large veins with fibrinolytically active plasma," Kardi-
    ologiya, 7, 111-116 (1973).
6. V. S. Savelyev, E. G. Jablokov, and A. I. Kirichenko,
    Thromboembolism of the Pulmonary Artery, Meditsina, Mos-
    cow (1979).
7. V. A. Shestakov, "The hemostasis system during throm-
    boses of large veins," Thesis, Pirogov Second Moscow
    Medical Institute, Moscow (1979).
8. G. Ya. Levin, "Hemocoagulating properties and clinical
    use of plasma and platelets from cadaver blood," Abstract,
    M. D. Thesis, Central Scientific Research Institute of
    Hematology and Transfusion, Moscow (1978).
9. I. V. Androzhskaya, "Fibronectin characteristics of post-
    mortem blood," Abstract, Ph.D. Thesis, Central Scientific
    Research Institute of Hematology and Transfusion, Moscow
    (1973).
10. I. V. Andozhskaya, S. V. Ryzhkov, V. T. Pleshakov, and
    B. N. Anuschenko, "Cadaver blood and fibrinolysin obtained
    from it: application for postoperative treatment of
    patients subjected to chest surgery," Probl. Gematol.
    Pereliv. Krovi, 9, 15-A (1972).
11. O. T. Zhukova and E. P. Pospelova, "Effect of fibrino-
    lytic blood on hemocoagulation system (used for the first

time in cardiovascular surgery)," Abstract, in: Fourth
All-Union Conference on the Blood Coagulation System
and Fibrinolysis, S. A. Georgieva (ed.), Ministry of
Public Health, Saratov (1975), Part 2, pp. 455-456.

12. V. M. Denisov, E. A. Efimova, A. S. Surkova, and A. I.
Virovlansky, "Transfusion of fibrinolytically active
plasma in treating infant pneumonia," in: Prophylaxis,
Diagnosis, and Treatment of Acute Infant Pneumonia, Moscow (1978), pp. 114-145.

13. V. S. Ilyin, "Fibrinogeno- and fibrinolytic factor (fibrinogenase) of postmortem blood," Biokhimiya, 14, No.
4, 354-360 (1949).

14. M. Ya. Golovanova, "Fibrinogenase," Biokhimiya, 15, No.
3, 256-266 (1950).

15. Z. A. Chaplygina, "Fibrinogenase action: experiments
on a whole organism," Abstract, Ph.D. Thesis, Leningrad
Scientific-Research Institute of Hematology and Transfusion, Leningrad (1953).

16. K. I. Zykova, "Prospects for storing cadaver blood with
high fibrinolytic activity," in: Transfusion of Postmortem Blood and Certain Aspects of Homoplasty, M. M.
Tarasov (ed.), Sklifosovsky Scientific-Research Institute of Emergency Aid, Moscow (1967), pp. 27-30.

17. G. A. Pafomov, "An installation for storing fibrinolytic
blood in flasks," Probl. Gematol. Pereliv. Krovi, 9,
56-58 (1962).

18. A. M. Bratchik and V. B. Levandovsky, "Experimental effect of thrombolytic agents on the blood coagulation
system," Trudy Krymsk. Med., Kharkov, 52, No. 5, 26-28
(1973).

19. V. B. Levandovsky, "Effect of thrombolytic preparations
on experimental mural thrombosis of the aorta," Abstract,
Ph.D. Thesis, Crimean Medical Institute, Simferopol
(1974).

20. G. A. Pafomov, V. B. Khvatov, O. A. Petrenko, and T. K.
Platonova, "Characteristics of fibrinolytic system com-
ponents in postmortem blood as guidelines for obtaining
fibrinolytic preparations," Abstract, in: Fourth All-
Union Conference on the Blood Coagulation System and
Fibrinolysis, S. A. Georgieva (ed.), Ministry of Public
Health, Saratov (1975), Part 1, pp. 112-113.

21. V. B. Khvatov and T. K. Platonova, "Biochemical substan-
tiation of the practicability of using euglobulin fraction from cadaver plasma for obtaining a fibrinolytic
preparation," Probl. Gematol. Pereliv. Krovi, 3, 38-42
(1976).

22. T. K. Platonova and V. B. Khvatov, "Comparative charac-
    teristics of fibrinolytically active fractions obtained
    from cadaver blood plasma," Probl. Gematol. Pereliv.
    Krovi, 6, 47-50 (1979).
23. O. A. Petrenko, "Purification and properties of plasmin-
    ogen activator from postmortem blood plasma," Biokhimiya,
    43, No. 8, 1438-1443 (1978).
24. G. A. Pafomov, V. B. Khvatov, V. M. Rusanov, T. K. Plat-
    onova, A. V. Sirotenko, G. P. Titova, M. V. Podolsky,
    and I. V. Maksimovich, "A thrombolytic preparation from
    the blood plasma of sudden death victims," in: Medical
    Preparations of Plasma Proteins, O. K. Gavrilov (ed.),
    Sborn. Nauchn. Rabot Tsentr. Inst. Gematol. Pereliv.
    Krovi, Moscow (1981), pp. 90-99.
25. G. A. Pafomov, V. M. Rusanov, T. A. Belova, V. B. Khvatov,
    V. P. Merzlov, and L. I. Simonova, "Prospects for obtain-
    ing an antiproteinase preparation on the basis of human
    α_2-macroglobulin," Probl. Gematol. Pereliv. Krovi, 4,
    23-27 (1982).
26. V. B. Khvatov, "Postmortem blood as a source of prepara-
    tions with directed action," Abstract, in: Blood Frac-
    tionation, All-Union Meeting on Blood Fractionation,
    O. K. Gavrilov (ed.), Central Scientific-Research Insti-
    tute of Hematology and Transfusion, Moscow (1981), pp.
    46-47.
27. V. B. Khvatov and T. K. Platonova, "A thrombolytic prep-
    aration with activator action from human blood plasma,"
    Abstract, in: International Conference on Antithrombotic
    Drugs in Clinical Practice, F. I. Komarov (ed.), Sechenov
    First Moscow Medical Institute, Moscow (1979), pp. 61-63.
28. O. A. Petrenko and V. B. Khvatov, "Isolation of plasmino-
    gen activator from the plasma of sudden death victims,"
    Probl. Gematol. Pereliv. Krovi, 12, 46-48 (1976).
29. T. K. Platonova and O. A. Petrenko, "Characteristics of
    human plasminogen activator and fibrinolytically active
    plasma fraction enriched by this enzyme," Vopr. Med.
    Khim., 25, No. 3, 302-307 (1979).
30. V. B. Khvatov, T. K. Platonova, and V. M. Rusanov, "A
    procedure for obtaining fibrinolytic preparation," Byull.
    Isobret., 35, 24 (1979).
31. G. V. Andreenko, Fibrinolysis (Biochemistry, Physiology,
    Pathology), Moscow State University (1979).
32. N. K. Permayakov, G. P. Titova, N. L. Sinyanskaya, V. B.
    Khvatov, and T. K. Platonova, "Morphology of clot lysis
    under the effect of plasmakinase," Arkh. Patol., 6, 31-
    36 (1982).

33. E. I. Chazov, L. S. Matveyeva, A. V. Mazaev, K. E.
    Sargin, G. A. Sadovskaya, and M. Ya. Ruda, "Intracoro-
    nary administration of fibrinolysin with heparin during
    acute myocardial infarction," Ter. Arkh., 4, 8-19 (1976).
34. V. A. Shestakov and V. N. Ilyin, Methods of Studying
    Hemostasis in Surgical Patients. Methodological Recom-
    mendations of the USSR Ministry of Health, Moscow (1976).
35. V. B. Khvatov and O. A. Petrenko, "Isolation of specific
    substrate for human plasminogen for revealing and quan-
    titative determination of plasminogen activator in ca-
    daver blood plasma," Probl. Gematol. Pereliv. Krovi, 6,
    57-66 (1975).
36. N. Axelsen, L. Krell, and B. Veeke, Manual of Quantita-
    tive Immunoelectrophoresis, Univ. Park Press (1975).
37. W. Ganz, N. Buchbinder, H. Marcus, A. Mondkar, and L. O.
    Conner, "Intracoronary thrombolysis in evolving myo-
    cardial infarction," Herz, 6, 37-43 (1981).
38. L. A. Bessolitsyna, A. V. Mazaev, R. A. Markosyan, A. V.
    Suvorova, V. P. Torchilin, and E. I. Chazov, "Effect of
    bioresolving microspheric preparations of immobilized
    fibrinolysin on the fibrinolysis system," Byull. Eksp.
    Biol. Med., 89, No. 1, 16-18 (1980).
39. W. Auerwald, B. Binder, and W. Doleschel, "Angiokinase
    — molecular weights of proteins representing perivascu-
    lar plasminogen activation," Thromb. Diath. Haemorrh.,
    26, 411-413 (1971).

INHIBITORS OF PLATELET AGGREGATION

K. M. Lakin and V. A. Makarov

Moscow Stomatologic Institute, Ministry of Public Health of the RSFSR, Moscow

The functional state of platelets and, in particular, their ability to aggregate and adhere, plays an important role in the regulation of fibrinolytic and thrombolytic efects. It has been found that blood platelets not only contain different inhibitors of fibrinolysis but also can indirectly influence plasma fibrinolytic activity by affecting the release of plasminogen activator from vessels. This can be accomplished under the effect of biogenic amines — serotonin, histamine, and others. In turn, a change in fibrinolysis can affect the aggregating ability of blood platelets. Accordingly, the fibrinogen and fibrin degradation products formed during fibrinolysis can significantly alter platelet aggregation and adhesion. Thus, there is a complex relationship between the function of blood platelets and the state of the fibrinolytic system in humans and animals.

Because of this complex relationship, drugs that inhibit platelet aggregation and adhesion can play a definite role in thrombolytic therapy. Throughout the investigation of aggregation inhibitors, interesting reviews have been published which contain a systematization of substances that decrease platelet aggregation [1-3]. In recent years, this information has been supplemented by new publications on the antiaggregating activity of prostaglandins and prostaglandinlike

compounds, including prostacyclin. New antiaggregating preparations from other groups of chemical compounds have also
been found. It is therefore timely to review these publications and the results of our own studies on the inhibitors
of blood platelet aggregation — an important component of
antithrombolytic therapy.

The aim of the present chapter is to evaluate briefly,
based on published data and our own study results, the different groups of aggregation inhibitors. Analysis of the
nature of platelet aggregation and the mechanism of action
of different substances is extremely important for both understanding the physiological bases of these reactions and for
detecting pathways of drug control of hemostasis. Because
of the limited length of this chapter, not all sources are
cited.

In recent years, antiaggregating properties have been
found in many pharmacologic preparations possessing vasodilatory, antiinflammatory, neuroleptic, antidepressive, and
other mechanisms of action. Among the vasodilators, dipyridamole and its derivatives CH-105, CH-123, RA 233, S_{H-889}, and
V_{K744} are, evidently, the most active antiaggregating agents.
A strong antiaggregating effect has been observed with papaverine and its derivatives [4], including trimetoquinol [5],
the R(+) of which possesses especially high activity. Trimetoquinol does not have a strong effect on adenylate cyclase
or on cyclic adenosine monophosphate (cAMP) or cyclic guanosine monophosphate (cGMP) phosphodiesterase. A powerful antiaggregating effect is observed in the phosphodiesterase selective inhibitor cAMP cilostamide, and in ticlopidine, 5-(o-
chlorobenzyl)-4,5,6,7-tetrahydrothienyl[3,2-C]pyridine hydrochloride, and 7-ethoxycarbonyl-6,8-dimethyl-4-hydroxymethyl-
(2H)-phthalazinone [6, 7].

It has been found that the calcium antagonists verapamil,
molsidomin, and nifedipine injected into an organism decrease
aggregation. This ability to decrease platelet aggregation
has also been found in the coronary dilator dilazep (As-0.5):
tetrahydro-1,4(1H)-diazepine-1,4(5H)-dipropanol-3,4,5-trimeth-
oxybenzoic acid diester.

Antiaggregational action has also been detected in a
number of xanthine derivatives [8]. In this series, caffeine
and aminophylline have manifested significant effectiveness;
consequently, the latter agent is used clinically to treat

Table 13.1. Effect of Some Xanthine Derivatives on Platelet Aggregating Ability

Substance	Concentration	Inhibition of platelet aggregation
	(µg/ml)	%
Control (ADP-aggregation)	-	0
8-Diethylaminomethyl-3,7-dimethyl-xanthine	400	63
1-Hexyl-8-diethylaminomethyl-3,7-dimethylxanthine chlorohydrate	400	0
Theobromine-8-acetic acid diethyl-aminoethylamide	400	35
6-Methylmercapto-2-amino-3,7-di-methylpurine chlorohydrate	400	29

thrombotic states. Among other xanthine derivatives, an anti-aggregating effect has been observed with aminophylline and pentoxifylline [9]. It was found that pentoxifylline enhances phosphorylation of thrombocytic membranes.

Our studies have revealed some peculiarities in the relationship between the chemical structure and antiaggregating activity of xanthine derivatives. A marked antiaggregating action was found with 8-diethylaminomethyl-3,7-dimethylxanthine. The placement of an aliphatic chain in the first position of this substance completely removed its antiaggregating action [8]. An antiaggregating effect was also observed with 5-methylmercapto-2-imino-3,7-dimethylpurine. But placement of a methylmercapto group in the second position did not cause a decrease in platelet aggregation. Replacement of the methylmercapto group in the sixth position by a dimethylamino group significantly decreased the antiaggregating action of the substance (Table 13.1). Another vasodilator, sulocti-dil, also decreases aggregation. It is especially effective in aggregation caused by collagen and thrombophax.

An antiaggregating effect has also been described for halidor. This drug decreases serotonin release from plate-lets and reduces its absorption. Halidor does not affect

cAMP phosphodiesterase. Nevertheless, its antiaggregating
effect increases in combination with theophylline. Combination with prostaglandin E_1 does not change the force of the
antiaggregating action. Interestingly, an overdose of halidor increases aggregation and disturbs the formation of plasma
and tissue thromboplastin. Aggregation is also reduced by
viquidil, a preparation that dilates vessels in the brain.
This drug is close to papaverine in force of action. Hydralazine and trapidil also have significant antiaggregating effects.

A certain antiaggregating effect was found with inten-
sain, tolazoline, inderal, benzofuran derivatives, and nitrite
and nitrate preparations [9]. Magnesium ions also possess an
antiaggregating effect.

According to many authors [10-12], prostaglandins (PG)
are one of the most effective physiologic inhibitors of aggregation. Prostaglandin I_2 (prostacyclin) is the most powerful natural antiaggregant, while prostaglandins D_2 and E_1 have
a lesser effect [13-15]. In ischemic heart disease, the prostaglandin content in the blood bed and tissues changes, i.e.,
the metabolism of prostaglandins E and A is enhanced [16].
The addition of PGE_1 against a background of aggregation caused
by adenosine diphosphate (ADP) leads to destruction of thrombocytic aggregates [10]. The inhibitory action of PGE_1 decreases with an increase in calcium ion concentration.

The relationship between the chemical structure and
pharmacologic effect in a series of PGE_1 derivatives is interesting to consider. By different modifications of the chemical structure it is possible to change the force of PGE_1
action. For instance, esterification (transformation into
the methyl ester) leads to some decrease in antiaggregating
action. Loss of a hydroxyl group in the second position observed in PGE_1 and PGB_1 decreases the effect. A decrease in
activity is observed with oxidation of the hydroxyl group in
position 15 (15-keto-PGE_1) and with removal of the keto group
in position 9 ($PGE_{1\alpha}$, $PGF_{1\beta}$). A similar effect is observed
when the carboxyl side chain is lengthened (α-homo-PGE and
α-dihomo-PGE) and the carboxyl chain shortened (α-nor-PGE_1
and α-dinor-PGE_1). Lengthening of the methyl side chain
(ω-homo-PGE_1) leads to enhancement of the preparation's activity. Administration of a 2,3-trans chain activates antiaggregational action, while 3,4-trans, 4,5-trans, 5,6-trans,
and 18,19-cis chains have inhibitory effects on prostaglandin
activity. Isomerism in positions 11 and 15 can have a sig-

nificant effect on the compound's activity. Thus, 11-15-epi-PGE$_1$ has lower activity than PGE$_1$ [17]. At the same time, 6-keto-PGE$_1$ exceeds the natural analog in antiaggregating activity.

Administration of a 5,6-cis coupled chain into PGE$_1$ leads to the formation of PGE$_2$. At the same time, the nature of the effect on platelets changes. It has been demonstrated that PGE$_2$ increases the aggregation of blood platelets. Shio et al. [17] report that, in small concentrations, PGE$_2$ partially removes the antiaggregating effect of PGE$_1$ and sometimes promotes the appearance of a second wave of aggregation. Lipoxidase and prostaglandin-synthetase blocking agents decrease platelet aggregation. In high concentrations, PGE$_2$ inhibits aggregation caused by ADP, affecting both the second and first aggregation waves. Under these conditions, PGE$_2$ activates platelet adenylate cyclase. In different concentrations, PGE$_2$ is able to stimulate, as well as inhibit, phosphodiesterase. At the same time, PGE$_2$ differently affects the content of cAMP in platelets.

Numerous studies have been devoted to investigating the mechanism of action of prostaglandin on platelet aggregation. It has been suggested that PGE$_1$ affects the binding of ADP with specific receptors on the thrombocytic membrane [18]. This view is based on studies of the binding of labeled ADP with blood platelets. PGE$_2$ decreased platelet radioactivity. Other authors believe that the effect of prostaglandin can be associated with activation and synthesis of intrathrombocytic cAMP since many inhibitors of aggregation increase the content of cAMP in platelets [19].

A correlation in the action of prostaglandin on platelet aggregation and adenylate cyclase has also been observed. Evidently, it is not necessary for prostaglandins to penetrate into the cytoplasm in order to produce an effect. For instance, PGE$_1$ immobilized on ω-NH$_2$-hexylagarose has an antiaggregating effect. Honohan et al. [20] have reported a prostaglandin preparation of peroral use with a prolonged effect. These authors refer to the substance 3-oxa-4,5,6-trinor-3,6-inter-m-phenylene-PGE$_1$-amide. The PGE$_1$ precursor, dihomo-γ-linolenate, decreases platelet aggregation, probably due to its transformation into PGE$_1$. The methyl ester of γ-linolenic acid possesses the same effect. It was found that linolenic acid disturbs the synthesis of endoperoxide due to the blockage of phospholipase A$_2$.

In contrast to prostaglandins E_1 and I_2, PGD_2 does not have a noticeable side effect on the cardiovascular system. However, in acute thromboses or hyperlipoproteinemia, platelet sensitivity to this or other prostaglandins drops sharply [21]. In considering the interaction of prostaglandin preparations with other antithrombotic agents, it was found that heparin and dextran sulfate inhibit the antiaggregating action of prostacyclin and other prostaglandins [22]. At the same time, inhibitors of cAMP phosphodiesterase enhance the prostaglandin effect [23].

A high rate of inactivation is a drawback with prostaglandin preparations. Some substances, particularly prostacyclin, are rapidly destroyed at room temperature or with a change in pH. Moreover, an increase in platelet aggregational activity is observed in the organism after cessation of prostacyclin injections. There is a suggestion that prostacyclin is transformed in plasma into a more stable analog which accounts for the antiaggregating effect at distant stages [24]. According to another concept, more prolonged preservation of prostacyclin in plasma at incubation is associated with its formation of a complex with albumins [25]. Modification of prostacyclin's chemical structure has led to the creation of a more stable analog — carbacyclin [26]. However, when injected into the organism, the preparation is also rapidly inactivated. An increase in prostacyclin stability with preservation of its antiaggregating properties has been observed with ethylation and immobilization of β-cyclodextrin.

Prostacyclin and prostaglandins E_1 and E_2 have common receptors on human platelet plasmatic membrane, and prostaglandin D_2 interacts with other receptors [19]. Data show that prostacyclin inhibits the binding of fibrinogen by platelets. There is a certain specificity in the action of prostaglandins on platelet aggregation. For example, prostaglandins D_2 and I_2 have an insignificant effect on the aggregating ability of platelets in chicken cells, while prostaglandin E_1 and E_2 have a strong antiaggregating effect. Prostacyclin and other prostaglandins selectively affect thrombocytic hemostasis and do not influence the plasmatic factors of blood coagulation. The prostacyclin metabolite 6-keto-PGE_1 also possesses an antiaggregating effect.

Antiaggregational activity is also found in a number of synthetic derivatives of prostacyclin. This property is observed in 13,14-didehydro-20-methylprostacyclin, in (16S)-13, 14- didehydro-methylprostacyclin, and in 20-methylprosta-

cyclin. Antiaggregational activity is also found in 6-keto-
$PGF_{1\alpha}$, 6β-PGI_1, 5-iodo-PGI_1 methyl ester, (4R, 5R)-4-iodo-
deoxy-5,9α-epoxy-PGF_1 methyl ester, (4Z)-9-deoxy-5,9α-epoxy-
PGF_1 methyl ester and free acid, and in (5R)-9-deoxy-
5,9α-epoxy-PGF_1 methyl ester and free acid [27, 28, 29]. An
antiaggregating effect comparable to prostacyclin is manifested by 9-deoxy-9α,6-nitrilo-PGF_1 and 9-deoxy-9α,5-nitrilo-
PGF_1. In certain concentrations, 15-methyl-$PGF_{2\alpha}$ and $PGF_{2\alpha}$
inhibit aggregation.

The relationship between the chemical structure and
pharmacologic activity in PGH_1 derivatives is interesting to
consider: 11.9-epoxymethano- and 11,9-epoxycarbonyl-PGH_1 induced
aggregation and its 9,11-isomers inhibit it. 5,8,11,14-Eico-
satetraenoic and 4,5,8,11,14,17-eicosapentaenoic acids decrease
blood platelet aggregation. Prostaglandins D_3 and H_3 inhibit
platelet aggregation. Antiaggregating properties are
also found in 17S-methyl-ω-homo-6,9α-nitrilo-PGI_1, 16,18-
ethano-ω-homo-6,9α-nitrilo-PGI , 17S-methyl-ω-homo -5E,Z-6,9-
methano-PGI_2, and 17S-methyl-ω-homo-trans-Δ^2-PGE_1 [30].

New prostaglandin derivatives containing fluorine at different molecular positions have been studied in our laboratory
[11]. The placement of a fluorine atom at position 15 of
prostaglandins E_1, E_2, and I_2 leads to the creation of substances that possess antiaggregating activity similar to natural analogs. A fluoridated analog of prostacyclin has a
higher stability than PGI_2. A fluoridated derivative of 6-
keto-$PGF_{1\alpha}$ also has antiaggregating activity. An antiaggregating effect has been detected with 11-deoxy-PGE_1 and is
especially marked in its methyl ester. The ethyl ester does
not have an antiaggregating effect.

Platelet aggregation decreases under the effect of sym-
patholytic vasodilators. For example, reserpine and oktadin
decrease adhesiveness and aggregation [1, 31]. In contrast
to oktadin, reserpine increases aggregation caused by ADP and
thrombin. At the same time, aggregation with collagen was
somewhat decreased.

An antiaggregating effect is also found with α-adrenolyt-
ic drugs: dihydroergotamine, redergin, nicergoline, dihydro-
ergotoxin, phentolamine, and others [31, 32].

The β-adrenolytic agents bupranolol and propranolol decrease aggregation and adhesion; bupranolol has the highest

activity. The β-adrenomimetic drug isoproterenol also decreases platelet aggregation, while propranolol blocks its
effect. The β-adrenomimetic agent nonachlasine also inhibits
aggregation.

Among the neurotropic drugs, an antiaggregating effect
is found with some serotonin antagonists. In our laboratory,
we observed strong inhibition under the effect of diamine
and indocarb. Similar properties have been described for
the preparations BO-148, methysergide, and metergoline [31-
33]. Dihydroergotamine and the diethylamide of bromolyser-
gic acid also decrease aggregation, but they do not affect
serotonin and adrenalin content in platelets [34-36].

Antihistamine drugs also decrease aggregation. The
antihistamine preparations dimedrol (benadryl), suprastin
(chloropyramine), and diazoline possess an antiaggregating
effect both in vitro and in vivo. It has been found that
dimedrol inhibits oxidative phosphorylation in platelets.
However, the preparation does not produce an effect when
2,4-dinitrophenol is added. The antihistamine drug diphen-
hydramine decreases aggregation in vitro, but its effect is
significantly weaker in vivo. Burimamide also possesses an
antiaggregating effect. Methylamide and cyproheptadine also
decrease aggregation both in vivo and in vitro [37].

In comparing the force of action, it was found that di-
phenhydramine has a weaker effect than aminazin (chlorpro-
mazine hydrochloride), but its effect is stronger than that
of cyclozine (cyclazocin) and chlorpheniramine. A liberator
of mast cells, substance 48/80, decreases aggregation in
vitro. Fluothane also decreases aggregation.

The effect of anticoagulants on platelet aggregation
is of special interest. Many authors report a decrease in
aggregation when heparin is injected at high doses, although
others have not found a decrease in platelet aggregation or
adhesiveness with this anticoagulant. A comparative study
demonstrated that low doses of heparin decrease adhesiveness
and high doses increase adhesiveness. Some observations show
that heparin increases aggregation caused by ADP but inhibits
the effect of collagen and norepinephrine [38, 39]. Salzman
et al. [40] believe that the proaggregating effect of heparin
is manifested against a deficiency of antithrombin III. Heparin and dextran sulfate inhibit the effect of PGE_1 on platelets. These substances suppress the activity of adenylate

cyclase in platelets. Heparin activates thromboxane A_2 synthesis [41]. Most authors agree that short-term use of heparin and indirect-acting anticoagulants do not significantly affect platelet function. A decrease in adhesiveness and aggregation has been observed only with long-term application of the indirect-acting anticoagulants from 4-hydroxycoumarin and indandione derivatives.

In our experiments, we found that heparin injected intravenously into animals decreases platelet aggregation and adhesiveness only at comparatively high doses — 1000, 1500, or more units per kg of rabbit weight [35, 36]. The effect of heparin on platelet aggregation is enhanced in combination with citrate—phosphate—dextrose or acid—citrate—dextrose. There is a report on the inhibitory effect of the anticoagulant hirudin on platelet aggregation induced by thrombin.

We have also studied new anticoagulants with indirect action: napharine, phepromarone, nitropharine, etc. Their high anticoagulant activity has been shown previously [35, 36]. However, a decrease in platelet aggregation and adhesiveness could be achieved only at doses significantly exceeding the therapeutic doses, and with long-term use. The heparin antagonist protamine sulfate increased platelet aggregation and adhesiveness. A similar effect has also been observed with the synthetic heparin antagonist polybrene [42]. Toluidine blue, also a heparin antagonist, suppresses platelet aggregation.

An antiadhesive effect in vitro has also been produced by thrombodym, an anticoagulant from the group of rare earth elements. This effect, however, was manifested when the drug was used at comparatively high concentrations.

Some heparinoids also decrease platelet aggregation. Hemoclar, Sp-54, possesses this effect, and ancrod, an anticoagulant causing defibrination, also decreases aggregation [43]. A correlation between the inhibition of aggregation and the production of fibrinogen degradation products has also been found.

Antithrombin activity of amidinophenylpyruvic acid has been revealed. In this connection, attempts have been made to search for antiaggregating drugs among amidine and guanidine derivatives [44]. It has been found that 4-

Table 13.2.   Effect of Some Amidine Derivatives on Platelet
Aggregating Ability

Substance	Concen- tration (mg/ml)	Platelet aggrega- tion (in % of fall in optic density)
Control (ADP-aggregation)	–	54 ± 5
3-Chloro-4-ethoxyphenylacetic acid amidine hydrochloride	150	27 ± 3
3-Chloro-4-isopropoxyphenylacetic acid amidine hydrochloride	150	30 ± 3
3-Chloro-4-isobutoxyphenylacetic acid amidine hydrochloride	150	0
3-Chloro-4-amyloxyphenylacetic acid amidine hydrochloride	150	0

guanidinebenzoic acid esters inhibit aggregation induced by
thrombin.   Meta-compounds appear to be more active than para-
compounds.   An especially strong effect is observed with benzamidine derivatives.   There are also other inhibitors of aggregation from the groups of amidine and guanidine derivatives
[45].

In order to obtain a high antiaggregating activity from
amidine derivatives, we have demonstrated in our laboratory
that it is very important to have a structure containing an
amidine base and a side chain represented by a phenyl ring
with an aliphatic chain in the para-position (Table 13.2).
We found that maximal activity was possessed by compounds
containing isobutoxyl and amyloxyl groups in the para-position of the phenyl ring.   Placement of a chlorine atom in
the para-position of the phenyl ring decreased the antiaggregating activity of the compounds.   Replacement of the chlorine atom by a bromine atom in the third position of the phenyl
radical enhanced antiaggregating activity, while replacement
with a nitro group in this position decreased antiaggregating
activity.   Incorporation of substituents in the amidine group
deprived the compounds of an antiaggregating effect and, in
some cases, potentiated the effect of ADP [45].

In studying compounds containing heterocycles (piperidyl and morpholine) instead of a phenyl ring and aliphatic side chain, we found that these compounds enhanced the proaggregating effect of ADP. Replacement of the amidine group by a guanidine insignificantly decreased antiaggregating activity. Compounds containing two guanidine groups weakly suppressed aggregation induced by ADP. Incorporation of a sulfur atom in the aliphatic chain of these compounds led to an enhanced effect of ADP. Guanidine derivatives containing an aromatic side chain represented by isonicotinyl, benzimidazole, and hydroxybenzylpyrimidine manifested moderate antiaggregating activity, with the exception of N-guanylbenzamidine which enhanced ADP action. N-Benzyl-N-propionylguanidine was the most active inhibitor of aggregation among the guanidine derivatives. Among the phenylamidine derivatives, the strongest antiaggregating effect was observed in 3-chloro-4-isobutoxyphenylacetic acid amidine hydrochloride, 3-chloro-4-amyloxyphenylacetic acid amidine hydrochloride, and 3-bromo-4-methoxyphenylacetic acid amidine hydrochloride.

Some authors try to detect antiaggregating agents by modifying the chemical structure of the anticoagulant. For example, it has been proven that platelet aggregation and adhesiveness decrease under the effect of dicumarol, bismacetic acid derivatives, compound K-47, and phenprocoumon [46]. Another group of researchers has studied the activity of 11-benzylidene coumaranone derivatives [47]. It has been found that aggregation inhibitors are characterized by the presence and position of chlorine, methyl, and methoxyl groups. Replacement of hydrogen by a methoxyl group increases the effect, although a direct association between the chemical structure and function has not been established. An antiaggregating effect and inhibition of thromboxane synthesis has been observed with 8-monochloro-3-β-diethylaminoethyl-4-methyl-7-ethoxycarboxylmethoxycoumarin [48].

Silicon compounds represent an interesting class of antithrombotic drugs. A marked antiaggregating effect has been observed with 1-(2-ethylsulfinyl)-silatrane. The activity of the derivatives of this group depends directly on the sulfur electron state. Thus, the higher the sulfur electron density, the stronger the inhibitory activity. The ability to decrease platelet aggregation has also been found in 1-(2'-methylsulfinylethyl)-silatrane, (1-silatranylmethyl)-2-thiouranyl, (silatranylmethyl)-2-mercaptoquinoline, and other compounds [49]. We have also demonstrated in in vitro experiments that

a silicon derivative of heparin — trimethylsilylheparin —
possesses an anticoagulant effect and can decrease platelet
aggregation [50].

Some authors have studied the effect of fibrinolytic
drugs on platelet function and have shown that small doses
of fibrinolysis activators increase platelet aggregation [51,
52]. Lüscher and Davey [53] have reported that trypsin decreases platelet adherence by collagen.

Increased platelet aggregation has been found immediately after the addition of streptokinase to plasma containing
platelets (before fibrinolysis activation). However, decreased
aggregation has also been observed after fibrinolysis activation occurring with more prolonged incubation and a higher
concentration of streptokinase, e.g., 1000 units/ml and more.

Urokinase does not affect aggregation in vitro and, after
infusion in vivo, decreases platelet aggregation. Similar results have been obtained in both clinical and experimental
conditions. Many authors explain these data by the fact that
the high-molecular-weight products of fibrinogen formed during
fibrinolysis inhibit platelet aggregation. Stachurska et al.
[54] has observed a decrease in platelet aggregation when induced by ADP and thrombin under the effect of increased fibrinolysis and fibrinogen products. It has been noted that
the decay products of antihemophilic globulin do not effect
aggregation. Chymotrypsin also does not effect aggregation
in high concentrations of fibrinogen.

Enzymes from the Aspergillus genus possess the ability
to decrease platelet aggregation while simultaneously enhancing fibrinolytic activity [55]. It has been found that plasmin inhibits aggregation but does not suppress the release
reaction caused by thrombin.

Thrombolytic therapy requires the combination of fibrinolytic drugs with aggregation inhibitors. Mamedov et al. [56]
suggest a combination of terrilytin and nicotinic acid. Decreased platelet aggregating ability has also been observed
with D-glucitol-hexanicotinate.

Platelet aggregation can be decreased by compounds affecting the nervous system. An antiaggregating effect has
been observed with a number of phenothiazine derivatives, such
as aminazin, triftazin, etc. [1, 57]. The effect of aminazin

is evidently associated with the fact that it decreases calcium input in a cell. Flupenthixol, close to the phenothi-
azines in chemical structure, also increases aggregation. At
the same time, the α-isomer is significantly more active than
the β-isomer [58].

Inhibitors of platelet aggregation have also been detected among the antidepressants. This group includes tricyclic
antidepressants and monoamine oxidase (MAO) inhibitors. Among
the tricyclic antidepressants, the highest activity is observed with imizin. It should be noted that imizin has a high
affinity for blood platelets. Among the MAO inhibitors, ni-
alamide and iprazid possess antiaggregating effects. However,
it is not clear whether the action of these substances is associated with the accumulation of biogenic amines or whether
there is another mechanism of action [1, 31].

Platelet aggregation is decreased under the effect of
ethanol. The effect of the alcohol derivatives 2-phenyleth-
anol, 3-phenylethanol, 4-phenylbutylamine, and phenylacetic
and 4-phenylbutyric acids on aggregation has been studied
[59], and it has been found that lengthening of the aliphatic
chain enhances the antiaggregating effect which is higher in
amines than in alcohols, and higher in alcohols than in acids.
The in vitro aggregating ability of blood platelets decreases
under the effect of phenobarbital. Decreased aggregation induced by ADP is also observed under the effect of lithium
ions. However, in low concentrations, lithium stimulates aggregation [60]. Decreased platelet aggregating ability is
also caused by nootropil [61].

Local anesthetics also inhibit aggregation. Dibucaine
hydrochloride, lidocaine hydrochloride, and novocaine possess
antiaggregating effects [62]. However, the most marked effect
has been observed with cocaine.

Platelet aggregation is inhibited by many butyric acid
derivatives, e.g., GABA, ethylchlorophenoxyisobutyrate, ethyl-
chloroisobutyrate, and β-benzoylbutyrate.

Hypocholesterolemic preparations contain a large group
of aggregation inhibitors. Clofibrate possesses a high antiaggregating effect [2, 63]. This preparation inhibits aggregation caused by ADP, thrombin, collagen, and adrenaline, and
the release of serotonin and β-glucuronidase from platelets.
Clofibrate suppresses the incorporation of labeled acetate

(but not glucose) into phospholipids, glycerides, and free
fatty acids.

Halofenate, another hypolipidemic preparation, also decreases platelet aggregation and is more active than clofibrate. Parmidin possesses antiaggregating action [64]. This
preparation more actively inhibits aggregation associated
with adrenalin and serotonin than with ADP. Hypocholesterolemic drugs from the group of nonsaturated fatty acids also
decrease aggregation. A Japanese preparation for treatment
of atherosclerosis — S-8527 — also has antiaggregating activity. This property is also observed in a new hypolipidemic
drug — 2[3-(2-thiazolylthio)-phenyl]propionic acid.

Marked antiaggregating action is found in nicotinic acid
[65]. Because of the antiaggregating action of nicotinic
acid, a similar activity has been looked for among its derivatives. Antiaggregating properties have been found in tetra-
nicotinoylfructofuranose and clonixin. Niflumic (trifluoro-
methyl-3-phenylalanine-2-nicotinic) acid decreases aggregation of both platelets and erythrocytes.

Many vitamins and their derivatives possess antiaggregating action (Table 13.3). This effect is found in phytin,
calcium pantothenate, pyridoxal phosphate, choline, ascorbic
acid, potassium orotate, cocarboxylase, mononucleotide ribo-
flavin, cyanocobalamin, calcium pangamate (calgam), rutin,
linaetol, tocopherol, vikasol, and folic acid [66, 67].

Study of the effect of hormonal preparations on platelets is of great practical importance. An antiaggregating
effect has been observed with androsterone in combination
with ethylchlorophenoxyisobutyrate.

Aggregation also changes under the effect of antidiabet-
ic drugs. A strong antiaggregating action has been found
with phenylbiguanide and with the sulfonamide derivative gly-
closide [34]. The latter suppresses the release reaction
caused by thrombin and collagen, and activates adenylate cyclase. Aggregation caused by ADP is decreased by insulin
[66].

Data also indicate that melatonin is able to suppress
aggregation. An in vivo antiaggregating effect has been observed with somatostatin. This effect, however, has not been
manifested in vitro.

Table 13.3. Effect of Some Vitamins on Platelet Aggregating
            Ability

Substance	Concen-tration (mg/ml)	Platelet aggrega-tion (in % of fall in optic density)
Control (ADP aggregation)	—	55 ± 6
Phytin	0.07	19 ± 2
Nicotinic acid	0.07	33 ± 4
Pyridoxal phosphate	0.01	25 ± 2
Choline chloride	0.4	10 ± 1

Savelieva [68] has found that retabolil inhibits platelet aggregation and increases the level of free heparin. Cor-
tisone decreases platelet adhesiveness in 7-10 days after injection. A similar effect has been observed with predniso-
lone and methylprednisolone. Dexamethasone and triamcinolone
do not affect aggregation.

Inhibition of aggregation has been observed with estradiol. After ovariectomy in rats, 17β-estradiol decreases aggregation. Injection of this substance into male rats after
castration leads to an increase in aggregation [69].

Many authors report decreased aggregation with enzymatic
preparations. Aggregation caused by epinephrine, collagen,
and thrombin is inhibited by phospholipase A_2 [70]. But data
also indicate that inhibitors of phospholipase A_2 — bromophenacyl, mepacrine hydrochloride, and others — prevent aggregation [71].

Some enzyme-substrate systems such as phosphoenolpyruvate—pyruvate kinase or creatine phosphate—creatine phosphate
kinase also possess antiaggregating effects [72].

The action of metabolic poisons is of great interest in
understanding the mechanism of platelet aggregation. It has
been found that platelet aggregation is inhibited by deoxy-
glucose, monoiodoacetate, cyanides, mercury compounds, and
others [73].

Fluoride possesses a peculiar action on platelets. Small concentrations do not affect aggregation and adhesion, but do decrease the ADP content in platelets. In high concentrations, fluoride blocks aggregation. Small concentrations of molecular iodine suppress aggregation; high concentrations activate aggregation [74]. Aggregation is suppressed by the cellular poison melfalan [75]. At the same time, this poison has little effect on platelet adhesion to glass. Hydroxyurea has no effect on aggregation, while high concentrations of urea inhibit aggregation and low concentrations activate it. N'-Morpholinomethylurea derivatives decrease aggregation [76].

The inhibitor of nucleic acid synthesis, azathioprine, suppresses platelet aggregation but does not affect adhesiveness. This action may be associated with the fact that azathioprine, in the organism, transforms into thioimidazole which suppresses aggregation.

An antitumoral preparation, dianhydrogalactitol, possesses antiaggregating action [77].

Decreased aggregation is also observed with the mitotic poisons colchicine and vinblastine [78].

Takano [79] found that low concentrations of vinblastine activate aggregation. Vincristine does not affect adhesiveness of blood platelets. However, it should be noted that some authors explain the decreased tolerance to aggregating agents, which occurs with injection of vincristine, by the effect on megakaryocytes [80].

Another inhibitor of tubulin synthesis, nocodazol, also suppresses aggregation.

Aggregational ability of blood platelets is decreased by other cytostatics, in particular by mitramycin [81].

The interferonogenase stimulator tilorone also decreases platelet adhesiveness.

Nonsteroidal anti-inflammatory drugs can be related to antiaggregational preparations which have been used clinically. Acetylsalicylic acid, indomethacin, phenylbutazone (butadion), and sulfinpyrazone are the most widely used [78, 82, 83].

The antithrombotic effects of acetylsalicylic acid, suprifen, indomethacin, and 5-methoxy-2-methyl-3-indoleacetic acid has been clearly shown in experimental models [84]. In clinical application acetylsalicylic acid potentiates the effect of heparin and the coumarin anticoagulants. At the same time, data indicate that acetylsalicylic acid is inferior to butadion in the force of antiaggregation [85].

Based on the effects of acetylsalicylic acid, new inhibitors of aggregation are being investigated. A good effect has been observed with colphyrine, aspirin-isopropylantipyrine, and sulindac [86]. The positive property of salicylates as antiaggregating agents lies in their enhancement of the action of dipyridamole, which is increased in the blood with combined injection. It has been found that salicylates suppress the activity of thrombocytic factors 3 and 4 [87]. Also, it has been noted that the effect of acetylsalicylic acid decreases when other salicylates are injected into the organism.

Antiaggregational action has also been observed with hydroxyphenbutazone, ibuprofen, flubiprofen, dipyrone, naproxen, compounds of AHR-5850 and AHR-6293, indobufen, tromal, alclofenac, ketoprofen, and 1-sulfophenyl-3-carboxy-5-hydroxypyrazole [88, 89].

We have conducted studies in our laboratory on the effect of a number of compounds related to the group of 3,5-dioxopyrazolidine derivatives. The strongest antiaggregating effect among these compounds was found with the sodium chloride of 1,2-diphenyl-4-[N-sulfophenyl]-hydrazono-3,5-dioxopyrazolidine, and with the sodium chloride of enol 1,2,4-tri-[N-sulfophenyl]-3,5-dioxopyrazolidine [90]. Tetraphenyl derivatives of 3,5-dioxopyrazolidine and diphenyl derivatives of diethylaminoethylpyrazolidine did not affect the aggregating ability of platelets.

We have also detected an antiaggregating effect with the new derivatives of anthranilic acid which activate fibrinolysis. It was found that flufenamic acid possesses higher antiaggregating activity than mefenamic acid. N-(4-difluoromethoxyphenyl)anthranilic acid also decreases aggregation, and its effect is stronger than the action of N-(3-difluoromethoxyphenyl)anthranilic acid. At the same time, N-(3-difluoromethylthiophenyl)anthranilic acid and N-(4-difluoromethylthiophenyl)anthranilic acid manifested similar anti-

aggregating activity. The lowest antiaggregating activity
was observed with N-(4-methoxyphenyl)-5-nitroanthranilic acid
(Table 13.4).

Among other compounds, the antiaggregational action of
chemotherapeutic preparations should be noted. This property
is characteristic of antimalarial drugs — quinine, quinidine,
chloroquine, hydroxychloroquine, proguanil hydrochloride [1],
and benzyl alcohol and phenol.

Nitrofuran derivatives also decrease platelet aggregation. A significant effect in this group has been observed
with nitrofurantoin [91]. 5-Nitro-2-furaldehyde dipropionate,
5-nitro-2-furaldehyde dibutyrate, and 5-nitro-2-furaldehyde
acetate are inactive. Adhesiveness and aggregation of blood
platelets also decrease under the effect of ditazol, 4-[4-
bromophenyl]-2-methyl-5-nitroxazole, and 6-methyl-2,3,4,5-
tetrahydroimidazo-[2,1]quinazolin-2-one monohydrate hydrochloride [92].

A similar effect is also caused by 6-[p-(4-phenylacetyl-
piperazin-1-yl)-phenyl]-4,5-dihydro-3(2H)-pyridazinone [93].

Inhibition of platelet aggregation is observed under
the effect of penicillin G and similar antibiotics — carben-
icillin, ticarcillin, oxacillin, tetracycline, and others [94].

Among diuretics, platelet aggregation is affected by
furosemide which suppresses aggregation and the release reaction. Another diuretic, spironolactone, has similar action
[95]. Analysis of antiaggregational activity of naphthal-
alkyllactam amides shows that the highest antiaggregating
activity is possessed by substances in which the substituent
radical is located on the carbon adjacent to the lactam amide
group. Low-molecular-weight dextrans, β-diethylaminoethyl-
diphenyl-propylacetate, 2-(p-chlorophenyl)-4-thiazole acrylic
acid, polyphloretine phosphate, and cephalothin [96, 97] are
notable among other compounds that decrease aggregation.

The search for antiaggregating preparations is actively
being conducted among substances that are close in structure
to aggregation inducers. An ability to inhibit platelet aggregation has been found in adenosine, AMP, and adenosine triphosphate (ATP) [98]. Among adenosine derivatives, the highest antiaggregating action has been observed with 2-chloro-
adenosine, 2-bromoadenosine, 2-azoadenosine, and 2-fluoroadeno-

Table 13.4. Effect of Some Anthranilic Acid Derivatives on
            Platelet Aggregating Ability

Substance	Concentration (mg/ml)	Platelet aggregating ability (in % of fall in optic density)
Control (ADP-aggregation)	—	71
N-(2,3-dimethylphenyl) anthranilic acid	2	16 ± 1
N-(3-trifluoromethylphenyl) anthranilic acid	2	9 ± 1
N-(4-difluoromethylthiophenyl) anthranilic acid	2	13 ± 1
N-(3-difluoromethoxyphenyl) anthranilic acid	2	26 ± 3
N-(4-difluoromethoxyphenyl) anthranilic acid	2	8 ± 1
N-(4-methoxyphenyl)-5-nitro- anthranilic acid	2	47 ± 2

sine. Aggregation is also decreased by 2-hydroxy-6-amino-purine-riboside, nicotinamide adenine dinucleotide (NAD), 6-methyl-aminopurine riboside, PAD, and others. Adenosine-1-N-oxide, 2-methylthioadenosine, 2-deoxyadenosine, and AMP have significantly lower effects. The antiaggregating action of these substances is blocked by 5-methyl-thioadenosine and 2,5-di-deoxyadenosine [99].

The metabolism of some adenosine analogs has been studied in platelets. It has been shown that fluoroadenosine mainly transforms into fluoro-ATP, which replaces ATP. By force of action, fluoroadenosine and N-phenyladenosine are inferior to 2-deoxyadenosine and arabinosyladenosine.

Cusack and Hourani [100] have found an antiaggregating effect with 5-N-ethylcarboxyaminoadenosine.

Polylysine derivatives also possess antiaggregating action. The effect of collagen is inhibited by a 1-amino-1-

Table 13.5. Effect of Some Aggregation Inhibitors on the Formation of Experimental Micro-thrombi Under Conditions of Anodic Stimulation (100 V)

Substance	Dose (mg/kg)	Time of onset (s) of primary thrombotic masses	Time of onset (s) of thrombus decay	Mean thrombus area (M^2)
Control	—	8.0 ± 0.9	7.0 ± 1.8	3430 ± 197
Amidine hydrochloride-3-chloro-4-isopropoxyphenyl-acetic acid	11	199 ± 13	1.3 ± 0.3 $p < 0.01$	293 ± 88
1,2-Diphenyl-4-p-sulfophenyl-hydrazono-3,5-dioxopyrazol-idine sodium chloride	13	over 300	$p < 0.01$	
8-Diethylaminomethyl-3,7-dimethylxanthine		over 300	$p < 0.01$	

Table 13.6. Effect of Certain Aggregation Inhibitors on the Amounts of Platelets when Disseminated Intravascular Coagulation (DIC) is Induced by Thrombin and Pituitrin

Substance	Dose (mg/kg)	Platelet amount (thousands/μliter)
Control I: healthy animals	—	552 ± 19
Control II: DIC syndrome	—	286 ± 65
Papaverine	2	514 ± 7
Amidine hydrochloride-3,4-iso-propoxyphenylacetic acid	11	420 ± 11 $p < 0.01$
1,2-Diphenyl-p-sulfophenyl-hydrazono-3,5-dihydroxypyrazol-idine sodium chloride	13	502 ± 25 $p < 0.01$

deoxy-2-ketopolylysine derivative, L-arginine, and a nonapeptide which is a collagen fragment [101, 102]. Decreased platelet aggregation is also observed under the effect of hypertonic solutions [103]. A similar effect has been found in sulfonates and carboxylates [104], α-1-acid glycoprotein [105], the oxidizer of the SH-group —diamine, N-acetylneuraminic acid [106], some carbamoylpiperidine derivatives [107], thiazole [108], some cytochrome P-450 inhibitors [109], and pyrrole and pyrrolidine derivatives [110].

Decreased aggregation also occurs under the effect of bile acids.

Kodairo et al. [111] isolated some aggregation inhibitors from Fraxinus japonica Blume; these were 3-methoxy-4-hydroxyphenylethanol, 2,6-dimethoxy-p-benzoquinone, and p-hydroxyphenylethanol.

In studying the antiaggregating properties of different chemical compounds, it is not sufficient, in our view, to study only the effect on aggregation in vitro and with injection into intact animals. Study of the effect in an experimental

model of thrombosis is of great importance. For projecting
the effectiveness of any substance in thrombolytic therapy
in our laboratory, we have used a model of thrombocytic micro-
thrombosis of the mesentery vessels, obtained by stimulation
of the vessel wall by 100 V of current for 20 μsec [112]. In
this model, the time of appearance of a white thrombus, the
onset of its degeneration, and the mean area of the thrombus
can serve as indices of the degree of thrombogenesis. Table
13.5 presents the results of our studies on the effect of
some antiaggregational drugs on the formation of experimental
thrombocytic microthrombus. These antiaggregational drugs
retard the formation of the microthrombus, activate its degradation, and decrease the mean area of the thrombus.

The ability to prevent platelet uptake is important in
projecting the effectiveness of preparations in cases of disseminated intravascular coagulation (DIC) [76]. Table 13.6
presents the action of certain new aggregation inhibitors
when the DIC syndrome was induced by injection of thrombin
(20 units/kg) and pituitrin (0.6 unit/kg). The aggregation
inhibitors in this study decreased platelet uptake with disseminated intravascular coagulation.

Different opinions are reported in the literature on
the nature of platelet aggregation and the mechanism of action
of antiaggregating preparations. A number of authors pay attention to the force of electrical resistance in platelet interaction [113, 114]. This resistance is confirmed by the
fact that ADP decreases and that certain antiaggregating
agents increase the force of this interaction [115]. Electrostatic interaction obviously plays an important role in
the aggregation of other cells. Thus, a number of the features
of aggregation and adhesion of cells can be explained by the
stability of colloid systems. This is confirmed by the fact
that polycations, for example polylysine, are able to activate platelet aggregation [116]. Treatment with neuramini-
dase also increases the aggregational and adhesive properties
of platelets and other cells, which is associated with the
removal of the negative charge of the plasmatic membrane.
Sialic acid also possesses antiaggregating action [117].

A number of hypotheses suggest that the formation of a
complex of ADP, calcium ions, von Willebrand factor, fibrinogen, and platelets is necessary for platelet aggregation
[118].

Inada et al. [119] pay great attention to the thrombocytic clumping of such proteins as albumin and aggrexons A and B.

Blomback and Blomback [120] believe that ADP can open receptor sites for fibrinogen contact with platelet membrane by disulfide links.

Hovig [121] has hypothesized that the mechanism of many aggregation inducers is associated with the release of ADP. This, however, is evidently not true for collagen since substances which block ADP aggregation do not affect aggregation caused by collagen [122].

A number of investigators attach importance to changes in the phospholipid content of thrombocytic membranes. It has been observed that some glycerophospholipids, in particular phosphatidylethanolamine, promote aggregation.

Valles et al. [123] have found that antiaggregating agents affect the membrane content of phospholipids. Acetylsalicylic acid decreases sphingomyelin content and increases phosphatidylcholine. It is also known that the antiaggregating action of bromophenacyl bromide and certain other preparations is accompanied by decreased degradation of phosphatidylcholine and phosphatidylinositol, and increased synthesis of phosphatidylethanolamine [124]. Aggregation is activated by cardiolipin and phosphatidic acid.

Glycoproteins, which are receptors of certain aggregation inducers, obviously play an important role in aggregation. The absence of different glycoproteins on the surface is, possibly, a leading pathogenic factor in Glanzmann's thrombasthenia, Bernard-Soulier syndrome, and other disorders. One should also not exclude the evidence that glycoproteins form interthrombocytic contacts [125].

A certain role in platelet aggregation belongs to the release of factor 4 on the surface, which interacts with fibrin soluble monomer. Factor 4 has two active centers, one of which interacts with heparin and the other which interacts with fibrin monomer.

In incubation with aggregation inducers, blood platelets can enter a refractory state accompanied by a decrease in their aggregational ability. Holme and Holmsen [126] explain this state by the blocking of external parts of the membrane due to their link with the inducer.

Taking into account a certain similarity in the metabolism of thrombocytes and muscles, some authors believe that the main role in initiating aggregation is played by an increase in the concentration of thrombocytes in ionized calcium [127-130]. The causes for the increase can differ. It can be associated either with the penetration of ions from the extracellular space or with the release of contact with some other intercellular structures. We believe that the latter reason makes more sense given biological expediency since, under these conditions, a cell always has a component necessary for aggregational reaction.

The proaggregating effect of ionophores such as A 23187 and others is associated with increased concentration of calcium ions inside the cells. Inhibitors of calcium permeability decrease aggregation. Also, the antiaggregational effect of local anesthetic and antiarrhythmic preparations, and derivatives of phenothiazine and other drugs, is connected with a disturbance of calcium intake in platelets [131].

Studies by Sneddon [132] and Miller et al. [133] demonstrate that thrombin induces aggregation in the absence of extracellular calcium, which can indirectly confirm the importance of intercellular calcium release for the organism. The calmodulin inhibitor N-(6-aminohexyl)-5-chloro-1-naphthalene sulfonamide inhibits platelet aggregation [134].

The level of cyclic nucleotides in platelets and the ratio of cGMP play a great role in regulating platelet aggregational ability. Substances that promote the accumulation of cAMP suppress aggregation. Prostaglandins E_1, D_2, and I_2, and adenosine, isoprenaline, activated arachidonic acid, epinephrine, norepinephrine, serotonin, and others inhibit adenylate cyclase [135].

Neuroleptic and vasodilatory drugs with myotropic action suppress cAMP phosphodiesterase. Also, the force of the anti-aggregating effect of these substances is proportional to the increase in cAMP content [136].

A decrease in cAMP concentration is noted under the effect of many aggregation inducers [137]. Some of these — ADP, serotonin, and collagen — increase cGMP content in plate-

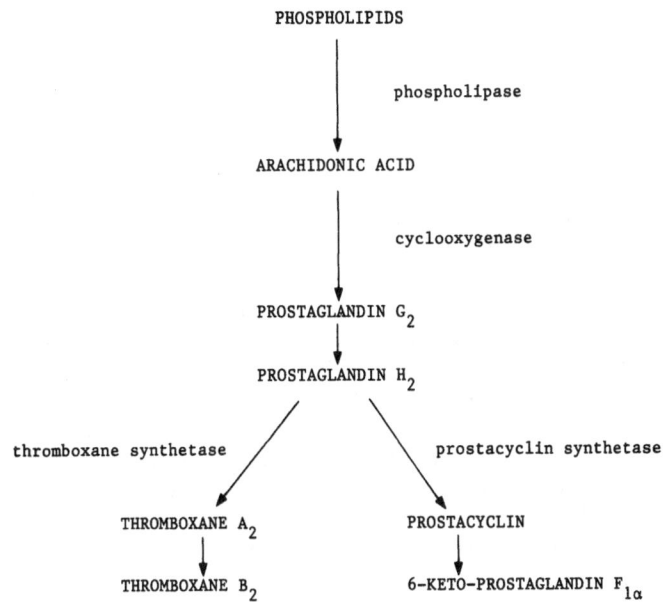

Fig. 13.1.   Thromboxane and prostacyclin synthesis.

lets.   Salicylates, however, which suppress aggregation, do not affect cGMP content.

Experiments by Nikulin [138], in our laboratory, demonstrate that the addition of cGMP to platelet-rich plasma causes insignificant platelet aggregation, but decreases the proaggregating action of ADP.

In recent years, the aggregating ability of platelets has been associated with a change in prostaglandin metabolism. This association was first reported by Silver et al. [139] and by Vargaftig and Zirinis [140], who demonstrated that arachidonic acid induces aggregation.   In platelets, arachidonic acid is first transformed into prostaglandin G_2 under the effect of cyclooxygenase (Fig. 13.1).   It then transforms into prostaglandin H_2, which, under the effect of thromboxane synthetase, is transformed into thromboxane A_2 and later metabolized into thromboxane B_2.   Prostaglandins G_2 and H_2 and thromboxane A_2 possess a higher proaggregating activity than arachidonic acid.   Thromboxane A_2 is the most powerful inducer of aggregation.

Pickett et al. [141] have observed increased concentration of thromboxane A_2 in platelet aggregation induced by
ADP, thrombin, and A 23187.

The antiaggregating action of vitamin E, butylated hy-
droxyanisole, and other antioxidants is accounted for by the
inhibition of autooxidation [142]. In this respect, a large
number of compounds have been studied that permit a change
in the antithrombotic activity of vessel wall. Acetylsalicylic acid decreases the aggregating ability of platelets
and simultaneously suppresses the antiaggregating and antiadhesive activity of vessel wall [143-146]. This effect is associated with the fact that this substance blocks cyclooxygenase activity in platelets and the vessel wall and disturbs
the synthesis of prostaglandins G_2 and H_2 which are precursors
of both thromboxane A_2 and prostacyclin. There is a dose-dependent relationship between acetylsalicylic acid and the
degree of inhibition of platelet aggregation and synthesis
of vessel wall prostacyclin: Low doses (3.5 mg/kg of body
weight) suppress thrombocytic cyclooxygenase and, therefore,
platelet aggregation, and insignificantly inhibit synthesis
of prostacyclin. Higher doses (5-10 mg/kg of body weight and
higher) insignificantly increase the antiaggregating effect,
but lead to complete inhibition of the synthesis of prostacyclin. There is also a time-dependent relationship in the
effect of acetylsalicylic acid on cyclooxygenase activity in
platelets and vessel wall: Inhibition of platelet cyclooxygenase activity and platelet aggregation after a single dose
of the preparation (4 mg/kg) is maintained for 4-5 days, and
inhibition of prostacyclin synthesis is maintained for about
24 h [143, 144]. The suppressive effect of acetylsalicylic
acid on the biosynthesis of prostacyclin has a different duration in the arteries and veins.

Similar to nonnarcotic analgesics, the synthesis of prostacyclin and thromboxane A_2 is disturbed by hydrocortisone.
It has been demonstrated in in vitro experiments that synthesis of prostacyclin in vessel wall is decreased with certain concentrations of glucose. At the same time, nicotinic
acid, complamine, and vitamin E, which have an antiaggregating
effect, do not disturb the synthesis of prostacyclin since
they block the synthesis of thromboxane A_2 and do not effect
cyclooxygenase activity [56, 146].

A number of antiaggregating and vasodilatory drugs stimulate synthesis of prostacyclin. This ability is possessed

by dipyridamole, trental, nicotinic acid, papaverine, cloni-
dine, propranolol, dihydrolazine-2-nicotinamidoethylnitrate,
nitroprussin, OKU-1581, nitroglycerin, trapidil, dihydroergotamine, and other agents [32, 146, 147].

We have demonstrated that nicotinic acid and complamine
not only inhibit platelet aggregation without suppressing
prostacyclin synthesis in the vessel wall, but also can prevent a decrease of prostacyclin synthesis in extreme states
and restore the antiaggregating properties of the vessel wall.
Nicotinic acid, as an antiaggregating agent, has some advantage over acetylsalicylic acid.

In order to decrease the unfavorable effect of acetyl-
salicyclic acid on vessel endothelium and to increase its
antiaggregating activity, the preparation is used in combination with dipyridamole. Theophylline does not affect the
synthesis of prostacyclin.

Imidazole, ascorbic acid, estradiol, angiotensins I and
II, and bradykinin are notable among other substances that
are able to increase the synthesis of prostacyclin [13]. It
should be noted that an inhibitor of protein synthesis, cyclohexamide, removes this effect of estradiol. Contraceptive
drugs containing a combination of estrogens and progesterones
activate the release of prostacyclin. This release increases
when the adrenergic nerves are stimulated under the effect of
sodium ions, ionophore A 23187, trypsin, or arachidonic acid.

Propranolol, pindolol, lidocaine, and aminazin do not
influence the pressor effects of angiotensin II, but do remove
its action on the synthesis of prostacyclin. Phentolamine,
atropine, and practolol do not affect the increase in prostacyclin generation caused by angiotensin II.

Inhibitors of thromboxane synthetase UK-37 and 248-01
do not affect synthesis of prostacyclin in the lungs but do
activate synthesis in the kidneys. Excretion of prostacyclin
from the uterus increases with oxytocin, ergometrine, carba-
chol, serotonin, and prostaglandin $F_{2\alpha}$, and decreases with
mepacrine hydrochloride, atropine, histamine, and relaxin.

As Radomski et al. [148] have shown, cholinomimetic preparations (acetylcholine, methacholine, and others) increase
the release of prostacyclin into the blood flow, while atropine decreases this effect.

The ability to activate synthesis and release of prostacyclin has also been observed with β-adrenergic blocking
agents (practolol, pindolol).

It has been found that prostacyclin synthesis in vessels
increases with such substances as HG-626, AL-122, MK-447, and
different thiazide derivatives [149], and decreases with
tryptophan.

Among other factors, the synthesis of prostacyclin is
affected by lipoproteins. Low-density lipoproteins stimulate
synthesis and high-density lipoproteins suppress synthesis.

Heparin neutralizes the antiaggregating effect of prostacyclin.

The coumarin derivative AD_6 increases synthesis of prostacyclin [150].

Alimentary factors also play an important role in prostacyclin synthesis. Beitz et al. [151] have found, in rats
fed a diet deficient in linoleic acid, that fatty acids,
phospholipids, and cholesterol esters isolated from their kidneys inhibit prostacyclin synthesis. When these substances
are isolated from the kidneys of rats fed linoleic acid, synthesis of prostacyclin is stimulated. Cholesterol ester with
a high concentration of oleic acid suppresses prostacyclin-
synthetase activity to a higher degree than cholesterol with
a high level of linoleic acid [151].

The effect of hemostatic preparations on prostacyclin
synthesis has not been studied sufficiently. In some vessels
thrombin stimulates synthesis and in other vessels it does
not.

Prostacyclin-synthetase activity decreases under the
effect of 15-hydroperoxyarachidonic acid. According to Nakao
et al. [152], testosterone decreases synthesis of prostacyclin
in the vessel wall.

The above data clearly show that it is necessary to
search for new inhibitors of platelet aggregation taking into
account the effect of the substance on the generation of prostacyclin.

The ability of substances to prevent the action of different agents on prostacyclin synthesis in the vessel wall
is very important in the treatment of thromboembolic diseases.
We have found in our laboratory that nicotinic acid, complamine, azethylnicotinate, nicomorpholin, isoptin, instenon, and
others prevent the suppressive effect of adrenalin on prostacyclin synthesis [146, 153]. In many laboratories, new antiaggregating preparations are being sought from among substances
that selectively inhibit thromboxane synthetase. Among these
substances, an antiaggregating effect has been found with
nictindol, benzydamine, 3-phenyl-5-[2,2,2-trifluoro-1-hydroxy-
1-(trifluoromethyl)-ethyl]indole-2-carbonitrile, 9,11-azo-13-
oxa-15-hydroxyprostanoic acid, (8R, 9S, 11R, 12S)-9α-homo-9,
1-epoxy-(5Z, 13E)-15-(3-phenyl-2-propenyl)-1H-imidazole, 4-
[2-1H-imidazole-1-ethoxy]hydrochloride benzoic acid, β-[4-(2-
carboxyl-1-propenyl) benzyl] hydrochloride pyridine, 2-ethyl-
pyridine, 5-ethyl-2-methylpyridine, diazepam, hydralazine,
9,11-azo-15-hydroprosta-5,13-dienic acid, 2-isopropyl-3-
nicotinylindole, sodium-p-benzyl-4-[1-oxo-2-(4-chlorobenzyl)-
3-phenylpropyl]phenyl phosphonate, 1-alkylimidazole derivatives, and a number of other inhibitors of this enzyme [154,
155].

The expediency of screening such substances lies in the
fact that, in order to obtain an antiaggregating effect, it
is necessary to suppress synthesis of thromboxane A_2 without
disturbing synthesis of prostacyclin in the vascular endothelium. Such antiaggregating agents as nicotinic acid and
its derivatives, and 9,11-aza-prosta-5,13-dienic acid, suppress thromboxane synthetase activity [156].

Imidazole derivatives comprise an interesting group of
thromboxane synthetase inhibitors and are potentially active
antiaggregating substances. According to Moncada et al.
[157], imidazole does not effect aggregation in inhibiting
thromboxane synthetase activity. These authors associate
this finding with the ability of imidazole to activate cAMP
phosphodiesterase. Although imidazole itself does not effect aggregation, it is possible to create the antiaggregant
KC-6141 and other compounds based on the chemical structure
of imidazole [158]. Anagrelide (6,7-dichloro-1,5-dihydro-
imidazo[2,1-b]-quinazolin-2(3H)-one monohydrochloride) has
turned out to be a strong inhibitor of aggregation. Its action has been manifested in both in vivo and in vitro experiments, and the substance has been found to be an active antithrombotic agent. Thromboxane synthesis is also inhibited

340 K. M. LAKIN ET AL.

by dipyridamole. A number of compounds, including 13-aza-prostanoic acid, possess the ability to block thromboxane A_2 receptors.

The antiaggregating effect of local anesthetic drugs — aminosyn, propranolol, clofibrate, cAMP, and others — is associated with the blockade of phospholipase A_2 and, thus, with a disturbance in prostaglandin synthesis [159].

Malmsten et al. [160] have found that cAMP suppresses platelet cyclooxygenase activity.

There is a relationship between the aggregating ability of platelets and the excitability of their α-adrenoreceptors [32]. Norepinephrine stimulates aggregation, while α-adreno-blockers suppress it [78, 161]. The effect of some compounds can be associated with a change in the permeability of thrombo-cytic membranes. An increase in permeability leads to the release from blood platelets of aggregation inducers and acti-vators of plasmatic hemostasis. Different phospholipases and proteolytic enzymes can activate aggregation in this way.

Study of the interaction between the aggregating abil-ity and energy metabolism of blood platelets is very impor-tant in developing a strategy for finding new inducers and inhibitors of aggregation. This problem has not been suffi-ciently addressed, although it has been shown that the ratio of ATP to ADP on a membrane is very significant for stability of the platelet suspension [34].

Studies by Mant [162] demonstrate that platelet aggrega-tion requires an expenditure of energy.

Bettrex-Galland and Lüscher [163] report that respira-tion is suppressed in blood platelet aggregation. At the same time, Hussain and Newcomb [164] have observed that res-piration is enhanced under the effect of physiological in-ducers of aggregation.

Data indicate that calcium ions induce aggregation and block glycolysis [165]. The anticoagulant EDTA inhibits ag-gregation and increases accumulation of pyruvate and lactate. The aggregation inducer ristocetin decreases the level of ATP in a cell.

At the same time, the blocking of glycolytic enzymes by chemical agents and a decrease in platelet glycolytic activity in different diseases are accompanied by a decrease in aggregating ability. ATP is known to be necessary for clot retraction [163] and, in particular, for thrombostenin contraction [166]. Many authors consider thrombostenin to be very important in the aggregation process. Odesskaya [166] thinks that suppression of thrombostenin M activity is the decisive factor in aggregation. Also, Booyse and Rafelson [167] comment on the necessity of thrombostenin relaxation for induction of aggregation of ADP and adrenalin. In addition, Puszkin et al. [168] report that ADP can increase thrombostenin activity.

Kirkpatrick et al. [169] think that platelet aggregation is independent of the contractile mechanism since cytochalasin B, a contractile protein of platelets, does not effect aggregation.

There are no convincing data in the literature to explain why both a deficit and an excess of ATP lead to decreased platelet aggregation — once again demonstrating the incompleteness of our understanding of platelet aggregation. Some authors explain the proaggregating action of ATP by the fact that it transforms into ADP.

Some authors hypothesize that a change in the water structure plays a role in platelet aggregation [170]. This hypothesis is very interesting; however, it has not been developed further in subsequent studies. At the same time, our studies demonstrate that changes in water structure can serve as a quantitative criterion for determining the concentration of formed elements in the blood and for evaluating the degree of platelet aggregation.

Some reports address the special nature of collagen-induced aggregation. Czenave et al. [171] associate this with interaction of the collagen glycosyl transferase located on the platelet membrane and collagen incomplete carbohydrate chains.

Until now, it has been supposed that the mechanism of action of many antiaggregating preparations has an extrathrombocytic genesis. Galactosidase, periiodate, dithiothreitol, p-hydroxymercuribenzoic acid, 2,4,6-trinitrobenzenesulfonic acid, and phospholipase C evidently influence the activity of von Willebrand factor and antihemophilic globulin.

Table 13.7. Effect of Some Inducers and Inhibitors of Aggregation on Pyruvate Kinase Activity and ATP Content in Platelets

Substance	Concentration (per ml)	Pyruvate kinase activity (μmole/min/mg protein)	ATP content (μmole/ml platelet suspension)
Control		0.41 ± 0.02	4.43 ± 0.02
Thrombin	0.4 unit	0.35 ± 0.02 $p < 0.05$	3.95 ± 0.12 $p < 0.05$
ADP	200 μg	0.33 ± 0.03 $p < 0.01$	3.99 ± 0.13 $p < 0.05$
cAMP	630 μg	0.48 ± 0.01 $p < 0.05$	4.93 ± 0.18 $p < 0.05$

Table 13.8. Effect of Some Inducers and Inhibitors of Aggregation on Calcium Binding to Thrombocytic Membranes

Substance	Concentration (per ml)	Chlorotetracycline fluorescence (in %)
Control	—	100
ADP	200 μg	74 ± 7
Thrombin	0.4 unit	55 ± 6
15-Fluoro-11,9-epoxy-methanoprostaglandin H_2	10 μg	60 ± 6
cAMP	630 μg	122 ± 3
Dipyridamole	4 μg	136 ± 15
Prostacyclin	10 μg	126 ± 15
ATP	200 μg	131 ± 5

According to Born [98], aminosyn, apyrase, and β-diethylaminoethyldiphenylpropylacetate can increase erythrocyte resistance and, thus, decrease the release of aggregation inducers.

Evidently, the high clinical effectiveness of dipyridamole, trental, and others is associated with the fact that
they inhibit aggregation by direct influence on platelets
and by activation of prostacyclin synthesis as well [146].
Prostacyclin, however, is not the only agent that prevents
platelet adhesion to the vessel wall.

Study of the action of aggregation inhibitors on the
conformation of platelet plasmatic membrane can be very important for understanding their mechanisms of action. It
has been shown that prostaglandins E and F induce differently
directed conformational changes in the protein phase of platelet plasmatic membrane [172].

Attention should be paid to studies on the relationship
of platelet aggregating ability and serotonin accumulation
in platelets. It is known that this biogenic amine induces
aggregation. A decreased aggregational effect of serotonin
is observed when dosage is increased [173]. Antagonism with
serotonin on different organs and tissues has been observed
with many antiaggregating preparations: sympathomimetic,
adrenolytic, local anesthetic, anti-inflammatory, tricyclic
antidepressants, and others. Acetylsalicylic acid, in particular, suppresses serotonin uptake and promotes serotonin
release from blood platelets. However, a distinct parallel-
ism has not been found between the antiaggregating activity
and the antiserotonin activity of preparations [31, 173].

Our studies demonstrate that the aggregational mechanism
of action of many compounds (xanthine and amidine derivatives,
cAMP stabilizers, and others) can be associated with glycolysis activation. In particular, inhibitors of aggregation
from the derivatives of pyrazolone, guanidine, and phenothiazine inhibited glucoso-6-phosphate dehydrogenase activity
and activated glyceraldehydrophosphate dehydrogenase and lactate dehydrogenase [45, 174]. The aggregation inducers ADP
and thrombin inhibit pyruvate kinase activity and decrease
ATP content in platelets, while cAMP activates this enzyme
and increases ATP content (Table 13.7). It is known that
glycolysis is a very important source of ATP synthesis. Consequently, glycolysis activators change the ATP/ADP ratio.

We have demonstrated, in our laboratory, the antiaggregating action of glycolysis intermediators such as fructose,
diphosphate, phosphoenol pyruvate, and others [175]. Phosphoenol pyruvate, in combination with pyruvate kinase, decreases

Table 13.9. Effect of Some Inducers and Inhibitors of Aggregation on the Morphometric Parameters of Platelets

Substance	Concentration (μg/ml)	% of degranulated platelets	Number of granules per platelet	Number of dense bodies per platelet
Control	–	11 ± 1	3.29 ± 0.32	1.86 ± 0.17
ADP	200	60 ± 7	1.76 ± 0.29	0.63 ± 0.13
Adrenalin	10	52 ± 6	1.96 ± 0.21	0.60 ± 0.10
Trifluoroperazine + ADP	200	15 ± 2	2.94 ± 0.26	0.01 ± 0.13
Papaverine + ADP	200	24 ± 3	3.09 ± 0.33	1.22 ± 0.10

Table 13.10. Effect of Heparin, Dipyridamole, Phytin, and Glutamic Acid on Some Hemostasis Parameters in Rabbits with Experimental DIC Syndrome

Indices studied	Control (injection of thrombin and pituitrin)	After injection of preparations and before injection of thrombin and pituitrin
Fibrinogen degradation products (mg%)	38.5 ± 2.5	12.8 ± 2.4 $p < 0.05$
Fibrinogen concentration (mg%)	2.54 ± 0.07	445 ± 25 $p < 0.05$
Antithrombin III (%)	55 ± 4	74 ± 3 $p < 0.05$
Platelet aggregation (%)	29 ± 2	38 ± 4 $p < 0.05$

platelet aggregating ability in vitro and in vivo [72]. Since pyruvate kinase activity depends on the concentration of calcium ions and since calcium is necessary for aggregation, we studied the effect of different platelet inducers and inhibitors on calcium binding to hydrophobic membrane structures. We found that inducers of aggregation such as ADP, thrombin, and 15-fluoro-11,9-ethoxymethanoprostaglandin H_2 decrease calcium binding to membrane structures (Table 13.8). Also, inhibitors of aggregation such as cAMP, prostacyclin, ATP, and others increase calcium binding to membranes [176]. These results have been confirmed by electron cytochemical observations of the dynamics of calcium localization in induction and inhibition of platelet aggregation. ADP decreases the number of electron-dense granules, while cAMP prevents this effect and promotes an increase in calcium deposits in electron-dense structures. Platelet aggregation is always accompanied by destruction of the granular apparatus of blood platelets.

Morphometric studies performed in our laboratory allow us to conclude that the degree of degranulation correlates

with the force of the aggregational effect of the substance
[176, 177]. Antiaggregating preparations prevent degranulation caused by inducers of aggregation (Table 13.9).

The results of the above studies confirm that calcium
release from electron-dense structures plays a significant
role in platelet adhesion. It is known that ionized calcium
is able to activate phospholipase A_2 and thus initiate the
cascade of arachidonic acid. Calcium also activates some
proteases. These calcium effects promote destruction of the
plasmatic membrane and the membranes of thrombocytic organ-
elles. Moreover, calcium activates thrombostenin. All of
these effects condition the release of various substances
(thrombocytic fibrinogen, antiheparin factor, etc.) from
platelets, thus promoting adhesion of the platelets among
themselves and adherence to a heterologous surface. Many inhibitors of aggregation prevent the release of calcium from
intercellular depots and thus prevent aggregation [176, 177].

However, not all inducers of aggregation promote the release of calcium from a bound state. Some substances in this
series (adrenalin, serotonin) increase the penetration of
calcium through plasmatic membrane. Inhibitors of aggregation
can therefore also be searched for among substances which inhibit calcium passage through the membrane. This is why antiaggregating action was observed with isoptin, phenothiazine
derivatives, local anesthetic preparations, and other substances.

Based on the data concerning the role of calcium in
platelet aggregation, we developed a combination of preparations that decrease platelet adhesion. This combination includes dipyridamole, phytin, and glutamic acid. Dipyridamole
increases cAMP content in platelets. Phytin and glutamic
acid activate ATP synthesis and are part of the cellular components that bind calcium. The ability to activate synthesis
of vessel wall prostacyclin is a positive property of this
combination.

Table 13.10 presents study results on the action of heparin, dipyridamole, phytin, and glutamic acid on some hemostasis indices. Administration of these substances decreases
the manifestation of experimental DIC syndrome. The combination of these preparations was used in complex therapy of
the DIC syndrome in patients with epilepsy, ischemic stroke,
and multiple sclerosis. With these substances, we observed

a normalization of the aggregating ability of platelets in
these patients. Based on our observations, inhibitors of aggregation can be used to restore the microcirculation in patients with postmastectomy edema and other states. Application of antiaggregating drugs potentiates the effect of
thrombolytic therapy and removes undesirable effects of certain thrombolytic preparations.

The data presented herein indicate that a large number
of pharmacologic preparations have been detected at present
that are able to decrease platelet aggregation and, if used
within complex therapy, increase the possibilities for preventing and treating thromboses. For the majority of these
substances, their effect on aggregation and thrombogenesis
is manifested against the background of other pharmacologic
drugs. Undoubtedly, prostacyclin and prostacyclin-like compounds are most preferable. However, they are insufficiently
stable, and thus have limited clinical use. Further research
on this and other directions seems to be a promising way of
finding antithrombotic and thrombolytic preparations.

Only a rational combination of thrombolytic drugs, anticoagulants, and aggregation inhibitors can lead to successful
treatment of thromboembolic states.

REFERENCES

1. K. M. Lakin, V. A. Feldbaum, and A. A. Lebedeva, "Sub-
    stances decreasing platelet aggregation and adhesiveness.
    Literature review," Farmakol. Toksikol., 34, 104-113
    (1971).
2. K. M. Lakin, V. A. Feldbaum, and M. S. Ovnatanova, "Ef-
    fect of some vasoactive substances on platelet aggrega-
    tion," Kardiologiya, 12, 21-25 (1972).
3. A. G. G. Turpie and J. Hirsh, "Platelet suppressive
    therapy," Br. Med. Bull., 34, 183-190 (1978).
4. L. M. Fuccella, "Clinical pharmacology of inhibitors
    of platelet aggregation," Pharmacol. Res. Commun., 11,
    825-852 (1979).
5. D. R. Feller, S. S. Navran, J. R. Mayro, et al., "Differ-
    ential stereoselective inhibition by trimetoquinol (TMQ)
    of platelet aggregation mediated by prostaglandin de-
    pendent and independent pathways," in: Eighth Inter-
    national Congress of Pharmacology, Tokyo (1981), Abstract,
    p. 585.

6. J. R. O'Brien, M. D. Etherington, and R. D. Shuttle-
    worth, "Ticlopidine — an antiplatelet drug: effects in
    human volunteers," Thromb. Res., 13, 245-254 (1978).
7. B. B. Vargaftig and J. Randon, "Inhibition by ticlopi-
    dine on the ex vivo aggregation of rat platelets: induc-
    tion of insurmountable anti-ADP activity in absence of
    plasma," in: Eighth International Congress of Pharma-
    cology, Tokyo (1981), Abstract, p. 458.
8. K. M. Lakin, M. S. Ovnatanova, M. A. Matyashova, and
    M. D. Mashkovsky, "Studies of effect of some xanthine
    derivatives on platelet aggregation and other indices
    of hemostasis," Byull. Eksp. Biol. Med., 2, 181-182
    (1976).
9. O. K. Gavrilov and A. M. Shislov, "BASR system in pa-
    tients with ischemic heart disease," in: Problems and
    Hypotheses in Studies of Blood Coagulability, O. K.
    Gavrilov (ed.), Meditsina, Moscow (1981), pp. 38-58.
10. K. M. Lakin, V. A. Makarov, M. S. Ovnatanova, and I. S.
    Azhgikhin, "Effect of Soviet preparation prostaglandin
    E_1 on platelet and erythrocyte aggregation," Farmakol.
    Toksikol., 39, 436-440 (1976).
11. K. M. Lakin, V. A. Makarov, V. V. Bezuglov, and L. D.
    Bergelson, "Prostaglandin effect on platelet aggrega-
    tion," in: Prostaglandins and Blood Circulation, V. V.
    Zakusov (ed.), Ministry of Health of the USSR, Erevan
    (1980), pp. 62-63.
12. J. R. Vane and S. Moncada, "Prostacyclin and its thera-
    peutic potential," in: Eighth International Congress
    of Pharmacology, Tokyo (1981); Advances in Pharmacology
    and Therapeutics II, Vol. 4, H. Yoshida, Y. Hagihara,
    and E. Ebashi (eds.), Pergamon Press, New York (1982),
    p. 215.
13. R. J. Gryglewski, "Regulation of prostacyclin (PGI_2) re-
    lease into circulation," in: Eighth International Con-
    gress of Pharmacology, Tokyo, 1981; Advances in Pharma-
    cology and Therapeutics II, Vol. 4, H. Yoshida, Y. Hagi-
    hara, and S. Ebashi (eds.), Pergamon Press, New York
    (1982), p. 235.
14. S. Moncada and J. R. Vane, "Discovery, biological sig-
    nificance, and therapeutic potential of prostacyclin,"
    in: Clinical Pharmacology of Prostacyclin, P. J. Lewis
    and J. O'Grady (eds.), Raven Press, New York (1981),
    pp. 1-8.
15. S. Moncada, "The pharmacology and clinical potential
    of prostacyclin," in: Ninth World Congress of Cardiology,
    E. I. Chazov (ed.), Sandos, Moscow (1982), No. 0050.

16. O. I. Aleshin, V. P. Zykova, and L. F. Nikolaeva, "Prostaglandins of blood flowing and outflowing from the heart in experimental myocardial infarction," in: Abstracts, USSR Symposium on Prostaglandins and Blood Circulation, E. S. Gabrielyan, Kh. M. Markov, and R. G. Boroyan (eds.), Ministry of Health of the USSR, Yerevan (1980), pp. 77-79.

17. H. Shio, P. Ramwell, and S. J. Jessup, "Prostaglandin E_2: effect on aggregation, shape change, and cyclic AMP of rat platelets," Prostaglandins, $\underline{1}$, 29-36 (1972).

18. D. J. Bouillin, A. R. Green, and K. S. Price, "The mechanism of adenosine diphosphate-induced platelet aggregation: binding to platelet receptors and inhibition of binding and aggregation by prostaglandin E_1," J. Physiol., $\underline{221}$, 415-420 (1972).

19. O. V. Miller and R. R. Gorman, "Evidence for distinct prostaglandin I_2 and D_2 receptors in human platelets," J. Pharmacol. Expt. Ther., $\underline{210}$, 134-140 (1979).

20. T. Honohan, F. A. Fitzpatrick, D. G. Booth, et al., "Hydrolysis of an orally active platelet inhibitory prostanoid amide in the plasma of several species," Prostaglandins, $\underline{19}$, 123-138 (1980).

21. H. Sinzinger, K. Silberbauer, A. K. Horsch, and A. Gall, "Decreased sensitivity of human platelets to PGI_2 during long-term intraarterial prostacyclin infusion in patients with peripheral vascular disease — a rebound phenomenon?" Prostaglandins, $\underline{21}$, 49-51 (1981).

22. D. E. MacIntyre, R. J. Handin, R. Rosenberg, and E. W. Salzman, "Heparin opposes prostanoid and nonprostanoid platelet inhibitors by direct enhancement of aggregation," Thromb. Res., $\underline{22}$, 167-175 (1981).

23. W. Rettkowski, K.-E. Blass, and W. Förster, "Influence of caffeine-sodium salicylate on PGI_2 mediated inhibition of ADP- and thrombin-induced platelet aggregation," in: Prostaglandins and Thromboxanes in the Cardiovascular System and Gynecology and Obstetrics, W. Förster (ed.), Pergamon Press, New York (1981), pp. 337-339.

24. M. F. Gimeno, L. Sterin-Borda, E. S. Borda, et al., "Human plasma transforms prostacyclin (PGI_2) into a platelet antiaggregatory substance which contracts isolated bovine coronary arteries," Prostaglandins, $\underline{19}$, 907-917 (1980).

25. M. A. Orchard and C. Robinson, "Stability of prostacyclin in human plasma and whole blood: studies on the protective effect of albumin," Thromb. Haemostas., $\underline{46}$, 645-647 (1981).

26. B. J. Whittle, S. Moncada, F. Whiting, and J. R. Vane,
     "Carbacyclin — a potent stable prostacyclin analog for
     the inhibition of platelet aggregation," Prostaglandins,
     19, 605-627 (1980).
27. R. A. Johnson and E. G. Nidi, "Synthesis and stereochem-
     istry of stable prostacyclin analogs," in: Chemistry,
     Biochemistry, and Pharmacological Activity of Prostanoids,
     S. M. Roberts and F. Scheinmann (eds.), Pergamon Press,
     New York (1979), pp. 71-78.
28. M. Lombroso, S. Nicosia, S. Moncada, et al., "PGI$_2$ and
     its stable analog 6β-PGI$_1$ bind to the same receptor
     coupled to adenylate cyclase in human platelets," in:
     Eighth International Congress of Pharmacology, Tokyo
     (1981), p. 738.
29. M. O. Whitaker, P. Needleman, and A. Wyche, "PGD$_3$ is the
     mediator of the antiaggregatory effects of the trienoic
     endoperoxide PGH$_3$," in: Advances in Prostaglandin and
     Thromboxane Research, Vol. 6, B. Samuelsson, P. W. Ram-
     well, and R. Paoletti (eds.), Raven Press, New York
     (1980), pp. 301-303.
30. T. Tsuboi, N. Natano, K. Nakafsuji, et al., "Pharmaco-
     logical evaluation of ONO 1206, a prostaglandin E$_1$ der-
     ivative, as antianginal agent," in: Advances in Pros-
     taglandin and Thromboxane Research, Vol. 6, B. Samuel-
     son, P. W. Ramwell, and R. Paoletti (eds.), Raven Press,
     New York (1980), pp. 347-349.
31. K. M. Lakin, V. A. Feldbaum, and A. A. Lebedeva, "Phar-
     macological regulation of platelet adhesiveness and ag-
     gregation," in: Aspects of Neurohumoral Regulation of
     the Blood Coagulation Process in Normal and Pathological
     Conditions, B. I. Kuznik (ed.), Academy of Pedagogical
     Sciences of the USSR, Chita (1971), pp. 83-88.
32. F. Markwardt and E. Glusa, "Pharmacological control of
     adrenergic effects on platelets," in: Physiologically
     Active Compounds in Medicine, V. V. Zakusov (ed.), Medi-
     tsina, Yerevan (1982), p. 336.
33. M. G. Doni, "Metergoline: a strong inhibitor of rat
     thrombocyte aggregation in vitro," Haematologica (Pavia),
     65, 717-724 (1980).
34. K. M. Lakin, Yu. F. Krylov, V. A. Makarov, et al., "Chem-
     ical composition, metabolism, and aggregation of plate-
     lets (review)," Farmakol. Toksikol., 14, 128-139 (1974).
35. E. I. Chazov and K. M. Lakin, Anticoagulants and Fibrin-
     olytics, Meditsina, Moscow (1977).
36. E. I. Chazov and K. M. Lakin, Anticoagulants and Fibrin-
     olytics, Yearbook Medical Publishers, Inc., Chicago—
     London (1980).

37. M. C. Scurutton, "Metiamide is not an α_2-adrenoreceptor antagonist in human platelets," Commun. J. Pharmacol., <u>32</u>, 438-444 (1980).

38. G.Kindness, W. F. Long, and F. B. Williamson, "Anticoagulant effects of sulphated polysaccharides in normal and antithrombin III-deficient plasmas," Br. J. Pharmacol., <u>69</u>, 675-677 (1980).

39. J. Morley, C. P. Page, and W. Paul, "The effect of heparin on platelet aggregation in vivo," Br. J. Pharmacol., <u>75</u>, 55 (1982).

40. E. W. Salzman, R. D. Rosenberg, H. Smith, et al., "Effect of heparin and heparin fractions on platelet aggregation," J. Clin. Invest., <u>65</u>, 64-73 (1980).

41. W. H. Anderson, S. F. Mohammad, H. J. K. Chuang, and R. G. Mason, "Heparin potentiates synthesis of thromboxane A_2 in human platelets," in: Advances in Prostaglandin and Thromboxane Research, Vol. 6, B. Samuelsson, P. W. Ramwell, and R. Paoletti (eds.), Raven Press, New York (1980), pp. 287-291.

42. B. S. Coller, "Polybrene-induced platelet aggregation and reduction in electrophoretic mobility: enhancement by von Willebrand factor and inhibition by vancomycin," Blood, <u>55</u>, 276-281 (1980).

43. C. L. Slade, W. A. Andes, and A. D. Mason, "Platelet aggregation following defibrination with ancrod," Thromb. Haemostas., <u>36</u>, 424-429 (1976).

44. E. Glusa, A. Hoffmann, and F. Markwardt, "Influence of benzamidine derivatives on thrombin-induced platelet reaction," Folia Haematol., <u>109</u>, 98-106 (1982).

45. K. M. Lakin, L. N. Lebydj, S. V. Gatash, and Yu. F. Krylov, "Effect of amidine and guanidine derivatives on platelet aggregation," Farmakol. Toksikol., <u>5</u>, 575-579 (1978).

46. K. Breddin, "Pro-kontra antikoagulantien oder Thrombozytenfunktionshemmer in der Prophylaxe der Herzinfarktes," Internist (Berlin), <u>21</u>, 394-396 (1980).

47. G. Castel, Darmanaden, A. M. Noel, et al., "Activité antiaggregante plaquenttaire de derivés de la benzylidene coumaranone," Trav. Soc. Pharm. Montpellier, <u>36</u>, 239-245 (1976).

48. C. Galli, E. Agradi, A. Petroni, and A. Socini, "Effects of 8-monochloro-3-β-diethylaminoethyl-4-methyl-7-ethoxy carboxyl methoxy coumarin (AD_6) on aggregation, arachidonic acid metabolism, and thromboxane B_2 formation in human platelets," Pharmacol. Res. Commun., <u>12</u>, 329-337 (1980).

49. V. B. Kazimirovskaya, L. N. Kholdeeva, L. V. Aksenova,
    et al., "Effect of 1-(organylthioalkyl)-silatranes and
    their S-oxides on platelet aggregational activity," in:
    Biologically active Compounds of Silicon, Germanium,
    Tin, and Lead, M. G. Voronkov and A. T. Platonova (eds.),
    Academy of Sciences of the USSR, Irkutsk (1980), pp.
    91-92.

50. K. M. Lakin, V. B. Kazimirovskaya, E. K. Vugmeister,
    et al., "Comparative study of physicochemical properties
    of heparin and its preparation with prolonged action
    of trimethylsilylheparin," in: Biologically Active Com-
    pounds of Silicon, Germanium, Tin, and Lead, M. G. Voron-
    kov and A. T. Platonova (eds.), Academy of Sciences of
    the USSR, Irkutsk (1980), pp. 87-88.

51. P. A. Wilson, G. P. McNicol, and A. C. Douglas, "Some
    effects of trypsin and streptokinase on platelet aggre-
    gation," Thromb. Diath. Haemorrh., 18, 66-75 (1967).

52. E. F. Lüscher and M. D. Davey, "The initiation of vis-
    cous metamorphosis of the blood platelets," in: Physiol-
    ogy of Hemostasis and Thrombosis, American Lectures in
    Hematology, No. 675, S. A. Johnson and W. H. Seegers
    (eds.), Thomas, Springfield, Illinois (1967), pp. 9-45.

53. M. Brochier and P. Griguer, "L'hyperaggregabilité
    plaquenttaire secondaire aux traitments thrombolytiques,"
    Acta Clin. Belgica, 30, 210-213 (1975).

54. J. Stachurska, S. Lopaciuk, B. Gerdin, et al., "Effect
    of proteolytic degradation products of human fibrinogen
    and of human factor VIII on platelet aggregation and
    vascular permeability," Thromb. Res., 15, 663-672 (1979).

55. S. V. Andreev, A. A. Kubatiev, I. D. Kobkova, et al.,
    "The new thrombolytic preparation terrilytin," in: Urgent
    Problems of Hemostasiology, B. V. Petrovsky, E. I. Chazov,
    and S. V. Andreev (eds.), Nauka, Moscow (1979), pp. 249-
    254.

56. Ya. Mamedov, D. Tagdisi, R. Reish, et al., "Control of
    hemo- and lymphocoagulation with terrilytin-nicotine,"
    in: Ninth World Congress of Cardiology, E. I. Chazov
    (ed.), Sandos, Moscow (1982), No. 1095.

57. K. M. Lakin and V. P. Baluda, "Pharmacological regulation
    of blood fluidity," in: Urgent Problems of Hemostasiol-
    ogy, B. V. Petrovsky, E. I. Chazov, and S. V. Andreev
    (eds.), Nauka, Moscow (1981), pp. 430-460.

58. D. J. Boullin, R. P. J. Grimes, and M. W. Orr, "The ac-
    tions of flupenthixol upon 5-hydroxytryptamine-induced
    aggregation and the uptake of 5-hydroxytryptamine and
    dopamine by human blood platelets," Br. J. Pharmacol.,
    55, 555-557 (1975).

59. R. Brossmer and H. Patscheke, "2-Phenylethanol and some of its amphiphilic derivatives as inhibitors of platelet aggregation. Structure-activity relationship," Arzneim. Forsch., 25, 1697-1702 (1975).

60. L. Imandt, D. Tyhuis, H. Wessels, et al., "Observations on ADP aggregation of lithium chloride incubated platelets in a variety of mammalian species," Haemostasis, 9, 276-287 (1980).

61. V. Skondia, "Antiplatelet activity of known central venous system drug piracetam (nootropil)," in: Ninth World Congress of Cardiology, Abstracts, Moscow (1982), No. 05119.

62. F. Gloss, H. Lippton, and P. J. Kadowitz, "Differential effects of local anesthetics and propranolol on arachidonic acid and adenosine S-diphosphate-induced aggregation in rabbit platelets," Prostaglandins Med., 5, 85-92 (1980).

63. J. R. O'Brien, M. D. Etherington, S. Jamieson, et al., "The effect of ICI 55, 897, and clofibrate on platelet function and other tests abnormal in atherosclerosis," Thromb. Haemostas., 40, 75-82 (1978).

64. M. D. Mashkovsky, K. M. Lakin, M. S. Obnatanova, and G. Ya. Schwartz, "Effect of parmidin (pyridinol-carbamate) on platelet aggregation, blood coagulation, and fibrinolysis," Byull. Eksp. Biol. Med., 3, 322-324 (1976).

65. V. A. Sheatakov, V. N. Ilyin, L. M. Danilova, et al., "Experimental and clinical studies of antithrombotic properties of nicotinic acid," Probl. Gematol. Pereliv. Krovi., 8, 29-35 (1977).

66. K. M. Lakin, J. F. Krylov, V. A. Makarov, et al., "Pharmacological effect on the blood platelet aggregation," in: Sixth International Congress of Pharmacology, M. K. Paasonen (ed.), University of Helsinki, Helsinki (1975), p. 625.

67. K. M. Lakin, V. A. Makarov, and G. I. Petruchina, "Die Wirkung von Antikoagulantien und Hemmstoffen der Thrombozytenaggregation auf die disseminierte intravaskulare Gerinnung (DIC)," Folia Haematol. (Leipz), 104, 816-818 (1977).

68. A. V. Savelieva, "Change in blood coagulability in long-term injection of retabolil," Farmakol. Toksikol., 3, 332-334 (1973).

69. M. A. Orchard and J. H. Botting, "The influence of sex hormone on rat platelet sensitivity to adenosine diphosphate," Thromb. Haemostas., 46, 496-499 (1981).

70. C. Ouyang, C. M. Teng, and T. T. Huang, "Classification of the purified venoms which affect blood coagulation and platelet aggregation," in: Eighth International Congress of Pharmacology, Tokyo (1981), Abstract, p. 456.

71. B. B. Vargaftig, "Carageenan and thrombin trigger prostaglandin synthetase-independent aggregation of rabbit platelets: inhibition by phospholipase A_2 inhibitors," J. Pharm. Pharmacol., 29, 222-228 (1977).

72. K. M. Lakin, M. A. Matyashova, F. B. Levin, and V. A. Makarov, "Effect of pyruvate kinase and phosphoenolpyruvate on formation of experimental thrombus," Farmakol. Toksikol., 41, 405-409 (1978).

73. A. J. Marcus and M. B. Zucker, The Physiology of Blood Platelets, Grune & Stratton, New York (1965).

74. K. Kikugawa, "The effect of molecular iodine on platelet aggregation in vitro," Thromb. Diath. Haemorrh., 31, 160-171 (1974).

75. P. Klener, A. Kubisz, and J. Suranova, "Influence of cytotoxic drugs on platelet functions and coagulation in vitro," Thromb. Haemostas., 37, 53-61 (1977).

76. J. M. Ribalta, J. J. Artus, L. Salvador, et al., "Synthese und pharmakologische Answertung von N-Morpholono-methylharnstoffderivaten mit Antithrombozytenaggregations-Aktivitat," Arzneim. Forsch., 31, 1782-1786 (1981).

77. P. Kubisz, F. Seghier, P. Klener, and S. Cronberg, "Influence of dianhydrogalactitol on some platelet functions in vitro," Acta Haematol. (Basel), 66, 27-30 (1981).

78. S. Cronberg, "Klinische und pharmakologische Untersuchungen über einige Plättchenhemmstoffe," Folia Haematol. (Leipzig), 106, 834-838 (1979).

79. S. Takano, "Difference in effects of vinblastine and vincristine on the dog platelet aggregation," Tohoku J. Exp. Med., 135, 79-85 (1981).

80. P. G. Steinherz, D. R. Miller, and H. W. Hilgartner, "Platelet dysfunction in vincristine-treated patients," Br. J. Haematol., 32, 439-450 (1976).

81. P. Kubisz, P. Klener, and S. Cronberg, "Influence of mitramycin on some platelet functions in vitro," Acta Haematol. (Basel), 63, 101-106 (1980).

82. V. Cepelak, H. Cepelakova, B. Brunova, et al., "Pyrazolidine derivatives; a comparative study of their effect on platelet aggregation," Folia Haematol. (Leipzig), 106, 839-848 (1979).

83. H. Vinazzner, "Klinisch-experimentelle Studien zur Hemmung der Thrombozytenfunktion mit Azetylsalizylsäure und mit Indobufen," Folia Haematol. (Leipzig), 106, 783-796 (1979).
84. E. Pogliani, G. Corvi, V. Mandelli, et al., "Preliminary human pharmacology studies on the inhibition of platelet aggregation by indobufen (K 3920)," Haematologica (Pavia), 66, 160-170 (1981).
85. T. R. Petrova and S. A. Pavlishchuk, "Ischemic heart disease and microcirculatory hemostasis," in: Vascular Pathology and Microcirculatory Hemostasis, T. R. Petrova (ed.), Ministry of Health of the RSFSR, Krasnodar (1975), pp. 41-51.
86. S. Aonuma, Y. Kohama, S. Fujimoto, and T. Makino, "Studies on aspirin derivatives with very little side effect. II. Potent platelet antiaggregant activity and no mutagenicity of aspirin-isopropylantipyrine (AIA)," J. Pharmacobiodyn., 4, 803-811 (1981).
87. D. Loew and H. Vinazzer, "Dose-dependent influence of acetylsalicylic acid on platelet functions and plasmatic coagulation factors," Haemostasis, 5, 239-249 (1976).
88. I. Weinberg, H. Jushua, J. Friedman, et al., "Inhibition of ADP-induced platelet aggregation of dipyrone in patients with acute myocardial infarction," Thromb. Haemostas., 42, 752-756 (1979).
89. W. J. Rosenblum and F. El-Sabban, "Use of AHR-5850 and AHR-6293 to distinguish the effect of antiplatelet aggregating drug properties from the effect of anti-inflammatory properties on an in vitro model of platelet aggregation," Microvasc. Res., 17, 309-313 (1979).
90. K. M. Lakin, Yu. F. Krylov, A. F. Malyugin, et al., "Effect of some pyrazolone derivatives on platelet aggregation," Farmakol. Toksikol., 2, 180-186 (1976).
91. E. C. Rossi, S. Nimer, and G. Louis, "Inhibition of platelet aggregation by 5-nitro-2-furaldehyde diacetate with observation on structure-activity relationships," J. Lab. Clin. Med., 87, 703-709 (1976).
92. R. L. Vigdahl, R. H. Ferber, and S. C. Parrish, "Comparison of sulfinpyrazone and BL-3459 with S-20344, a potent new antithrombotic agent," Thromb. Res., 1, 547-555 (1981).
93. E. M. Griffett, S. M. Kinnon, A. Kumar, et al., "Effects of 6-[p-(4-phenylacetylpiperazin-1-yl)phenyl]-4,5-dihydro-(CCI 17810) and aspirin on platelet aggregation and adhesiveness," Br. J. Pharmacol., 72, 697-705 (1981).

94. M.J. Genua, J. Giraldez, E. Rocha, and A. Mouge, "Ef-
     fects of antibiotics on platelet functions in human
     plasma in vitro and plasma in vivo," J. Pharmacol. Sci.,
     69, 1282-1284 (1980).
95. M. Oka and M. J. Manku, "Spironolactone decreases the
     formation of prostaglandins and related substances from
     arachidonic acid in rat vascular rings and human plate-
     lets," Prostaglandins Med., 4, 193-203 (1980).
96. M. Aberg and S. Bergentz, "The effect of dextran on
     the platelet distribution and lysability of ex vivo
     thrombi in dog," Eur. Surg. Res., 11, 282-288 (1979).
97. P. Olsson, L. Schalin, H. Lagergren, et al., "Polyphlor-
     etine phosphate as an inhibitor of platelet adhesion.
     Comparison with acetylsalicylic acid and dipyridamole,"
     Thromb. Res., 10, 349-364 (1977).
98. G. V. Born, "Platelets in haemostasis and thrombosis,"
     in: Platelets. Cellular Response, Mechanisms, and
     Their Biological Significance, A. Rotman, F. A. Meyer,
     C. Gitler, and A. Silberberg (eds.), John Wiley and
     Sons, New York (1980), pp. 1-15.
99. K. C. Agarwal and R. E. Parks, Jr., "5-Methylthioaden-
     osine and 2',5'-dideoxyadenosine blockade of the in-
     hibitory effects of adenosine on ADP-induced platelet
     aggregation by different mechanisms," Biochem. Pharm-
     acol., 29, 2529-2532 (1980).
100. N. J. Cusack and S. M. Hourani, "Specific but noncom-
     petitive inhibition by 2-alkylthio analogs of adenosine
     5-monophosphate and adenosine 5-triphosphate of human
     platelet aggregation induced by adenosine 5-diphosphate,"
     Br. J. Pharmacol., 75, 2 (1982).
101. J. A. Brown, S. A. Jimenez, and R. W. Colman, "Collagen-
     induced platelet shape change. The role of collagen qua-
     ternary structure," J. Lab. Clin. Med., 95, 90-98 (1980).
102. H. J. Messmore, J. Fareed, and A. J. Karczmar, "Studies
     of low-molecular-weight peptide inhibitors of platelet
     function," in: Eighth International Congress of Pharm-
     acology, Tokyo (1981), Abstract, p. 375.
103. M. B. Zucker and R. A. Grant, "Hypertonic solutions
     decrease light transmission of platelet suspensions,"
     Thromb. Haemostas., 42, 1062-1063 (1979).
104. E. R. Anderson and J. G. Foulks, "The effect of small
     organic anions on aggregation and shape change of rab-
     bit platelets," Thromb. Haemostas., 40, 43-60 (1978).
105. P. Andersen and C. T. Eika, "Thrombin-, epinephrine-,
     and collagen-induced platelet aggregation inhibited by
     α_1-acid glycoprotein. Influence of heparin and anti-
     thrombin III," Scand. J. Haematol., 24, 365-372 (1980).

106. H. Patscheke, "Selektive Aggregationshemmung Ein neues
     Konrept für Hemmstoffe der Thrombozytenfunktion," Klin.
     Wochenschr., 59, 451-457 (1981).
107. R. P. Quintana, A. Lasslo, M. Dugdale, and L. L. Goodin,
     "Relationships between the chemical constitution of
     carbamoylpiperidines and related compounds, and their
     inhibition of ADP-induced human blood platelet aggrega-
     tion," Thromb. Res., 22, 665-680 (1981).
108. R. H. Rynbrandt, E. E. Nishizawa, D. P. Balogoyen, et
     al., "Synthesis and platelet aggregation inhibitory
     activity of 4,5-bis(aryl)-2-substituted-thiazoles,"
     J. Med. Chem., 24, 1507-1510 (1981).
109. M.J. Parnham, P. C. Bragt, A. Bast, and F. J. Zijlstra,
     "Comparison of the effects of inhibitors of cytochrome
     P-450-mediated reactions on human platelet aggregation
     and arachidonic acid metabolism," Biochim. Biophys.
     Acta, 677, 165-173 (1981).
110. S. H. S. Makoni and J. K. Sugden, "Some pyrrolidine
     and pyrrolizine derivatives as inhibitors of blood plate-
     let aggregation," Arzneim. Forsch., 30, 1135-1137 (1980).
111. H. Kodaira, M. Ishikawa, Y. Komoda, and T. Nakajima,
     "Antiplatelet aggregation principles from the bark of
     Fraxinus japonica Blume," Chem. Pharm. Bull. (Tokyo),
     29, 2391-2393 (1981).
112. K. M. Lakin, V. A. Makarov, M. A. Matyashova, et al.,
     "Pharmacological effect on experimental thrombi of the
     mesentery microvascular system," in: Problems of Micro-
     circulation (Function and Structure), A. M. Chernukh
     (ed.), Academy of Medical Sciences of the USSR, Moscow
     (1977), pp. 222-223.
113. A. K. Chepurov and G. M. Yelchaninov, "ATP and ADP ef-
     fect on platelet ζ-potential," Byull. Eksp. Biol. Med.,
     3, 14-16 (1970).
114. Yu. Shuteu, T. Bendyle, A. Kafrits, et al., Shock,
     Military Publ., Bucharest (1981).
115. J. R. Hampton and J. R. A. Mitchell, "Effect of glass
     contact on the electrophoretic mobility of human blood
     platelets," Nature, 209, 470-472 (1966).
116. S. F. Mohammas, H. J. K. Chuang, P. E. Crowter, et al.,
     "Interactions of poly(L-lysine) with human platelets.
     Correlation of binding with induction of platelet func-
     tion," Thromb. Res., 15, 781-791 (1979).
117. I. B. Kovacs and P. Görög, "Inhibition by N-acetyl neur-
     aminic (sialic) acid of platelet aggregation induced
     by different stimuli," Thromb. Haemostas., 42, 1187-1192
     (1979).

118. J. McPherson and M. B. Zucker, "Platelet retention in glass bead columns: adhesion to glass and subsequent platelet-platelet interactions," Blood, 47, 55-67 (1976).

119. J. Inada, M. Okada, and J. Saito, "Cooperativity of albumin and aggrexans A and B for ADP-induced aggregation of platelets," Thromb. Haemostas., 42, 1557-1560 (1979).

120. B.Blomback and M. Blomback, "Molecular defects and variants of fibrinogen," Nuov. Rev. Fr. Hematol., 10, 671-678 (1970).

121. T. Hovig, "Release of a platelet-aggregating substance (adenosine diphosphate) from rabbit blood platelets induced by saline 'extract' of tendons," Thromb. Diath. Haemorrh., 9, 264 (1963).

122. B. Nunn, "Collagen-induced platelet aggregation: evidence against the essential role of platelet adenosine diphosphate," Thromb. Haemostas., 42, 1193-1206 (1979).

123. J. Valles, J. Aznar, and M. T. Santos, "Effect of aspirin on platelet phospholipids," Thromb. Haemostas., 36, 628-633 (1976).

124. B. B. Vargaftig, F. Fougue, and M. Chignard, "Interference of bromphenacyl bromide with platelet phospholipase A_2 activity induced by thrombin and by the ionophore A 23187," Thromb. Res., 17, 91-102 (1980).

125. D. R. Phillips, L. K. Jennings, M. C. Berndt, et al., "Platelet membrane glycoproteins as thrombin and aggregation receptors," in: Platelets. Cellular Response, Mechanisms, and Their Biological Significance, A. Rotman, F. A. Meyer, C. Gitler, and A. Silberberg (eds.), Wiley, New York (1980), pp. 131-141.

126. S. Holme and H. Holmsen, "ADP-induced refractory state of platelets in vitro. I. Methodological studies on aggregation in platelet-rich plasma," Scand. J. Haematol., 15, 96-103 (1975).

127. E. F. Lüscher, P. Massini, and R. Käser-Glanzmann, "The role of calcium ions in the induction of platelet activities," in: Platelets. Cellular Response, Mechanisms, and Their Biological Significance, A. Rotman, F. A. Meyer, C. Gitler, and A. Silberberg (eds.), Wiley, New York (1980), pp. 67-77.

128. N. B. Chernyak, "Biochemical aspects of platelet functional properties in the BASR system," in: Problems and Hypotheses in Studies of Blood Coagulability, O. K. Gavrilov (ed.), Meditsina, Moscow (1981), pp. 38-58.

129. E. Ya. Pozin, E. G. Popov, Z. A. Gabbasov, et al., "Study of the redistribution of intracellular calcium

in platelets in the process of aggregation and release reaction," Byull. Vses. Kardiol. Nauchn. Tsentra AMN SSSR, 2, 55-57 (1981).

130. A. V. Suvorov, "Dependence of aggregational ability of platelets on the content of intrathrombocytic Ca^{2+}," Byull. Kardiol. Vses. Nauchn. Tsentra AMN SSSR, 2, 52-54 (1981).

131. E. R. Anderson, J. G. Foulks, and D. V. Godin, "The effect of local anesthetics and antiarrhythmic agents on the responses of rabbit platelets to ADP and thrombin," Thromb. Haemostas., 45, 18-23 (1981).

132. J. M. Sneddon, "Divalent cations and the blood platelet release reaction," Nature, 236, 103-105 (1972).

133. J. L. Miller, A. J. Katz, and M. B. Feinstein, "Plasmin inhibition of thrombin-induced platelet aggregation," Thromb. Diath. Haemorrh., 33, 286-289 (1975).

134. G. Suda and V. Aoki, "Inhibition of platelet function by a calmodulin interacting agent, W-7," Thromb. Res., 21, 447-453 (1981).

135. G. J. Johnson, G. H. R. Rao, L. A. Leis, et al., "Effect of agents that alter cyclic AMP on arachidonate-induced platelet aggregation in the dog," Blood, 55, 722-729 (1980).

136. J. E. Tateson, S. Moncada, and J. R. Vane, "Effects of prostacyclin (PGX) on cyclic AMP concentrations in human platelets," Prostaglandins, 13, 389-399 (1977).

137. N. B. Chernyak, "Platelet biochemistry," in: Normal Hemopoiesis and Its Regulation, N. A. Fedorov (ed.), Meditsina, Moscow (1976), pp. 275-295.

138. A. A. Nikulin, "Effect of calcium, thrombin, and nucleotides (ADP, cAMP, cGMP) on glycolysis and energy metabolism of blood platelets," Farmikol. Toksikol., 5, 585-590 (1980).

139. M. J. Silver, J. B. Smith, C. Ingerman, et al., "Arachidonic acid-induced human platelet aggregation and prostaglandin formation," Prostaglandins, 4, 863-869 (1973).

140. B. B. Vargaftig and P. Zirinis, "Platelet aggregation induced by arachidonic acid is accompanied by release of potential inflammatory mediators distinct from PGE_2 and PGF_2," Nature, 244, 114-116 (1973).

141. W. C. Pickett, R. L. Jesse, and P. Cohen, "Initiation of phospholipase A_2 activity in human platelets by the calcium ion ionophore A 23187," Biochim. Biophys. Acta, 486, 209-218 (1977).

142. E. Agradi, A. Petroni, and A. Solini, "In vitro effect of synthetic antioxidants and vitamin E on arachidonic acid metabolism and thromboxane formation in human

360   K. M. LAKIN ET AL.

platelets and on platelet aggregation," Prostaglandins,
22, 255-266 (1981).
143. G. Masotti, L. Poggesi, G. Galanti, et al., "Stimula-
      tion of prostacyclin by dipyridamole," Lancet, 1, 1412-
      1414 (1979).
144. K. M. Lakin, V. P. Baluda, T. I. Lukoyanova, et al.,
      "Effect of acetylsalicylic and nicotinic acids and
      complamine on formation of prostacyclin-like substance
      by aortic wall," Farmakol. Toksikol., 5, 581-585 (1980).
145. W. Förster, W. Rettkowski, and H.-U. Block, "The anti-
      aggregatory effect of nitroglycerine (N.G.), in com-
      bination with prostacyclin and PGE, and its effect on
      prostaglandins (PGS) and TXA_2 biosynthesis," in: Pros-
      taglandins and Thromboxanes in the Cardiovascular Sys-
      tem and in Gynecology and Obstetrics, W. Förster (ed.),
      Pergamon Press, New York (1981), pp. 275-285.
146. K. M. Lakin, V. P. Baluda, T. I. Lukoyanova, and V. A.
      Makarov, "Effect of some inductors and inhibitors of
      thrombocyte aggregation on vascular wall prostacyclin
      generating function," in: Eighth International Con-
      gress of Pharmacology, Tokyo (1981), Abstract.
147. K. Pönicke and W. Förster, "Influence of antianginal
      drugs on the biosynthesis of prostaglandins and thrombox-
      ane A_2 in rat lung spleen and aorta in vitro," in: Pros-
      taglandins and Thromboxanes in the Cardiovascular Sys-
      tem and in Gynecology and Obstetrics, W. Förster (ed.),
      Pergamon Press, New York (1981), pp. 29-302.
148. M. Radomski, J. Swies, A. Korbut, et al., "Regulation
      of PGI_2 release by mediators of autonomic nervous sys-
      tem," in: Seventh Congress of the Polish Pharmacology
      Society, A. Chodera (ed.), Academy of Medicine, Poznan
      (1980), p. 195.
149. Y. Harada, K. Tanaka, and M. Katori, "Acceleration of
      PGJ_2 generation from isolated rat aortas by MK-447 and
      AI-122," in: Eighth International Congress of Pharma-
      cology, Tokyo (1981), Abstract, p. 743.
150. A. Socini, E. Agradi, and C. Galli, "The coumarin deriv-
      ative AD_6 reduces platelet aggregation and thromboxane
      formation and enhances aortic prostacyclin production
      in normal and hyperlipaemic rabbits," in: Eighth Inter-
      National Congress of Pharmacology, Tokyo (1981), p.
      742.
151. J. Beitz, P. Hoffmann, and W. Förster, "Influence of
      renal lipids extracted from rats fed pre- and postnatal-
      ly a linoleic acid-rich or -deficient diet on the in
      vitro biosynthesis of prostaglandin I_2 (PGI_2)," Pharm.
      Pharmacother. Lab. Diagnos., 119, 1149-1150 (1980).

152. J. Nakao, W.-C. Chang, S.-J. Murota, and H. Orima,
     "Testosterone inhibits prostacyclin production by rat
     aortic smooth muscle cells in culture," Atherosclerosis,
     $\underline{39}$, 203-209 (1981).
153. K. M. Lakin, "Pharmacological effect on biosynthesis
     of thromboxane A_2 and prostacyclin in an organism," in:
     Physiologically Active Compounds in Medicine, Meditsina,
     Yerevan (1982), pp. 167-168.
154. R. J. Gryglewski, "Prostaglandin and thromboxane bio-
     synthesis inhibitors," Naunyn Schmiedebergs Arch. Phar-
     macol., $\underline{297}$, 85-91 (1977).
155. S. Moncada, P. Needleman, S. Hunting, et al., "Prosta-
     glandin endoperoxide and thromboxane generating systems
     and their selective inhibition," Prostaglandins, $\underline{12}$,
     323-334 (1976).
156. P. Needleman, H. Spreeher, M. O. Whitaker, and A. Wyche,
     "Mechanism underlying the inhibition of platelet aggre-
     gation by eicosapentaenoic acid and its metabolites,"
     in: Advances in Prostaglandin and Thromboxane Research,
     Vol. 6, B. Samuelsson, P. W. Ramwell, and R. Paoletti
     (eds.), Raven, New York (1980), pp. 61-68.
157. S. Moncada, S. Bunting, K. Mullane, et al., "Imidazole:
     a selective inhibitor of thromboxane synthetase," Pros-
     taglandins, $\underline{13}$, 611-620 (1977).
158. T. Umetsu and T. Kato, "Effect of 1-methyl-2-mercapto-
     5-(3-pyridyl)-imidazole (KC-6141) on rabbit platelet
     aggregation in vitro and rat platelet retention,"
     Thromb. Haemostas., $\underline{39}$, 167-176 (1978).
159. E. Vallee, J. Gougat, J. Navarro, et al., "Anti-inflam-
     matory and platelet anti-aggregant activity of phospho-
     lipase-A_2 inhibitors," J. Pharm. Pharmacol., $\underline{31}$, 588-
     592 (1979).
160. C. Malsten, E. Granström, and B. Samuelsson, "Cyclic
     AMP inhibits synthesis of prostaglandin endoperoxide
     (PGG_2) in human platelets," Biochem. Biophys. Res.
     Commun., $\underline{68}$, 569-574 (1976).
161. J. R. O'Brien, "Some effects of adrenaline and anti-
     adrenaline compounds on platelets in vitro and in vivo,"
     Nature, $\underline{200}$, 763 (1963).
162. M. J. Mant, "Platelet adherence to collagen: metabolic
     energy requirement," Thromb. Res., $\underline{17}$, 729-736 (1980).
163. M. Bettrex-Galland and E. F. Lüscher, "Studies on the
     metabolism of human blood platelets in relation to clot
     retraction," Thromb. Diath. Haemorrh., $\underline{4}$, 178-195 (1960).
164. Q. Z. Hussain and T. F. Newcomb, "Thrombin stimulation
     of platelet oxygen consumption rate," J. Appl. Physiol.,
     $\underline{19}$, 297-300 (1964).

165. B. K. Kim and M. G. Baldini, "Glycolytic intermediates and adenine nucleotides of human platelet. I. The influence of ACD and EDTA anticoagulants," Haematologia (Budap), 6, 447-457 (1972).

166. T. A. Odesskaya, "Thrombostenin — contractile protein of blood platelets — an important factor in dynamic transformations of platelets," Probl. Gematol. Pereliv. Krovi, 21, 37-40 (1976).

167. F. M. Booyse and M. E. Rafelson, Jr., "Human platelet contractile proteins: location, properties, and function," Ser. Haematol., 4, 152-174 (1971).

168. S. Puszkin, E. Puszkin, M. Katz, et al., "Control of platelet actomyosin activity: effect of ADP on super-precipitation and ATP-ase activity of human platelet actomyosin," Biochim. Biophys. Acta, 34, 102-112 (1974).

169. J. P. Kirkpatrick, L. V. McIntire, et al., "Differential effects of cytochalasin B on platelet release, aggregation, and contractility: evidence against a contractile mechanism for the release of platelet granular contents," Thromb. Haemostas., 42, 1483-1489 (1979).

170. Y. H. Abdulla, "Effect of water structure on platelet aggregation in vitro. A theoretical and experimental study," J. Atheroscl. Res., 7, 415-423 (1967).

171. J. P. Cazenave, H. J. Reimers, M. A. Packham, and J. F. Mustard, "Effects of reserpine on rabbit platelet aggregation and adherence to collagen or injured rabbit aorta," Biochem. Pharmacol., 26, 149-157 (1977).

172. E. S. Gabrielyan, S. E. Akopov, E. A. Amroyan, and A. G. Aivazyan, "Prostaglandin regulation of the functional state of the vascular wall-platelet system," in: Physiologically Active Compounds in Medicine, V. V. Zakusov (ed.), Ministry of Health of the USSR, Yerevan (1982), pp. 73-74.

173. K. M. Lakin, V. A. Feldbaum, and M. S. Ovnatanova, "Effect of some indole derivatives on platelet aggregation," Farmakol. Toksikol., 35, 711-714 (1972).

174. K. M. Lakin, Yu. F. Krylov, and V. D. Nepsha, "Effect of some aggregation inhibitors on the activity of thrombocytic enzymes," Farmakol. Toksikol., 41, 284-288 (1978).

175. K. M. Lakin, V. A. Makarov, L. N. Bobrova, et al., "Effect of fructose derivatives on platelet aggregational ability," Farmakol. Toksikol., 6, 692-694 (1981).

176. V. A. Makarov, N. V. Novikova, M. S. Ovnatanova, et al., "Pharmacological regulation of platelet aggregational ability," in: Physiologically Active Compounds in Medicine, V. V. Zakusov (ed.), Ministry of Health of the USSR, Yerevan (1982), pp. 181-182.

177. K. M. Lakin, V. A. Makarov, and A. G. Mular, "Der Einfluss einiger Induktoren und Inhibitoren der Aggregation auf die funktionellen Eigenschaften und die Ultrastruktur der Blutplättchen," Folia Haematol. Leipzig, 106, 849-852 (1979).

PLATELET SPREADING AND THE FORMATION OF THROMBI-LIKE AGGREGATES ON A COLLAGEN SUBSTRATE. EFFECT OF SOLUBLE INDUCERS OF PLATELET ACTIVITY UNDER DIFFERENT HYDRODYNAMIC CONDITIONS

A. V. Mazurov, V. L. Leytin, V. S. Repin, V. N. Smirnov

Laboratory of Cell Culture, Cardiology Research Center
Academy of Medical Sciences of the USSR, Moscow

W. Forster

Department of Pharmacology and Toxicology
Martin Luther University of Halle-Wittenberg, GDR

ABSTRACT

Soluble plasma inducers and inhibitors of platelet activity and fluid dynamics of the bloodstream are effective modulators of platelet–vessel wall interaction. The effects of two inducers of platelet activity, arachidonic acid (AA) and a stable prostaglandin endoperoxide analog (U46619), on platelet deposition on the bottom of multiwell tissue culture plates coated with fibrillar calf skin collagen (CSC) were studied by scanning electron microscopy. Both agents stimulate platelet spreading and the formation of large surface-bound, multilayer (thrombi-like) aggregates on a CSC substrate. The effects of AA and U46619 on spreading and thrombi-like aggregate formation depend on the speed of platelet suspension shaking during platelet deposition on the surface. In the absence of shaking, both inducers mostly stimulate the

spreading of platelets: Spread platelets fuse and form widespread sheets covering up to 50% of the CSC-coated surface.
An increase in the shaking speed leads to a decrease in platelet spreading, while the number of surface-bound thrombi-like
aggregates grows, reaching a maximum at a shaking speed of
40 back-and-forth cycles per minute.   The thrombi-like aggregates consist mainly of fused platelets and always contain
the basal sheet of spread platelets, suggesting that the latter participates in the attachment of aggregates to the surface.   Large aggregates are absent in the population of nonadherent platelets.   The data obtained indicate that AA metabolites participate in platelet spreading and thrombi-like aggregate formation — processes that are specific for platelet—
surface interactions.   Use of the proposed model for in vitro
study of platelet spreading and the formation of mural thrombi and for screening of antithrombotic and thrombolytic drugs
is discussed.

INTRODUCTION

Platelet interaction with damaged areas of vessel wall
is determined by three main factors: (a) composition and
contour of the surface which is exposed to the lumen after
injury and which plays an important role in primary attachment and surface-induced activation of adherent platelets;
(b) soluble inducers and inhibitors of platelet activity,
which can be liberated from circulating cells after injury or
lysis, or be released locally from adherent platelets and
vessel wall cells at the site of injury; and (c) blood flow
pattern at the luminal surface, which affects both the interaction of platelets with an exposed surface and soluble factors [1-6].

In order to investigate the role of exposed surface composition in the attachment and activation of adherent platelets, we previously analyzed the deposition of platelets on
surfaces coated with human collagen types I, III, IV, and V,
and with fibrillar calf skin collagen (CSC) [7-9].   We established that intensive platelet spreading and formation of
large, multilayer (thrombi-like) aggregates take place on the
surfaces coated with human collagen types I and III.   On surfaces coated with human collagen types IV and V and with CSC,
we observed only a weak activation of adherent platelets:
transformation of discoid platelets to spherical ones, formation of surface projections (pseudopods), and moderate spread-

ing. Thrombi-like aggregates are not formed on these collagenous substrates.

Many authors [10-13] have investigated, using a photometric method, the effects of various soluble platelet inducers [arachidonic acid (AA) and its derivatives, adenosine
diphosphate (ADP), Ca-ionophore A23187, and thrombin] on
change in platelet shape and aggregation of platelets in suspension. This technique, however, does not allow for the
study of the effects of inducers on processes that are specific for platelet—surface interaction, i.e., platelet spreading and formation of the substrate-bound platelet aggregates.
In the present study, we investigate by scanning electron
microscopy (SEM) the role of two soluble inducers of platelet activity (AA and U46619) in the interaction of platelets
with a nonthrombogenic substrate — a CSC-coated surface. The
effects of these agents were studied under various hydrodynamic conditions: incubation of platelets with the CSC
substrate without shaking or with shaking at different speeds.
It was shown that both inducers stimulate platelet spreading
and formation of surface-bound, thrombi-like aggregates. By
altering the hydrodynamic conditions, one can stimulate primarily platelet spreading (without shaking) or formation of
thrombi-like aggregates (horizontal shaking of a multiwell
at a speed of 40 back-and-forth cycles per min).

MATERIALS AND METHODS

Reagents

AA and U46619 [(15S-hydroxy-11α,9α)epoxymethano(5Z,13E)-
prostadienic acid] were a generous gift from The Upjohn Com-
pany. Other reagents were purchased from Sigma. AA and
U46619 were kept in 96% ethanol at −20°C under nitrogen. Immediately before use, an aliquot of AA was neutralized with
an equimolar amount of NaOH; ethanol was vacuum-evaporated;
and AA was dissolved in 0.15 M NaCl. U46619 was dissolved in
0.15 M NaCl.

Isolation of Platelets

Blood was obtained from the cubital vein of healthy donors
and was anticoagulated with 1/6 volume acid—citrate—dextrose—
apyrase (65 mM citric acid, 85 mm trisodium citrate, 2% dextrose, 2 mg/ml apyrase) (1.9 units of ADPase activity/mg protein).

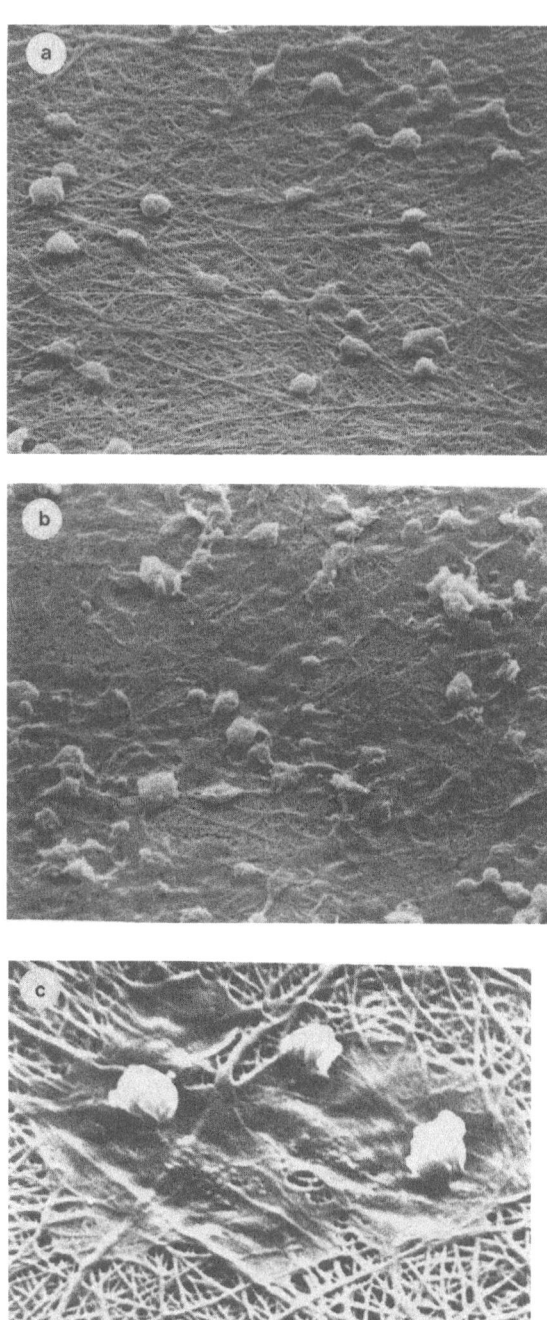

Fig. 14.1. Platelet spreading on CSC substrate in the ab-
            sence of soluble inducers of platelet activity
            (a), and stimulated by AA (b) or U46619 (c).
            Platelets were incubated in CSC-coated wells for
            40 min at 37°C without shaking; the samples were
            studied by SEM (see "Materials and Methods").
            a) Platelets adherent to fibrillar CSC are main-
            ly represented by single, unspread platelets of
            a spherical or discoid shape. Top right: local
            sheet of fused spread platelets with adherent
            unspread platelets; b) 200 µM of AA was added
            to the wells. Massive spreading of platelets on
            the CSC substrate was observed. Spread platelets
            are fused, forming vast sheets inside which the
            areas coated with CSC fibrils are observed.
            Single, unspread platelets and microaggregates
            are adherent to the upper surface of the sheets.
            The CSC-coated areas contain significantly less
            adherent unspread platelets than the sheets of
            spread platelets; c) 1 µM of U46619 was added
            to the wells. A group of three partly fused,
            spread platelets is seen on fibrillar CSC. Two
            have adherent spherical platelets on their upper
            surface.

Platelet-rich plasma was obtained by centrifugation at 200 g for 12.5 min at 20°C. Platelets were isolated from plasma by gel filtration on Sepharose 2B according to Tangen et al. [14]. The modified Tyrode solution, without Ca^{2+} and Mg^{2+}, containing 0.35% bovine serum albumin was used as an eluting buffer. The platelets were counted in a Coulter model 2B1 cell counter (Coultronics).

Preparation of CSC-Coated Surface

Acid-soluble CSC (Sigma, C-3511) was immobilized in fibrillar form on the bottom of 16-mm wells of multiwell tissue culture plates (Falcon) as described elsewhere [7, 8].

Platelet Deposition on CSC Substrate [7, 8]

$CaCl_2$ and $MgCl_2$ were added to the platelet suspension just before the incubation to final concentrations of 2 mM and 1 mM, respectively. Two hundred μliters of platelet suspension ($2-4 \cdot 10^7$ cells) were added into CSC-coated wells and incubated for 40 min at 37°C in a Dubnoff metabolic shaking incubator (GCA, Precision Scientific) without shaking or with shaking at the indicated back-and-forth cycles per min. AA and U46619 were added into the wells immediately prior to incubation.

Preparation of Samples for SEM

After incubation with the CSC substrate, nonadherent platelets were removed and nonadherent and adherent platelets were fixed with 2.5% glutaraldehyde in 0.1 M phosphate buffer (pH 7.3) for 1.5-2 h at 37°C. The nonadherent platelets were applied to 2.5-cm Nucleopore filters with a pore size of 0.4 μm. The fixed platelets were dried and coated with platinum-palladium or gold [7, 8].

Morphometric Analysis of Adherent and Nonadherent Platelets

Analysis was carried out in a Philips PSEM-500X scanning electron microscope. The number of adherent platelets was counted on 41 scanning fields (×5000) moving from the edge to the center of the well with a spacing of 0.2 mm. The area covered with spread platelets was measured using a MOP-3 automatic system (Reichert—Jung) and was expressed as the percentage of total area viewed. The number of thrombi-like aggre-

TABLE 14.1. Proportion of Spread and Unspread Platelets
Adherent to CSC Substrate in the Absence of
Soluble Platelet Inducers

Adherent platelets	Percentage of total adherent platelets	
	Shaking, cycles per min	
	0	40
Unspread		
Single	52.5 ± 5.5	63.2 ± 8.8
In microaggregates	31.5 ± 8.3	23.1 ± 6.9
In thrombi-like aggregates	0	0
Spread	16.0 ± 3.1	13.7 ± 6.6

0.2 ml of gel-filtered platelet suspension (2-$4 \cdot 10^7$ cells
per well) was added to the CSC-coated wells and incubated
for 40 min at 37°C without shaking (n = 4) or with shaking at 40 back-and-forth cycles per min (n = 5). Means ±
standard error of the means are given. The differences
between the means for the two groups (0 and 40 cycles per
min) are not significant.

gates was counted at ×640 and ×1250 continuously moving from
the edge to the center of the well. (Multilayer, surface-
bound aggregates consisting of more than 15 platelets per aggregate were classified as thrombi-like aggregates.) Morphometry of nonadherent platelets was performed by counting not
less than 100 cells.

Statistical Analysis

The significance of differences between the means (p)
was calculated using Student's t-test. The differences of
p > 0.05 were considered insignificant.

RESULTS

Interaction of Platelets with CSC Substrate
in the Absence of Soluble Inducers
of Platelet Activity

Five stages of platelet interaction with CSC substrate
were distinguished by SEM: (a) primary attachment of platelets to the substrate; (b) transformation of discoid platelets
into spherical ones, and formation of surface extrusions or
pseudopods; (c) spreading of platelets on the substrate; (d)
attachment of platelets from the suspension to the upper surface of the spread platelets; and (e) formation of small aggregates (microaggregates) consisting of two to seven platelets on the upper surface of the spread platelets and collagen-coated areas of the substrate [7, 8].

According to previous data [7, 8], CSC substrate has a
relatively small ability to stimulate platelet spreading in
the absence of soluble inducers (Fig. 14.1a). Spread platelets account for approximately 15% of the total number of adherent platelets (Table 14.1) and cover 2-6% of the collagenous substrate area (Table 14.2). Most of the adherent platelets remain in an unspread state: 50-60% are single spherical or discoid platelets; 20-30% are organized as microaggregates. In the absence of soluble inducers, thrombi-like aggregates are never formed on CSC substrate (Fig. 14.1; Table
14.1), i.e., the CSC substrate itself is not thrombogenic.
The platelet count in the different subpopulations of adherent
platelets depends weakly on the speed of shaking of the platelet suspension (Table 14.1).

Stimulation of Platelet Spreading
on CSC Substrate by AA and U46619

AA and U46619 stimulate an intensive spreading of platelets on CSC substrate. In the absence of shaking, spread
platelets usually fuse and form widespread cellular sheets
alternating with zones of free collagen (Fig. 14.1b). Separate groups consisting of several spread platelets are observed
relatively seldom (Fig. 14.1c). Massive spreading usually
makes the boundaries between the individual spread platelets
scarcely distinguishable. It is impossible, therefore, to
conduct morphometric analysis by directly counting the number
of spread platelets. For this reason, the quantitation of
platelet spreading consists of measuring the substrate area
covered with spread platelets.

TABLE 14.2.   AA- and U46619-Stimulated Spreading of
              Platelets on CSC Substrate

| Inducer | Surface covered with spread platelets, % | | |
| | Shaking, cycles per min | | |
	0	40	P
None	6.1 ± 4.0	2.2 ± 1.2	n.s.
AA	45.8 ± 7.2	20.5 ± 3.4	p < 0.01
	p < 0.005	p < 0.001	
U46619	43.7 ± 4.3	17.5 ± 4.5	p < 0.01
	p < 0.001	p < 0.01	

Prior to the incubation of platelets with CSC substrate,
0.15 M NaCl, 200 mM AA, or 1 mM U46619 was added to the
CSC-coated wells. The significance of differences between the means (p) for the groups without an inducer
and in the presence of AA or U46619 is given in the last
column, and the significance of differences between the
group shaken for 0 cycles per min (n = 4) and the group
shaken for 40 cycles per min (n = 5) is given in the
first two columns.
n.s. = Not significant.

In the absence of shaking, spread platelets cover up to
40-50% of the CSC substrate area (Fig. 14.1b; Table 14.2).
An increase in the shaking speed leads to a decrease in AA-
and U46619-stimulated spreading (Fig. 14.2a). At 40 cycles
per min, the area covered with spread platelets is 2- to 2.5-
fold smaller than that without shaking (Table 14.2).

AA- and U46619-Induced Formation of the
CSC-Bound Substrate Thrombi-like Aggregates

AA and U46619 stimulate the formation of surface-bound,
large, multilayer (thrombi-like) aggregates (Fig. 14.3) which
are never formed on the CSC substrate in the absence of soluble platelet inducers (Fig. 14.1a; Table 14.1). Thrombi-
like aggregates consist mainly of fused platelets (Fig. 14.3a
and b) and exhibit "viscous metamorphosis" specific for a

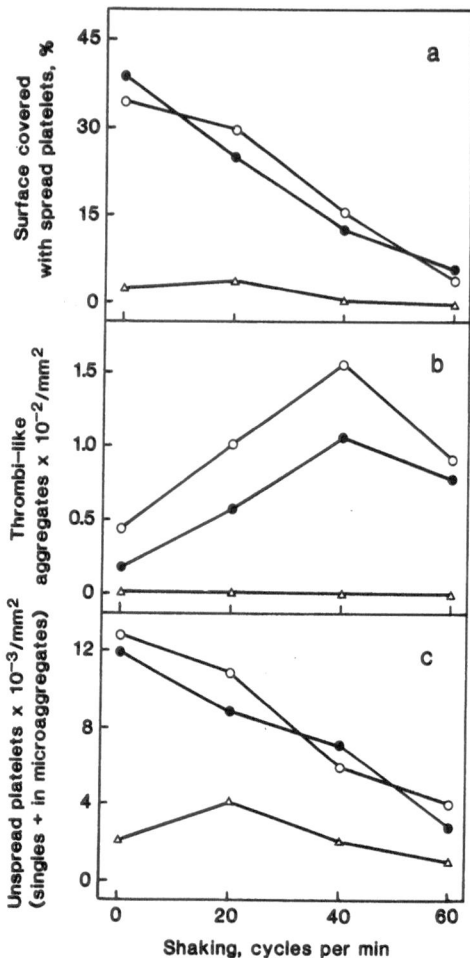

Fig. 14.2.   Effects of shaking the platelet suspension on
             platelet deposition on the CSC substrate. a)
             Spreading of platelets; b) formation of substrate-
             bound, thrombi-like aggregates; c) adhesion of
             single, unspread platelets and microaggregates.
             A suspension of gel-filtered platelets was added
             into the CSC-coated wells with subsequent addi-
             tion of 0.15 M NaCl (Δ), 200 μM of AA (o), or 1
             μM of U46619 (o). The platelets were incubated
             with CSC substrate for 30 min at 37°C without
             shaking or with shaking at 20, 40, and 60 back-
             and-forth cycles per min. Platelet deposits were
             quantified by SEM. The means of two experiments
             are presented.

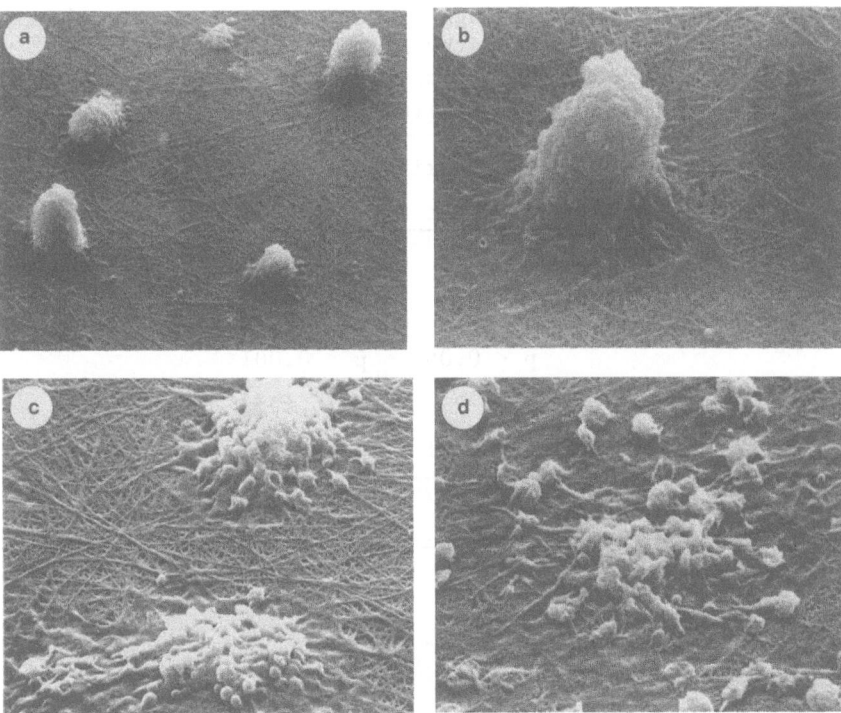

Fig. 14.3.   Thrombi-like aggregates formed on the CSC substrate in the presence of AA and U46619 at different speeds of shaking of the platelet suspension. a, b) The platelet inducer is 1 μM of
U46619, shaking speed 40 cycles per min. The
multilayer, thrombi-like aggregates are distributed discretely over the fibrillar CSC substrate.
They consist of fused platelets and contain local
sheets of spread platelets at the base. c) The
inducer is 1 μM of U46619, shaking speed 20 cycles
per min. The height of the thrombi-like aggregates is decreased. Fusion of the platelets is
less marked. Local sheets of spread platelets
underlie the thrombi-like aggregates. d) The inducer is 200 μM of AA, without shaking. A larger
part of the CSC substrate is covered with sheets
of spread platelets with adherent, low, thrombi-
like aggregates and microaggregates. Bottom
right: Two small areas of fibrillar CSC without
spread platelets.

TABLE 14.3. Thrombi-Like Aggregate Formation on CSC Substrate Induced by AA and U46619

Inducer	Thrombi-like aggregates per mm^2		
	Shaking, cycles per min		
	0	40	p
None	0	0	
AA	53 ± 22	197 ± 29	$p < 0.01$
	$p < 0.05$	$p < 0.001$	
U46619	35 ± 9	216 ± 53	$p < 0.01$
	$p < 0.01$	$p < 0.005$	

See legend for Table 14.2.

thrombus structure [15]. A layer of spread platelets always lies at the base of thrombi-like aggregates (Fig. 14.3a-d). Thrombi-like aggregates do not fuse, leaving most of the substrate area free (Fig. 14.3a). This mode of arrangement determines the method for quantifying the thrombi-like aggregates: the number of aggregates per square unit of substrate is counted.

As with spreading, the formation of thrombi-like aggregates on the CSC substrate depends largely on the speed of shaking of the platelet suspension. The number of AA- and U46619-induced thrombi-like aggregates grows linearly with an increase in shaking speed to 40 cycles per min; it then decreases at 60 cycles per min (Fig. 14.2b). At 40 cycles per min, the number of thrombi-like aggregates exceeds that without shaking by four- to sixfold (Table 14.3). Structural organization of thrombi-like aggregates is also affected by shaking: The aggregates formed in the absence of shaking (Fig. 14.3d) or at 20 cycles per min (Fig. 14.3c) are smaller in size, and the fusion of platelets is less pronounced compared to the aggregates that are formed at 40 cycles per min (Fig. 14.3a and b).

TABLE 14.4. Adhesion of Single, Unspread Platelets and Platelets in Microaggregates to CSC Substrate

Inducer	Unspread platelets $\cdot$ 10^{-3}/mm^2 (single + in microaggregates)		
	Shaking, cycles per min		
	0	40	p
None	5.4 ± 3.1	2.7 ± 0.7	n.s.
AA	20.3 ± 4.0	4.4 ± 1.0	p < 0.01
	p < 0.05	n.s.	
U46619	12.9 ± 1.1	4.5 ± 1.0	p < 0.02
	p < 0.05	n.s.	

See legend for Table 2.
n.s. = Not significant.

Adhesion of Single, Unspread Platelets on Microaggregates in the Presence of AA and U46619

Part of the platelets adherent to the substrate in the presence of AA and U46619 neither spread nor participate in the formation of thrombi-like aggregates. This fraction consists of single, unspread platelets and microaggregates (Figs. 14.1b and 14.3d). AA- and U46619-induced activation of platelets augments the total level of adhesion of single, unspread platelets and platelets in microaggregates to the CSC-coated wells. However, statistically significant differences are observed only in the absence of shaking (Table 14.4). After the addition of inducers, the number of adherent, unspread platelets decreases as the shaking speed increases (Fig. 14.2c). Without shaking, the level of adhesion of unspread platelets is three to four times higher than at 40 cycles per min (Table 14.4).

We also studied the effects of AA and U46619 on the distribution of platelets from the suspension between two types of platelet-binding sites: (a) the upper surface of platelets spread on CSC, and (b) the areas coated with CSC only (Table 14.5).

TABLE 14.5. Distribution of Adherent Unspread Platelets
Between Two Substrates: Upper Surface of
Platelets Spread out on CSC, and CSC-Coated
Areas Free from Spread Platelets

Inducer	Unspread platelets, adherent to spread platelets, %*	
	Shaking, cycles per min	
	0	40
None	45.3 ± 11.4	38.1 ± 13.7
AA	86.6 ± 6.3	75.8 ± 7.5
	$p < 0.01$	$p < 0.02$
U46619	85.2 ± 5.5	86.0 ± 3.1
	$p < 0.01$	$p < 0.01$

*We used as a 100% value the total number of unspread
platelets (single + in microaggregates) adherent to the
upper surface of platelets spread out on CSC and to the
areas of the well bottoms which were coated with CSC
only. The differences between the two groups (0 and 40
cycles/min) were not significant.

In the absence of the inducers, about 40% of the unspread platelets concentrate on the surface of the spread
ones, which occupy only 2-6% of the CSC substrate area (Table
14.2); about 60% are bound to the area coated with CSC only
(Table 14.5). Recalculation per square unit shows dramatic
differences in the density of binding: $82.3 \pm 23.8 \cdot 10^3$ platelets per mm^2 of spread platelet area and $2.0 \pm 0.7 \cdot 10^3$ platelets per mm^2 of free CSC-coated area (n = 9, $p < 0.001$). In
the presence of AA and U46619, 80-85% of the unspread platelets, which are not involved in the formation of thrombi-like
aggregates, bind to spread platelets. Distribution of unspread platelets between the two types of substrates does not
depend on the nature of the inducer or the conditions of shaking (Table 14.5).

As shown in Figure 14.4, both inducers cause: (a) a decrease in the percentage of single, unspread platelets; (b)
an increase in the percentage of platelets in microaggregates;
and (c) an increase in the size of microaggregates.

TABLE 14.6. Aggregates in the Population of Platelets Nonadherent to the CSC Substrate

Platelets*	No inducer		AA or U46619	
	Shaking, cycles per min			
	0	40	0	40
In thrombi-like aggregates, %	0	0	0	0
In microaggregates, %	10.9 ± 5.6	15.8 ± 2.6	10.3 ± 2.4	11.9 ± 1.7
Average number per microaggregate	2.0 ± 0.03	2.2 ± 0.12	2.3 ± 0.09	2.3 ± 0.10

Following interaction of the platelets with CSC-coated wells, aliquots of nonadherent platelet suspension were applied to Nucleopore filters (pore size 0.4 μM), and not less than 100 cells were counted by SEM.
*The total number of nonadherent platelets counted was used as the 100% value. The differences between the means of all types of groups (with and without an inducer; 0 and 40 cycles per min) were not significant (n = 4).

Microaggregates in Population of Platelets
Nonadherent to CSC Substrate

Interaction of platelets with the CSC substrate in the
presence of AA and U46619 induces the formation of surface-
bound, thrombi-like aggregates (Figs. 14.2b and 14.3; Table
14.4). However, no large aggregates were observed in the
population of platelets that was nonadherent to the CSC substrate following the action of the inducers (Table 14.6).
Moreover, independent of the shaking conditions, AA and U46619
neither stimulate the formation of nonadherent microaggregates
nor enlarge their size (Table 14.6), which is not the case
with adherent microaggregates (Fig. 14.4). These data demonstrate that practically all AA- and U46619-activated platelets bind to the CSC substrate.

DISCUSSION

This study established three main facts, as follows:

1. In the absence of soluble inducers of platelet activity, platelet–CSC substrate interaction is characterized by
a low level of spreading and a small number of surface-bound
microaggregates which mostly consist of two to three platelets.
CSC per se is a nonthrombogenic substrate, i.e., it is unable
to induce the formation of surface-bound,thrombi-like aggregates. A low level of activation of the platelets attached
to the substrate suggests that platelet deposition on the CSC
substrate in the absence of soluble inducers mainly reflects
the process of primary attachment of platelets to the surface.

2. AA and U46619, soluble inducers of platelet activity,
stimulate massive spreading of platelets and the formation of
thrombi-like aggregates on the CSC substrate.

3. The effects of AA and U46619 on the interaction of
platelets with CSC substrate depend on fluid dynamics. Alterations in the speed of shaking of the platelet suspension
during platelet deposition make it possible to stimulate selectively either platelet spreading or the formation of thrombi-like aggregates.

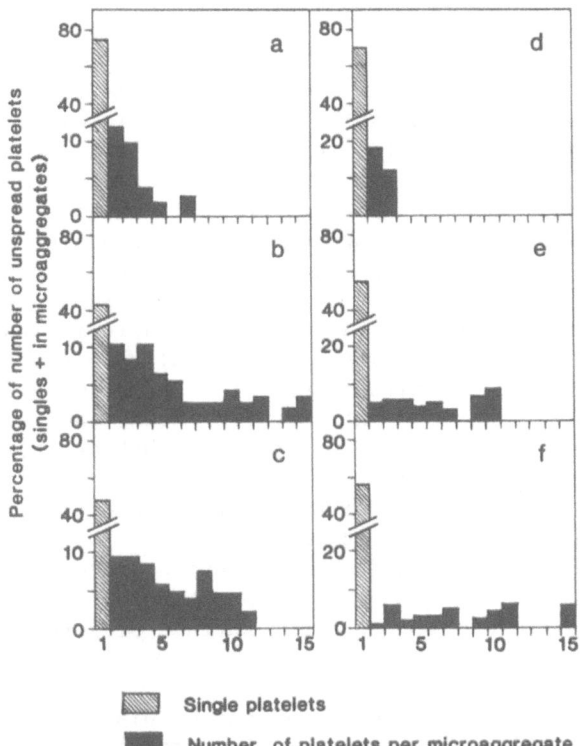

Single platelets

Number of platelets per microaggregate

Fig. 14.4. Histograms of the distribution of adherent, unspread platelets between the classes of single
platelets and microaggregates of different size
in the absence of soluble inducers (a, d) and in
the presence of 200 µM of AA (b, e) or 1 µM of
U46619 (c, f). Platelets were incubated with the
CSC substrate without shaking (a-c) or with shaking at 40 cycles per min (d-f). The means of
four experiments are presented. The number of
single, unspread platelets + the number of platelets in microaggregates was used as the 100% value.

Platelet Shape Change and Platelet–Platelet
Interactions in Suspension and on Solid Substrate

The above data were obtained on the borderline between
a platelet suspension and a solid substrate. This fact calls
for special comment. The shape change during platelet contact with the substrate and the platelet–platelet interaction

on the substrate differ considerably in a number of parameters from analogous processes in platelet suspension.

First, two types of platelet shape change take place in
suspension: (a) transformation of discoid platelets to a
more spherical shape and (b) formation of surface extrusions,
mostly long filiform projections (pseudopods). On a solid
substrate, platelets can undergo, in addition to these processes, a third type of morphologic transformation, i.e., the
formation of a flattened lamellar cytoplasm (spreading). This
process is likely a stronger morphologic manifestation of
platelet activation than a disc—sphere transformation and
ejection of surface extrusions. Kinetically, platelet spreading always follows the disc—sphere transformation and pseudo-
pod formation [7, 8], and intensive spreading takes place
only on highly adhesive or thrombogenic substrates [9].

Second, the character of platelet—platelet interaction
in suspension and on solid surfaces differs. Only one type
of interaction, i.e., cohesion of unspread platelets (aggregation), occur in suspension. Another type of interaction,
adhesion of platelets to the upper surface of spread platelets,
is observed on the substrate.

Thus, the specific manifestations of platelet interactions with a solid substrate are: (a) platelet spreading on
the substrate, (b) adhesion of unspread platelets from suspension on the upper surface of spread platelets, and (c)
formation of thrombi-like aggregates bound to spread platelets.

Soluble Inducers of Platelet Activity Stimulate
Spreading of Platelets on Substrate: A Shape
Change Specific for Platelet—Substrate Interaction

The spreading of platelets on a substrate is a necessary
stage in the growth of untransformed, nucleated cells in culture. To stimulate this process, highly adhesive substrates
are used and the adhesive properties of the substrates are
enhanced by pretreatment of the substrate with various spreading and growth factors. Spreading of platelets (unnucleated
cells) also depends on the adhesive characteristics of the
substrate. Platelets intensively spread on natural substrates
formed of thrombogenic human collagens (types I and III) and
on the artificial substrate, activated Falcon plastic. On
nonthrombogenic collagens (type IV human collagen and CSC),
platelet spreading is decreased substantially [7-9]. The

capability of the substrate to bind and spread platelets can
be increased by treating it with human plasma fibronectin
[16-18].

The present study has demonstrated a basically new way
to stimulate platelet spreading on the substrate by activation of platelets in suspension with soluble inducers. AA
and U46619 were used for this purpose. According to our data,
thrombin also actively stimulates platelet spreading on CSC
substrate, at a concentration of 0.5-5 units/ml. Further
studies will show whether this is a common feature of all
soluble inducers of platelet activity or whether there are
inducers which can activate platelets in suspension but which
are unable to activate their spreading on substrate.

It has been recently found that AA, thrombin, and ADP
stimulate the spreading of megakaryocytes, the cultured nucleated platelet precursors [19]. At present, it is still unclear whether the stimulating effects of the inducers are
limited to the cells of the platelet line or whether these
agents have a wider sprectrum of action.

Effect of Hydrodynamic Conditions on Formation of Surface-Bound, Thrombi-Like Aggregates and Platelet Spreading

In the model used, the shaking speed of the platelet
suspension is the main hydrodynamic factor in modulating the
occurrence of the two competitive processes: (a) platelet—
platelet interaction in suspension and (b) attachment of
platelets from suspension to the substrate.

Domination of platelet spreading in the absence of shaking is likely due to a high efficiency of the platelet—substrate interaction under these conditions. An increase in
the shaking speed leads to an increase in the frequency of
platelet—platelet contacts in suspension and, therefore, to
an increase in the ratio of effective platelet—platelet interactions to platelet—substrate interactions. The change in
this ratio accounts for (a) an increased number of thrombi-
like aggregates, (b) a decreased level of spreading, and (c)
a decreased number of single, adherent, unspread platelets
and platelets in microaggregates.

Regardless of the shaking conditions, a sheet of spread
platelets lies at the base of thrombi-like aggregates, leading

to the assumption that spread platelets participate in the
attachment of aggregates to the surface. However, the differences in the structure of thrombi-like aggregates following
changes in the hydrodynamic conditions suggest that these aggregates develop by different mechanisms. It is likely that,
in the absence of shaking and at a shaking speed of 20 cycles
per min, the growth of "low" thrombi-like aggregates may occur by a gradual buildup of a platelet mass on the preformed
surface of sheets of spread platelets. At more rapid shaking
(40 and 60 cycles per min), the thrombi-like aggregates are
formed first in suspension and then they are attached to the
substrate via the spreading of basal platelets. The absence
of thrombi-like aggregates in the population of nonadherent
platelets suggests that all aggregates preformed in suspension
effectively bind to the substrate. The possibility that surface-bound, thrombi-like aggregates may be formed by this
mechanism are confirmed by the data of Chervionke et al. [20].
These investigators have shown that platelet aggregates preformed in an aggregometer cuvette effectively bind to a plastic surface and to sheets of smooth muscle cells and fibroblasts.

Application of the Model

The use of soluble inducers of platelet activity substantially contributes to the development of methods for
studying thrombi formation in vitro. This approach makes it
possible to create conditions for massive spreading of platelets and/or the formation of surface-bound, thrombi-like aggregates. With this approach, one can study the (a) role of
various metabolic pathways of platelet activation in platelet
formation, (b) alterations of the cytoskeleton during platelet spreading, (c) topography of receptor distribution on the
surface of spread platelets, and (d) mechanism of mural thrombi formation via the deposition of platelet mass on the sheets
of spread platelets. The proposed model is a convenient method for screening antithrombotic and thrombolytic drugs, and
for studying their mechanisms of action.

REFERENCES

1. J. F. Mustard and M. A. Packham, "Platelets and throm-
    bosis in the development of atherosclerosis and its com-
    plications," in: Thrombosis: Animal and Clinical Models,
    H. J. Day, B. A. Molony, E. E. Nishizawa, and R. H. Ryn-
    brandt (eds.), Plenum Press, New York (1978), p. 7.

2. R. M. Jaffe, "Interaction of platelets with connective
    tissue," in: Platelets in Biology and Pathology, J. L.
    Gorman (ed.), North-Holland Publishing Company, Amster-
    dam (1976), p. 23.
3. H. R. Baumgartner, R. Muggli, T. B. Tschopp, and V. T.
    Turrito, "Platelet adhesion, release, and aggregation
    in flowing blood: effects of surface properties and
    platelet function," Thromb. Haemostas., 35, 124 (1976).
4. G. V. R. Born and M. A. A. Kratzer, "Endogenous agents
    in platelet thrombosis," Acta Med. Scand. [Suppl.], 651,
    85-90 (1981).
5. E. F. Grabowski, "Roles of blood flow in platelet ad-
    hesion and aggregation," in: Thrombosis: Animal and
    Clinical Models, H. J. Day, B. A. Molony, E. E. Nishi-
    zawa, and R. H. Rynbrandt (eds.), Plenum Press, New
    York (1978), p. 73.
6. M. B. Stemerman, "Interaction of vessel wall with formed
    blood elements," in: Contemporary Hematology/Oncology,
    J. Lo Bue, A. S. Gordon, R. Sieber, and P. M. Muggia
    (eds.), Plenum Medical Book Company, New York (1979),
    p. 47.
7. V. L. Leytin, E. V. Lyubimova, D. D. Sviridov, V. S.
    Ripin, and V. N. Smirnov, "Time-response changes in the
    thrombogenicity of platelets spread on a collagen-coated
    surface," Thromb. Res., 20, 335 (1980).
8. V. L. Leytin and D. D. Sviridov, "A model for studying
    platelet interaction with cellular and macromolecular
    constituents of the vessel wall in vitro," in: Vessel
    Wall in Athero- and Thrombogenesis. Studies in the USSR,
    V. N. Smirnov and E. I. Chazov (eds.), Springer-Verlag,
    Berlin (1982), p. 173.
9. V. L. Leytin, S. P. Domogatsky, V. E. Koteliansky, A. V.
    Mazurov, F. Misselvitz, O. A. Merzlikina, E. A. Podrez,
    Ch. Taube, and W. Förster, "Platelet spreading and throm-
    bi formation in vitro," in: Ninth World Congress of
    Cardiology, E. I. Chazov (ed.), Plenum, New York (in
    press).
10. G. V. Born, "Aggregation of blood platelets by adenosine
    diphosphate and its reversal," Nature, 194, 927 (1962).
11. R. L. Kinlugh-Rathbone, H. J. Reimers, J. F. Mustard,
    and M. A. Packham, "Sodium arachidonate can induce plate-
    let shape change and aggregation which are independent
    of the release reaction," Science, 192, 1011-1012 (1976).
12. R. L. Kinlough-Rathbone, M. A. Packham, H. J. Reimers,
    J. P. Cazenave, and J. F. Mustard, "Mechanisms of plate-
    let shape change, aggregation, and release induced by

collagen, thrombin, or A23187," J. Lab. Clin. Med., 90, 707 (1977).

13. M. J. Silver, J. B. Smith, C. M. Ingerman, and J. Kocsis, "Arachidonic acid-induced human platelet aggregation and prostaglandin formation," Prostaglandins, 4, 863 (1973).

14. O. Tangen, H. J. Berman, and P. Marfey, "Gel filtration: a new technique for separation of platelets from plasma," Thromb. Diath. Haemorrh., 25, 268 (1971).

15. J. G. White, "Platelet morphology," in: The Circulating Platelet, S. A. Johnson (ed.), Academic Press, New York (1971), p. 45.

16. V. E. Koteliansky, V. L. Leytin, D. D. Sviridov, V. S. Repin, and V. N. Smirnov, "Human plasma fibronectin promotes the adhesion and spreading of platelets on surfaces coated with fibrillar collagen," FEBS Lett., 123, 59 (1981).

17. R. O. Hydes, I. U. Ali, A. T. Destree, V. Maunter, M. E. Perkins, D. R. Senger, D. D. Wagner, and K. K. Smith, "A large glycoprotein lost from the surfaces of transformed cells," Ann. N. Y. Acad. Sci., 312, 317-342 (1978).

18. R. Grinnell, M. Feld, and W. Snell, "The influence of cold insoluble globulin on platelet morphological response to substrate," Cell Biol. Int. Rep., 3, 585-592 (1979).

19. R. M. Leven and V. T. Nashmias, "Cultured megakaryocytes: changes in the cytoskeleton after ADF-induced spreading," J. Cell Biol., 92, 313 (1982).

20. R. L. Chervionke, J. C. Hoak, and G. L. Fry, "Effect of aspirin on thrombin-induced adherence of platelets to cultured cells from the blood vessel wall," J. Clin. Invest., 62, 847 (1978).

IMPAIRED PLATELET ADHESION TO COLLAGEN-COATED SURFACE IN PATIENTS WITH BLEEDING DUODENAL ULCER (DIRECT MEASUREMENT OF PLATELET ADHESION IN A COULTER-TYPE CELL COUNTER FOLLOWING ELUTION OF ADHERENT PLATELETS)

V. L. Leytin, F. Misselwitz, A. G. Yumashkina,*
M. F. Bondarenko,** E. V. Lyubimova,
Yu. A. Sharova,* V. S. Repin, and V. N. Smirnov

*Cell Culture Laboratory, Cardiology Research Center, Academy
of Medical Sciences of the USSR, Moscow*

ABSTRACT

Platelet adhesion to a surface coated with fibrillar calf
skin collagen was examined to study the possible role of platelets in the pathogenesis of hemorrhagic complications caused
by peptic ulcer. The adhesion was measured using a specially
developed procedure based on quantitative elution of adherent
platelets with 0.25% trypsin solution and subsequent counting
of eluted platelets in a Coulter-type cell counter. By this
method, platelet adhesion to the collagen-coated surface was
measured in 8 patients with uncomplicated peptic ulcer, 6 patients suffering from peptic ulcer complicated by hemorrhages,
and 10 healthy subjects. The total amount of adhesion in the
patients with bleeding ulcer was 5620 ± 720 platelets/mm² of

*Department of Gastroenterology, Central Institute for Advanced
Medical Training, Moscow.
**Department of Pathophysiology, Central Institute for Ad-
vanced Medical Training, Moscow.

the collagen-coated surface, which is 1.6-fold lower than the healthy donors (8910 ± 580, p < 0.001). Platelets of the ulcerous patients without bleeding showed an intermediate adhesiveness (7660 ± 550) that is significantly higher than the patients with bleeding ulcers (p < 0.01), but similar to the healthy subjects. Different morphologic subpopulations of adherent platelets, reflecting successive stages of surface-induced activation, were studied by scanning electron microscopy. It was established that the healthy donors and ulcerous patients have the same percentage of platelets from different morphologic subpopulations. The results suggest that an impaired interaction of platelets with collagenous substrate in bleeding duodenal ulcer is involved in the stage of initial attachment of platelets to the substrate, while subsequent stages of surface-induced activation remain intact.

INTRODUCTION

Adhesion of platelets to collagenous components of vascular subendothelium is one of the triggering events in the formation of hemostatic plugs and in thrombogenesis. As a link in the platelet-vessel wall interaction and coagulation mechanisms of hemostasis, platelets may participate in the pathogenesis of inherited or acquired disorders of blood coagulation. Analysis of platelet aggregation and release, and of platelet interaction with an injured vessel wall or artificial surface in vitro, is often used as a diagnostic test [1-6].

From 10 to 30% of patients with peptic ulcer suffer from hemorrhagic complications, yet the mechanism by which these complications develop remains mostly obscure. Certain studies point to the possibility that platelets participate in the pathogenesis of these hemorrhages [7, 8]. In routine clinical diagnosis of ulcers, the blood coagulation system is evaluated only by global tests (bleeding time, clot observation test, partial thromboplastin time, thromboelastography, etc.) and by the content of plasma clotting factors. Seldom is platelet aggregability studied using aggregometry or the platelet retention test on glass bead columns [8-11]. There is an obvious shortage of simple techniques for quantifying platelet adhesion in clinical practice.

The procedures that have been described for the measurement of platelet adhesion can be divided into two main groups.

The first group of procedures is based on direct counting of
adherent platelets using light or electron microscopy [4,
12]. These methods are laborious and time-consuming. The
second group of procedures involves radioactive labeling of
platelets and subsequent measuring of the substrate-bound
radioactivity [13, 14], which requires preliminary separation
of platelets from the excess radioactive label in the incubation mixture. One can also count adherent platelets by calculating the difference between the number of adhered and
nonadhered platelets. This technique, however, is only possible when the percentage of adhered platelets is relatively
substantial ($\geq 50\%$).

In this chapter, we suggest a new procedure for measuring platelet adhesion to a surface coated with fibrillar collagen in the presence of plasma. The method is based on quantitative elution of adherent platelets with a 0.25% trypsin
solution, and subsequent counting of eluted platelets in a
Coulter-type cell counter. The technique gives reproducible
results which agree well with platelet-adhesion measurements
using the radioisotopic procedure, but it does not involve
preliminary separation of platelets from plasma [16].

Using this technique, we studied platelet adhesion in
patients suffering from peptic ulcer. A significant decrease
in the number of adherent platelets is found in patients with
duodenal ulcer and a history of hemorrhagic episodes compared
to both healthy donors and patients without hemorrhages. To
determine which stage of platelet—collagen interaction is impaired during peptic ulcer, the population of adherent platelets was studied by scanning electron microscopy. It was
shown that adhesion-induced activation (change in shape of
adherent platelets) does not differ between patients suffering
from peptic ulcer with hemorrhagic complications and healthy
donors. These results suggest that a defect of the platelet hemostatic function is localized in the stage of initial
attachment to the collagenous substrate.

MATERIALS AND METHODS

Isolation of Platelets

We collected 10-12 ml of blood into plastic tubes containing 1/7 final volume of ACD and anticoagulant (acid—citrate—
dextrose and apyrase). Platelet-rich plasma (PRP) was obtained

by centrifugation for 10 min at 200g and 37°C [17, 18]. In
the experiments in which the two types of measurements for
platelet adhesion (radionuclide technique and electron resistive-pulse-sensing procedure for platelet counting in a Coulter cell counter were compared), PRP was divided into two
portions. Platelet-free plasma (PFP) was obtained from one
portion by centrifuging PRP for 30 min at 3000g with subsequent filtration of the supernatant through a Nucleopore filter (pore size 0.4 μm). $Na_2^{51}CrO_4$ (100-400 mCi/mg chromium,
Amersham, 100 μCi/ml of PRP) was added to another portion
of PRP, and ^{51}Cr-labeled platelets were isolated by gel filtration on Sepharose 2B (Pharmacia) [19]. Elution of platelets from the column was carried out with a modified Ca^{2+}-
and Mg^{2+}-free Tyrode solution containing apyrase (Sigma, 0.2
mg/ml) and bovine serum albumin (Sigma, 3.5 mg/ml). Platelets were counted with a Coulter-type cell counter [20]
(Platelet Counter, PL-100, TOA Medical Electronics, Japan);
radioactivity was measured on a gamma-spectrometer (Searle
1176z, USA).

Measurement of Platelet Adhesion
with the Coulter-Type Cell Counter

In the experiments in which platelet adhesion was measured by the radionuclide and cell-counter technique, we used
PRP obtained by mixing one volume of ^{51}Cr-labeled platelets
with one to two volumes of unlabeled autologous PFP (for a
final concentration of 1-$2 \cdot 10^8$ platelets/ml PRP). Platelet
adhesion was carried out in multiwell tissue culture plates
(Falcon) coated with fibrillar calf skin collagen (Sigma, 150-
200 μg protein/cm²) [17, 18]. Into each well we placed 200
μliters of PRP, and the multiwell was rotated in a horizontal
incubator-shaker (Lab-Line, USA) for 40 min at 36 rpm and 37
The suspension of nonadherent platelets was removed and
the adherent platelets were washed off the surface by stepwise elution. For this purpose, the wells were washed three
times with 200 μliters of 0.25% trypsin (GIBCO) in 0.1 M
phosphate buffered saline (each washing continued for 20 min
at 37°C and 36 rpm, if not stated otherwise).

The eluates were combined, and the number of platelets
eluted from the surface was counted in the cell counter (Platelet Counter, PL-100) or by radioactivity of the eluate and
specific activity of PRP expressed in cpm per 10^6 platelets.
The number of platelets remaining on the surface after one,
two, or three subsequent elutions was also estimated; for

this, the multiwell bottoms were excised and the surface-ad-
sorbed radioactivity was measured. Free (not bound to platelets) ^{51}Cr in PRP was determined in the supernatant after
pelleting the platelets by centrifugation.

Unlabeled PRP (1.5·10^8 platelets/ml) was used in the
experiments to measure platelet adhesion in healthy donors
and ulcerous patients. Adherent platelets were eluted three
times for 20 min each at 37°C and 36 rpm; the adhesion was
measured in the cell counter in four aliquots. The fifth
well was used for scanning electron microscopy. The total
amount of adhesion was expressed as the number of adherent
platelets per mm^2 of the collagen-coated surface. The data
represent the means ± SEM.

Scanning Electron Microscopy

Adherent platelets were fixed with 25% glutaraldehyde
(Sigma), dehydrated in graded ethanol (40-100%), dried, and
coated with platinum-palladium (Jeol Fine Coat Sputter-1100,
Japan). The adherent platelets were analyzed in a scanning
electron microscope (Philips, PSEM-500×) [17, 18].

Selection of Donors and Patients

We examined 10 healthy male donors who had taken no
drugs for 2 weeks preceding the experiment, 6 patients (5
males and 1 female, aged 35-66 years) who were suffering from
duodenal bulb ulceration and with a history of bleeding, and
8 patients (6 males and 2 females, aged 25-73 years) who were
suffering from duodenal bulb ulcer without hemorrhage. In all
of the patients, the presence of ulcer in the duodenal cap
was corroborated by endoscopy.

                              RESULTS

Development of Procedure for Measuring
Platelet Adhesion to the Surface Via Elution
and Subsequent Counting in the Cell Counter

Platelet adhesion directly from PRP was measured in the
walls of multiwell tissue culture plates coated with fibrillar calf skin collagen [17, 18]. The amount of platelet adhesion to the surface was determined by elution of adherent
platelets and counting in the Coulter-type cell counter [16].

Elution was carried out with 0.25% trypsin solution. Radioisotopic technique, i.e., measurement of ^{51}Cr-labeled adherent platelets, was used for comparison. Toward this end,
the platelets were purified of labeled plasma components by
gel filtration, and the PRP was reconstituted from ^{51}Cr-label-
ed gel-filtered platelets and unlabeled autologous PFP. The
efficacy of elution was evaluated by the number of adherent
platelets remaining on the surface after one, two, or three
elutions by trypsin solution (Figure 15.1). The results demonstrate that washing of the wells three times at 37°C with
moderate shaking (three times for 20 min each at 36 rpm) en-
sures quantitative elution of adherent platelets. Less prolonged elution (three times for 10 min each) without shaking
at room temperature results in the elution of about half of
the adherent platelets (Figure 15.1).

The number of platelets eluted from the surface was measured using the radionuclide technique and in the Coulter-type
cell counter. The results obtained by both procedures correlated well in the dose interval of $0.5-3.0 \cdot 10^7$ platelets
per one well (Figure 15.2). Elution does not lead to lysis
of adherent platelets; the amount of free ^{51}Cr in the eluted
fractions does not exceed 3-8% compared to 1-3% in the initial platelet suspensions.

Platelet Adhesion in Patients
with Peptic Ulcer

We studied the adhesion of platelets from PRP to the collagen-coated surface in 10 healthy donors (control group), 6
ulcerous patients with a history of bleeding, and 8 patients
suffering from ulcer without hemorrhages. The amount of adhesion was determined using the elution-counting procedure
described above.

Platelet adhesion to the collagen substrate in patients
suffering from hemorrhages was 5620 ± 720 platelets/mm², which
is 1.6-fold lower than the control group (8910 ± 580) (p <
0.001). Platelets of patients suffering from an ulcer without
bleeding showed an intermediate adhesiveness (7660 ± 550 plate-
lets/mm²), which is significantly higher than the patients
with a bleeding ulcer (p < 0.01), but practically similar to
the controls (Table 15.1).

To determine which stage of platelet-collagen substrate
interaction is impaired in peptic ulcer, the population of

Table 15.1. Platelet Adhesion to a Collagen-Coated Surface
in Healthy Donors and Patients Suffering from
Peptic Ulcer With and Without Hemorrhages

Group	No.	Adherent platelets per 1 mm^2	
Healthy donors	10	8910 ± 580	$p < 0.001$
Patients suffering from peptic ulcer:			
with hemorrages	6	5620 ± 720	$p < 0.01$
without hemorrhages	8	7670 ± 550	

The number of adherent platelets (three aliquots for each
subject) was measured in a PL-100 platelet counter (TOA Medical Electronics, Japan). The values represent the means ±
SEM; statistical significance was calculated using Student's
t-test for two means.

adherent platelets was studied by scanning electron microscopy. We detected the following morphologic forms of adherent
platelets, which reflect successive stages of platelet activation during interaction with the substrate: (a) nonactivated platelets—discs, (b) discoid platelets with pseudopods,
(c) spheroid platelets without pseudopods or (d) with pseudopods, (e) bipolar sticklike platelets, (f) spread platelets,
and (g) platelets in microaggregates [17, 18, 21]. The proportion of native, nonactivated platelets to different types
of activated platelets did not differ significantly between
the healthy donors and the patients suffering from an ulcer
(Figure 15.3).

DISCUSSION

The pathogenesis of hemorrhagic complications of peptic
ulcer remains mostly obscure. As yet, it is unclear which
links in hemostasis are impaired in cases of gastroenteric
bleeding. The cause of these hemorrhages is likely to be un-
related to inherited or acquired defects in the intrinsic
blood clotting system [9, 11]. Some authors have tried to
link the bleeding syndrome in ulcerous patients to a reduced
platelet count [7, 11]. The tendency for moderate thrombo-
cytopenia, however, is insignificant and can hardly account
for acute gastroenteric hemorrhages.

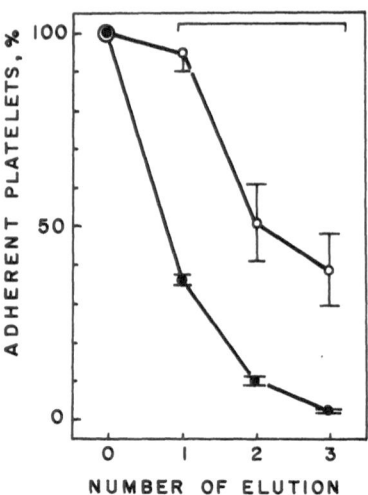

Fig. 15.1. The efficacy of elution of adherent platelets
            from a collagen-coated surface under differ-
            ent conditions. The number of ^{51}Cr-labeled
            platelets remaining on the surface after one,
            two, and three consecutive elutions was esti-
            mated by the surface-adsorbed radioactivity.
            The initial number of adherent platelets
            after the removal of nonadherent platelet sus-
            pension was used as the 100% value (⊙) . Elu-
            tion was performed with 0.25% trypsin solu-
            tion for 10 min at 22°C without shaking (o—o)
            or with shaking at 36 rpm for 20 min at 37°C
            (●—●). The values represent the means ± SEM,
            n = 4.

     Various functional tests are the most informative way
for studying the platelets of patients with peptic ulcer. An
impaired platelet aggregation in suspension has been found
in patients suffering from bleeding ulcer [8-11]. Estimation
of platelet adhesiveness is one of the most important tests
for characterizing the participation of platelets in thrombogenesis and formation of blood clots. A decreased platelet
retention on glass bead columns has been demonstrated in patients with peptic ulcer [11]. This test, however, mostly
characterizes not the platelet adhesion per se, but their subsequent massive aggregation [15]. There is a series of adequate methods to study platelet adhesion in vitro [12-14].

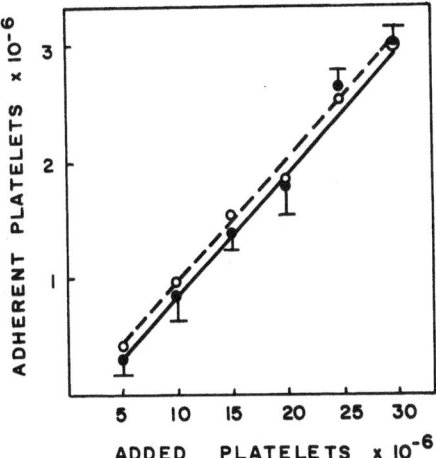

Fig. 15.2.   Dose-response of platelet adhesion to a col-
            lagen-coated surface. The number of adher-
            ent platelets was determined in the pooled
            eluate following three consecutive elutions
            with 0.25% trypsin solution (20 min, 37°C,
            36 rpm). o—o: Counting in a Coulter-type
            cell counter (PL-100, Platelet Counter),
            means, n = 2. ●—●: Measurement of ^{51}Cr-
            eluted radioactivity, means ± SEM, n = 4.

Baumgartner's perfusion chamber is based on the morphometry
of platelets adherent to the subendothelium of everted de-
endothelialized vessels by light or electron microscopy [4,
12]. This procedure, however, is time-consuming and laborious, which considerably limits its application in routine
clinical practice for examining large groups of patients.
Other methods [13, 14] involve the measurement of ^{51}Cr-labeled
platelet adhesion by surface-bound radioactivity. This technique requires preliminary separation of the labeled platelets
from labeled plasma components by gel filtration or a washing
procedure and, consequently, does not allow measurements in a
medium containing plasma coagulation factors.

    In vivo platelet interaction with damaged vessel wall
occurs in the presence of plasma, and the fraction of adherent platelets is negligible compared to the total number of
circulating platelets. Therefore, the amount of platelet adhesion cannot be determined by calculating the difference between the number of initial and nonadherent platelets.

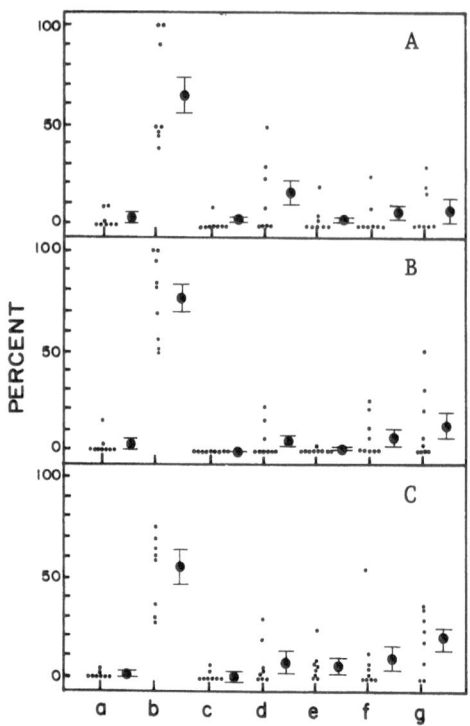

Fig. 15.3. Scatter diagrams reflecting the relation between
            the different forms of platelets adherent to a
            collagen-coated surface. The total number of ad-
            herent platelets was used as the 100% value. The
            differences between the percentage of platelets
            in the different subpopulations (a-g) of adherent
            platelets in the group of healthy donors (A, n =
            10) and groups of ulcerous patients without hem-
            orrhages (B, n = 8) and with hemorrhages (C, n =
            6) are not statistically significant according
            to Student's t-test for two means. a) Discs with-
            out pseudopods; b) discs with pseudopods; c)
            spheres without pseudopods; d) spheres with
            pseudopods; e) bipolar sticklike platelets; f)
            spread platelets; g) platelets in microaggregates.

    To study platelet adhesiveness in patients suffering
from peptic ulcer, we developed a method based on quantitative
elution of adherent platelets and subsequent counting by a
resistive-pulse-sensing technique in the Coulter-type cell
counter [20]. Fibrillar calf skin collagen immobilized on

the well bottom of multiwell tissue culture plates was used as a substrate. This method makes it possible to study (a) platelet adhesion on a natural substrate in the presence of plasma under standard and reproducible conditions, and (b) initial events of adhesion, which are not accompanied by massive aggregation. This method is both simple and inexpensive. It is not time-consuming and it allows mass measurements of platelet adhesion in the presence of plasma [16]. Application of this technique makes it possible to reveal a significant decrease in platelet adhesion to the collagen-coated surface in patients with peptic ulcer complicated by hemorrhages.

To determine which stage of platelet—surface interaction is impaired, the proportion of different forms of activated platelets to nonactivated ones was studied by scanning electron microscopy. Morphologic subpopulations of adherent platelets [17, 18, 21], which reflect successive stages of surface-induced platelet activation, were distinguished. It was shown that the ratios of various forms of activated platelets to each other and the ratio of activated platelets to nonactivated ones do not differ significantly between healthy donors and patients with ulcer. These data likely indicate that the defect is in the initial attachment of platelets to the collagen substrate. The platelets which come into primary contact with the collagenous surface are then normally activated and undergo a change in shape.

The data obtained suggest that platelets participate in the pathogenesis of gastroenteric hemorrhages caused by peptic ulcer. Experiments with platelet-rich plasma reconstituted from platelets and platelet-free plasma of ulcerous patients and healthy donors are likely to show whether this defect is localized in platelets themselves or in plasma factors which probably mediate the platelet—collagen interaction.

REFERENCES

1. J. L. Gordon and A. J. Milner, "Blood platelets as multifunctional cells," in: Platelets in Biology and Pathology, J. L. Gordon (ed.), Elsevier/North-Holland Biomedical Press, Amsterdam (1976), pp. 3-22.
2. A. L. Copley, "Roles of platelets in physiological defense mechanisms and pathological conditions," Folia Haematol. (Leipzig), 106, 732-764 (1979).

3. M. I. Barnhart, "Platelet responses in health and disease," Mol. Cell. Biochem., 22, 113-137 (1978).
4. H. R. Baumgartner, T. B. Tschopp, and H. J. Weiss, "Platelet interaction with collagen fibrils in flowing blood, II. Impaired adhesion-aggregation in bleeding disorders," Thromb. Haemostas., 37, 17-28 (1977).
5. G. V. R. Born, "Aggregation of blood platelets by ADP and its reversal," Nature, 194, 927 (1962).
6. A. Hellem, "Platelet adhesiveness in von Willebrand's disease. A study with a new modification of the glass-filter method," Scand. J. Haematol., 7, 374-382 (1970).
7. S. Langer and R. Stauch, "Frühe postoperative Rezidiv-blutung," in: Das Komplizierte Gastroduodenale Ulkus, Hrsg., R. Haring (ed.), Georg Thieme Verlag, Stuttgart, (1978), pp. 53-61.
8. F. W. Green, M. M. Kaplan, L. E. Curtis, and P. H. Levine, "Effect of acid and pepsin on blood coagulation and platelet aggregation: A possible contributor to prolonged gastroduodenal mucosal hemorrhage," Gastro-enterology, 74, 38-43 (1978).
9. A. S. Belousov, A. G. Yumashkina, N. Ya. Lagutina, G. A. Papikyan, G. D. Grigoryan, and S. A. Verkhovod, "Prediction of hemorrhagic complications of peptic ulcer," Klin. Med., 3, 40-44 (1979).
10. A. S. Belousov, G. A. Papikyan, N. Ya. Lagutina, G. V. Leontieva, and M. A. Tumanyan, "Tissue fibrinolysis in the mechanisms of gastroduodenal hemorrhage in patients with peptic ulcer," Klin. Med., 2, 23-28 (1981).
11. A. G. Yumashkina, Yu. A. Sharova, and E. M. Sharshukova, "Platelet-related hemostasis in patients with peptic ulcer complicated by hemorrhages," Klin. Med., 11, 41-45 (1980).
12. H. R. Baumgartner and R. Muggli, "Adhesion and aggregation: morphological demonstration and quantitation in vivo and in vitro," in: Platelets in Biology and Pathology, J. L. Gordon (ed.), Elsevier/North-Holland Biomedical Press, Amsterdam (1976), pp. 23-60.
13. J. P. Cazenave, D. Blandowska, M. Richardson, R. L. Kinlough-Rathbone, M. A. Packham, and J. F. Mustard, "Quantitative radioisotopic measurement and scanning electron microscopic study of platelet adherence to a collagen-coated surface and to subendothelium with a rotating probe device," J. Clin. Lab. Med., 93, 60-66 (1979).
14. L. F. Brass, D. Daile, and H. B. Bensusan, "Direct measurement of the platelet—collagen interaction by affinity chromatography on collagen—Sepharose," J. Lab. Clin. Med., 87, 525-534 (1976).

15. J. McPherson and M. B. Zucker, "Platelet retention in glass bead columns: Adhesion to glass and subsequent platelet—platelet interactions," Blood, 47, 55-64 (1976).

16. V. L. Leytin, F. Misselwitz, E. V. Lyubimova, V. S. Repin, and V. N. Smirnov, "Automated technique for measuring platelet adhesion to a surface in the presence of plasma," Thromb. Res., 23, 201-205 (1981).

17. V. L. Leytin, E. V. Lyubimova, D. D. Sviridov, O. S. Zakharova, V. S. Repin, and V. N. Smirnov, "Endothelial cell cultures on fibrillar collagen: Partial reconstruction of the vessel wall and platelet adhesion," Thromb. Res., 20, 509-521 (1980).

18. V. L. Leytin and D. D. Sviridov, "A model for studying platelet interaction with cellular and macromolecular constituents of the vessel wall in vitro," in: Vessel Wall in Athero- and Thrombogenesis, E. I. Chazov and V. N. Smirnov (eds.), Springer-Verlag, New York (1982), pp. 173-194.

19. O. Tangen, H. J. Berman, and P. Marfey, "Gel filtration. A new method of separation of blood platelets from plasma," Thromb. Diath. Haemorrh., 25, 268-278 (1971).

20. W. H. Coulter, "High-speed automatic blood cell counter and cell size analyzer," Proc. Natl. Electr. Conf., 12, 1034-1040 (1956).

21. R. D. Allen, L. R. Zacharski, S. T. Widirstky, R. Rosenstein, L. M. Zaitlin, and D. R. Burgess, "Transformation and motility of human platelets," J. Cell. Biol., 83, 126-142 (1979).

BIOLOGICAL PROPERTIES AND
THROMBOLYTIC ACTIVITY OF
PROTEINASE FROM *Aspergillus terricola*

A. A. Kubatiev, S. V. Andreev,
and Ya. D. Mamedov

Institute for Advanced Training of Physicians,
Ministry of Public Health of the USSR, Moscow

INTRODUCTION

The present stage in the progress of medicine is characterized by a growing interest among researchers in the pharmacotherapy of thromboses [1-5]. This interest is related, above all, to the ever-increasing risk of thromboembolic diseases and the mortality resulting therefrom, caused in part by the rapid industrialization of society. The intense rhythm of everyday life, accompanied by constant emotional stress, hypodynamia, and changes in the composition and quality of the diet, has lately raised mortality from acute thromboses and emboli in some countries to a tremendously high level.

Over the last 30 years, in England for example, mortality from pulmonary embolism increased sixfold [6]. The percentage of fatal outcomes from acute arterial thromboses was also high during the same period in other economically developed countries [7]. According to [6, 7], the frequency of thromboembolic complications in surgical clinics in many countries throughout the world recently averaged 20-25% in all postoperative complications.

401

Such an avalanche of growth in the percentage of thromboembolic complications and the frequency of fatal outcomes has
stimulated studies on the development of effective drugs that
have targeted fibrinolytic and thrombolytic action. Many
preparations with these properties have been obtained so far.
The most active among them are fibrinolysin, a natural fibrinolytic agent (mainly a prolonged one), and its specific activators such as streptokinase and urokinase [2, 8], which are
effectively used in the treatment of thromboembolic diseases.

Unfortunately, the use of plasminogen activator is somewhat limited. Possessing pronounced therapeutic properties
in early thromboses, these agents are not always effective
against massive, organized thrombi, which are followed by
a stable decrease in endogenous resources of the fibrinolytic
protective mechanism [1, 2]. Complex therapy based on the
combined administration of anticoagulant, disaggregating agent,
plasminogen activator, and direct-acting thrombolytic agent
has been shown to be a more reliable means of conservative
treatment against thromboses and embolism [2, 3]. It is good
practice to use proteolytic enzymes as anticoagulants, including brinase isolated from the culture fluid of Aspergillus
oryzae [9], as well as proteinase from Aspergillus ochraceus
[10] and other producers of fungal and microbic origin [11-14].

In the Soviet Union, active proteinase from Aspergillus
terricola (later termed terrilytin) was first isolated in
1965 [15-19]. Terrilytin is a lyophilized white powder that
has no taste or odor and is readily soluble in water and 0.25%
novocain solution. Its specific proteolytic activity is not
less than 2 PU (proteolytic units)/mg of preparation, and
protein concentration does not exceed 3%.

Experiments at various laboratories have shown that terrilytin possesses both proteolytic and high fibrinolytic
activities [20, 21]. This chapter reviews some aspects in
obtaining terrilytin, and its physicochemical, pharmacodyna-
mic, and thrombolytic properties.

TERRILYTIN AND ITS PHYSICOCHEMICAL PROPERTIES

The first preparations of terrilytin were obtained under
laboratory conditions by Academician Imshenetsky et al. [15].
In studying the fibrinolytic activity of 122 museum strains
of fungi from the genus Aspergillus, the enzymic complex from

Aspergillus terricola (strain 3-374) was found to possess a pronounced capability to lyse fibrinous clots. Further analytic selection permitted the isolation of strain 5 from this fungus, which was characterized by stable morphocultural and enzymic properties, and which could be used as a producer of terrilytin.

By ultraviolet (UV) irradiation of the initial strain, a series of mutants were obtained which synthesized proteases with more pronounced fibrinolytic properties [16]. Bactericidal lamps with a wavelength of 253.7 nm were used as a source of UV rays. The radiation dose was $5.4-12 \cdot 10^3$ erg/mm^2. Mutants were selected on a medium of milk agar with tetrazolium blue (0.015%) in addition to the usual wort agar. This allowed the differentiation of mutant colonies according to their color and milk casein proteolysis zones [22].

The proteolytic activity of modified cultures was studied [18] in their depth cultivation on saccharose-mineral medium composed of 0.72% saccharose, 0.36% glucose, 1.3% KH_2PO_4, 0.20% KN_3, 0.12% $MgSO_4$, and 2 ml of microelement solutions containing, per liter, 723 mg $FeSO_4$, 440 mg $ZnSO_4$, and 200 mg $MnSO_4$. The pH of the medium was 6.8.

During selection, hundreds of colonies cultured at various doses of UV irradiation were studied. From these colonies, two mutants were isolated which possessed the strongest fibrinolytic activity.

Conidia from the initial strain were exposed to a single irradiation of $12 \cdot 10^3$ erg/mm^2 to produce mutant H-20; mutant 103-B was obtained using the same single reirradiation of mutant H-20 conidia. Mutant 103-B appeared to be more active when comparing the fibrinolytic activity of both strains. Protein yield was 70%, and proteolytic activity, studied by the method of Kunitz [45], was 6.5 PU.

Chromatographic studies using various ion exchangers and sorbents demonstrated the heterogeneity of the preparation [23]. In addition to neutral proteinase, terrilytin contains negligible amounts of acid amylase and protease autolysates. These components are well pronounced in chromatography on DEAE-cellulose (Fig. 16.1).

Individual proteinase obtained by gel chromatography is homogeneous in ultracentrifugation and electrophoresis [21].

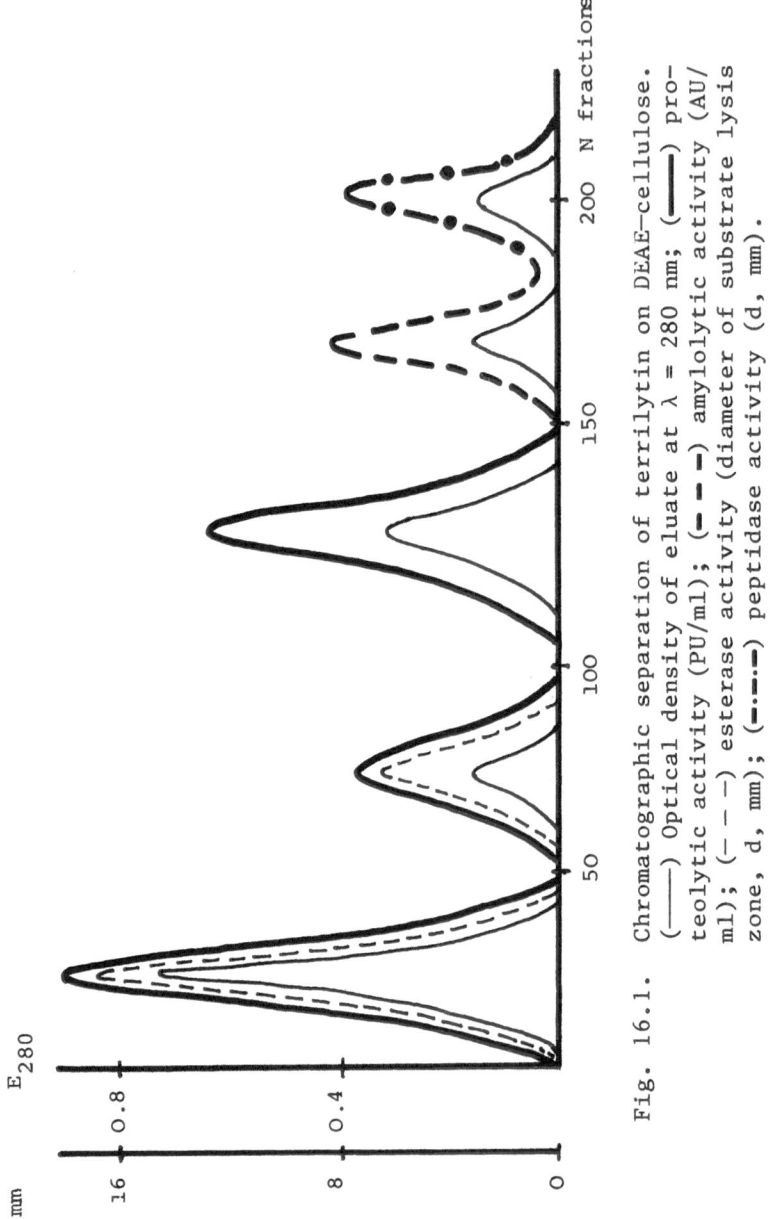

Fig. 16.1. Chromatographic separation of terrilytin on DEAE–cellulose.
(———) Optical density of eluate at λ = 280 nm; (———) proteolytic activity (PU/ml); (— — —) amylolytic activity (AU/
ml); (— — —) esterase activity (diameter of substrate lysis
zone, d, mm); (—·—·—) peptidase activity (d, mm).

The sedimentation coefficient (S) of the protease is $3 \cdot 10^{-13}$, while the diffusion coefficient is $10.4 \cdot 10^{-7}$ [23]. The molecular weight of the proteinase calculated from these parameters is 26,400 [21]; the isoelectric point, measured by polyacrylamide gel electrophoresis (pH ranging from 3.5 to 8.0 at 10°C), is 4.6; the optimum temperature is 45°C.

Terrilytin was autolyzed to form inactive products. Gel chromatography of the autolysate and the kinetic study of enzyme activity in solution versus time show that, during terrilytin autolysis, inactive impurities are formed, which eliminates the opportunity for identifying the N-terminal amino acid. Alanine, serine, glycine, asparagine, and glutamic acids were determined as N-terminal amino acids in autolysates, while monosaccharides such as mannose, glucose, and galactose were found in acidic hydrolysates of terrilytin analyzed by paper chromatography [24].

Analysis of the amino acid composition of the hydrolysate establishes the following general formula for terrilytin: $Lys_{13}His_4Arg_3Asp_{25}Trp_{13}Ser_{23}Glu_{16}Pro_4Gly_{25}Ala_{29}Val_{15}Met_1Iso$-$leu_{10}Leu_{12}Tyr_7Phe_6Thr_3H_3(amidic)$-17.

Protein isolated on biogel P-150 was separated further by Shatayeva et al. [21] by chromatography on carboxylmethyl-cellulose (CMC) using gradient elution. The choice of ionic strength gradient and pH values for NaH_2PO_4 between 0.003 M and 0.2 M permitted the authors to obtain two peaks on the elution curve in which peak 1 represents the polysaccharide and peak 2 represents the active protein with high, specific proteolytic and fibrinolytic activities.

In studying the composition of terrilytin by chromatography on DEAE-cellulose, five terrilytin components were isolated by Selezneva et al. [25, 26]. The first three peaks (i.e., proteases 1, 2, and 3) indicated the capability of hydrolyzing casein, hemoglobin, and tributyrin; peak 4 showed amylase activity and peak 5 showed peptidase activity. The rates of fibrin and fibrinogen hydrolysis of all the proteolytic components of terrilytin were comparable. Initial rates of fibrin hydrolysis (by these components) were higher than the initial rates of fibrinogen hydrolysis, indicating a specific affinity of the three proteolytic components for fibrin.

Proteinase and amylase activity depend essentially on the substrate pH. The enzyme was found to hydrolyze casein

most actively in an alkaline medium with a pH of 7.5-8.5.
Starch was hydrolyzed by amylase in an acidic medium, and
the maximum rate of substrate hydrolysis occurred at a pH
of 4.5-5.0.

INFLUENCE OF TERRILYTIN ON PLASMIC HEMOSTASIS

In small concentrations, terrilytin converts prothrombin
to thrombin [16]. A similar effect also seems to be produced
with large doses of terrilytin [21]. Among the various components of the blood coagulation system, fibrin, rather than
thrombin, is hydrolyzed most actively by terrilytin.

In an attempt to determine the amount of peptide chains
in thrombin, fibrin, and fibrinogen molecules which are capable of being cleft by terrilytin, Shatayeva et al. [21] calculated the K_M value during incubation of the preparation with
these components. The reaction rate was measured in μmole/ml·
sec of amino nitrogen in the concentration (μmole/ml). It
was impossible to determine the K_M value for the terrilytin—
thrombin complex. This indicates that no model can describe
this reaction based on the formation of an enzyme—substrate
complex, and that the increase in amino nitrogen concentration
in the system depends on thrombin dissociation.

The K_M value appeared to be three orders less for fibrin
than for fibrinogen. Provided that the chemical composition
and the amount of peptide bonds affected by terrilytin per
unit of body weight are similar for both fibrinogen and fibrin,
the affinity of terrilytin for the peptide bonds of the latter
is a hundred times as high as for the peptide bonds of fibrinogen.

· The components of plasmic hemostasis that are affected
by terrilytin were found by Mamedov et al. [27] to assume
peculiar conformations. In studying the physicochemical properties of prothrombin, thrombin, and fibrin(ogen) in various
combinations with terrilytin, the authors found substantial
differences in their spectra. The infrared spectrum of terrilytin features a high intensity of absorption bands within
800-1200 cm^{-1} where valence and deformational skeleton vibra-
tions, corresponding to such bonds as C—C (600-1500 cm^{-1}),
C—O (1000-1300 cm^{-1}), and C—N (1000-1200 cm^{-1}), are seen along
with peculiar amide absorption bands. This results from the
mere nature of terrilytin as a protein—carbohydrate complex.

At the same time, pure terrilytin is characterized by a spread spectrum and predominance of the integral intensity of the net absorption band over that at 1650 cm^{-1} [27]. The infrared spectrum is also characterized by the medium intensity band within 2450-2500 cm^{-1}, which is attributed to the valence vibration of the —OH bond with the —H bond. In the treatment of prothrombin with terrilytin, the spectral characteristics of prothrombin remained unaffected, while thrombin was markedly modified. In this instance, the relation between the intensities of amide I and amide II bands changed under the influence of low concentrations of terrilytin (Figure 16.2). The authors consider the behavior of the terrilytin—thrombin complex to be indicative of the destructive effect of the enzyme, proven by the broad spectrum absorption bands within 800-1200 cm^{-1}, the intensity of which is practically independent of the dose of terrilytin.

In the terrilytin—fibrinogen system, even minute quantities of the enzyme distorted the entire spectrum with the appearance of bands within 900-1200 cm^{-1}, which are peculiar to denatured proteins. Such behavior permits the assumption that fibrinogen is a substrate available for the preparation. Marked changes that are typical of denatured proteins were also found in the infrared spectra of the terrilytin—fibrin system (Fig. 2). Thus, both the proteolytic effects of terrilytin and its affinity for fibrinogen and fibrin are borne out by molecular spectroscopic data.

In in vitro experiments [28] we studied the influence of various terrilytin concentrations (5, 10, 50, 100, and 500 µg) on the lysis of fibrinous clots obtained from whole blood, plasma, and fibrinogen through the addition of equal volumes (0.1 ml) of thrombin solution. We simultaneously followed the behavior of fibrinous clots placed in equal volumes of native plasma and incubated with terrilytin in plasma as well. At 50-100 µg of the preparation, the time of clot lysis averaged 8.9 ± 2.0 h, while at 500 µg complete lysis occurred within 2.6 ± 0.4 h [28]. In 500 µg of the terrilytin in solution, lysis time decreased to 2.1 ± 0.3 h. Fibrinogen preincubation with terrilytin (5-10 µg in 0.05 ml) slightly inhibited clot formation to increase further lysis (averaging 1.4 ± 0.3 h at 500 µg).

There were no indications of clot lysis for 24 h. During clot incubation in the plasma and terrilytin mixture, 5-10 µg of the preparation were ineffective for clot lysis; at

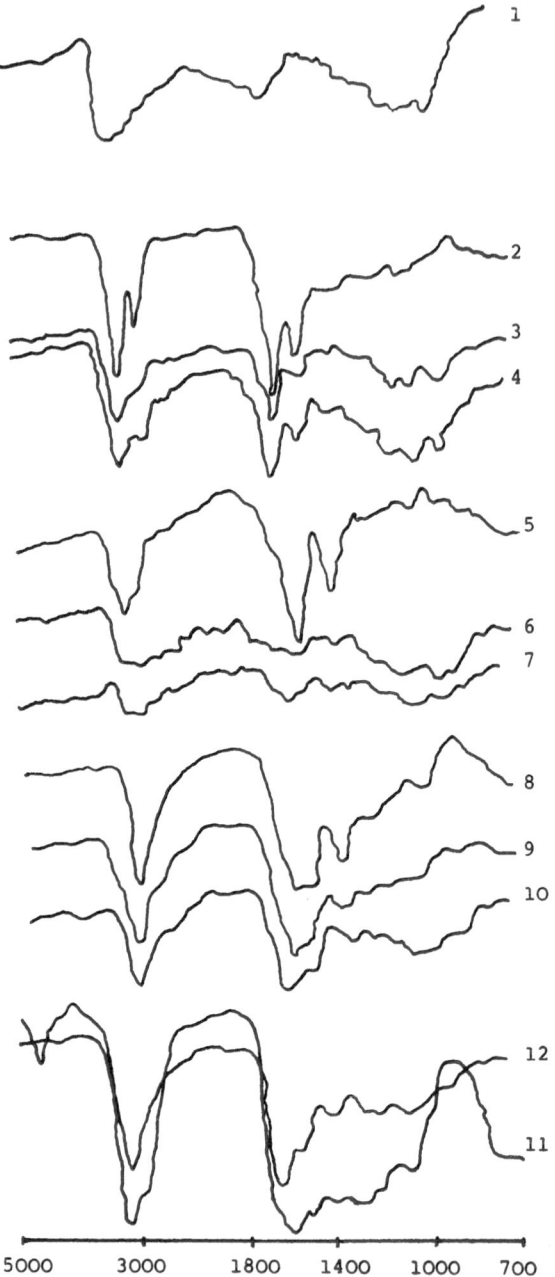

100 µg, the clot was lysed in 12.3 ± 1.1 h, and at 500 µg, in 3.7 ± 0.9 h.

In vivo study of terrilytin (in dogs and rabbits) demonstrated that the preparation, within effective lytic doses, significantly lengthened recalcification time, reduced thromboplastin and fibrinase activity, and increased plasma fibrinolytic activity. In addition, a decrease in plasma tolerance toward heparin and an increase in thrombin time, especially at 500 µg of terrilytin, were observed. Doses lower than 100 µg were less effective (Figure 16.3).

Recent research results indicate that it is possible to intensify the specific activity of terrilytin when combined with anticoagulants and plasminogen activators [29-31]. Administration of a mixture of terrilytin and heparin (60 PU/kg and 150 PU/kg, respectively) [29] decreased coagulation and accelerated activation of the fibrinolytic system. Thus, the combined mixture is more effective than injection of the enzyme alone.

Thirty minutes after administration of the mixture, the reaction time, R, on the thromboelastogram increased from 127.9 ± 10.3 sec to 175.3 ± 16.2 sec; clotting time formation, K, also increased, from 117.3 ± 8.3 sec to 632 ± 19.8 sec, while the specific coagulation constant, t, decreased from 764 ± 16 to 676 ± 14.2 sec. The solidification constant, S, reflected the deficiency of thromboplastin and thrombin formation due to an increase in the fibrinogen coagulation phase. The total coagulation constant was 2485 ± 10 sec compared to the norm of 967 ± 21 sec.

Fig. 16.2.   Effects of terrilytin on the infrared spectra of prothrombin, thrombin, fibrinogen, and fibrin. 1) Terrilytin; 2) prothrombin; 3) prothrombin treated with 1% terrilytin solution; 4) prothrombin treated with 15% terrilytin solution; 5) thrombin; 6) thrombin treated with 1% terrilytin solution; 7) thrombin treated with 15% terrilytin solution; 8) fibrinogen; 9) fibrinogen treated with 1% terrilytin solution; 10) fibrinogen treated with 15% terrilytin solution; 11) fibrin; 12) fibrin treated with 15% terrilytin solution.

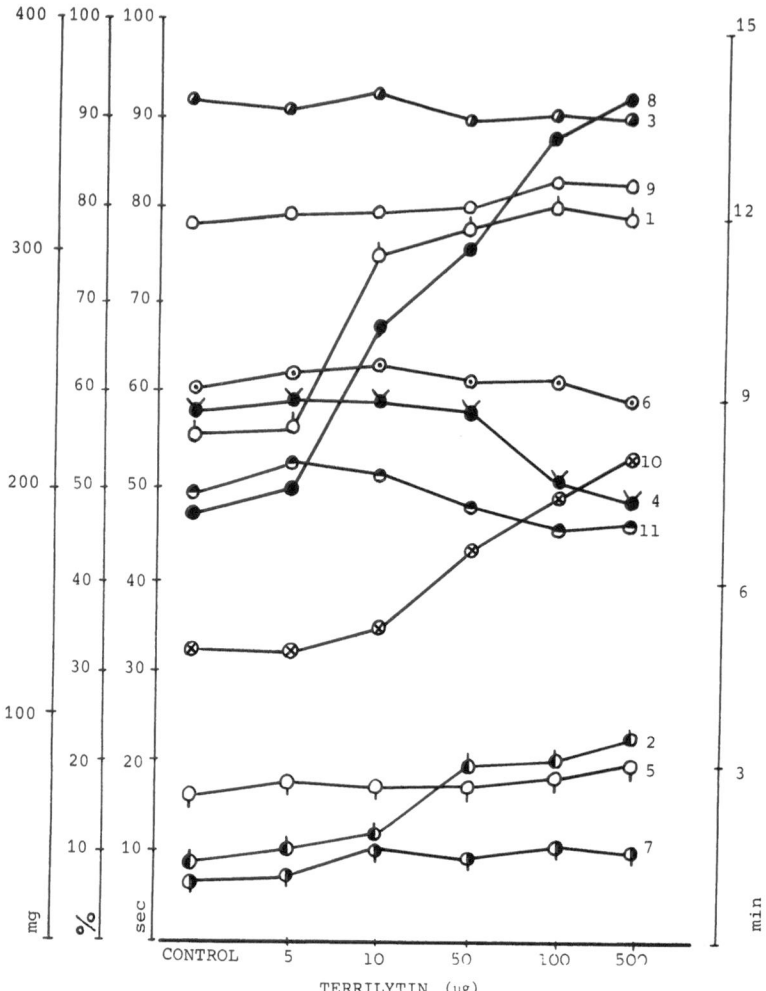

Fig. 16.3. Effect of terrilytin on coagulation parameters.
1) Thromboplastin activity (sec); 2) recalcification time (sec); 3) prothrombin index (%); 4)
activity of stable factors of prothrombin complex
(%); 5) thrombin time (sec); 6) heparin time
(sec); 7) free heparin (sec); 8) heparin tolerance test (min); 9) fibrinogen (g/liter); 10) fibrinolytic activity (%); 11) fibrinase (%).

        One hour after administration of the mixture, the reac-
tion time decreased. A simultaneous decrease in clotting

time and specific coagulation constant time was also observed.
The constant S showed an increase in the fibrinogen coagulation phase.

At 1 and 3 days, the activity of the prothrombin complex
continued to increase. By the end of the first day, the clotting time approached the norm, and exceeded it by the end
of the third day. On the second day, the specific coagulation constant was increased, and on the third day tended toward
a decrease. The duration of the fibrinogen coagulation phase
pointed to a deficiency in thromboplastin and thrombin formation: At 1 day it was 1554 ± 16 sec, and at 3 days 1400 ±
18 sec. The total coagulation constant for the same days
reached a high level (1599 ± 18 and 1480 ± 17 sec, respectively). All of the above indicate that fibrinolytic activation
induced by the mixture is more pronounced than by terrilytin
alone during the first days after injection [20].

Similar results have been obtained with the administration of terrilytin and sodium salicylate to animals [30].
One hour after injection, the thromboelastogram showed a hypo-
coagulable state. Maximum amplitude was 2.5 ± 0.82 mm (norm
is 0.8 ± 1.4 mm). Simultaneously, a sharp increase in R, K,
and t was recorded, indicating a decrease in activity of the
prothrombin complex factors and a marked deficiency in the
formation of thrombin and thromboplastin. At 3 hours, the
activity of the prothrombin complex remained below the norm,
the deficiency in thrombin and thromboplastin formation was
preserved, and fibrinolytic activity increased an average of
70%. By the end of the day, most of the indices had reached
the normal level, although the reaction time still exceeded
the initial level by a factor of 1.7.

A pronounced hypocoagulative effect was observed with
combined administration of terrilytin and nicotinic acid [31].
A single injection of these preparations in animals caused a
stable (within 24 h) inhibition of prothrombin activity in
the blood, with lengthening of the coagulation constant and
segment K, which is indicative of the inhibition of blood
coagulation due to a slowdown in fibrin coagulation and decreased clot elasticity.

The biological effect of proteolytic enzymes on the blood
coagulation system is known to be limited largely by natural
inhibitors occurring in the blood which reduce the fibrino-
and thrombolytic activity of proteinase [32-33].

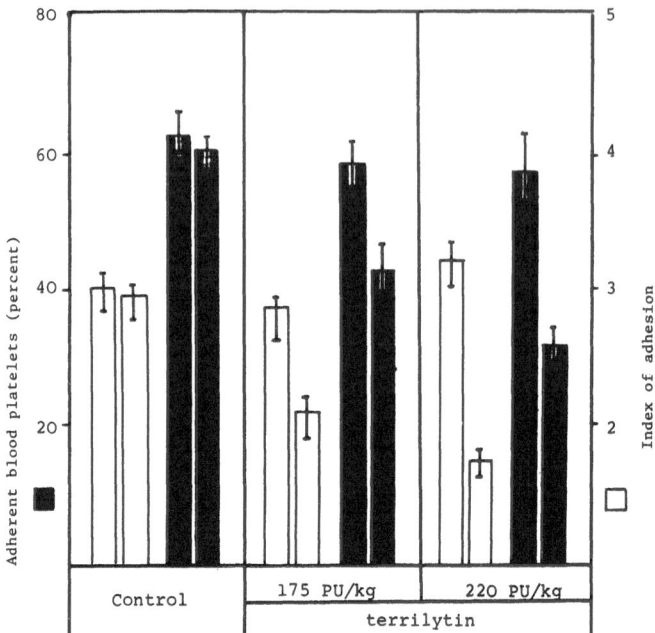

Fig. 16.4.   Effect of terrilytin on the adhesive properties
             of platelets.

        In comparing the inhibitory effect produced by blood
serum on caseinolytic activity of various proteolytic enzymes,
including Astra 1652 (Pharmacia), trypsin, thrombolytin, and
terrilytin, Shatayeva et al. [21] found that the inhibitory
effect is directed mainly at thrombolytin and trypsin; Astra
1652 is less affected and terrilytin remains virtually intact.
At the same time, the capability of these enzymes for splitting casein (per mg/protein) is the highest for terrilytin,
and is less pronounced for Astra 1652; it is minimal for trypsin and thrombolytin.

        Beattie et al. [32] have succeeded in isolating three
protein fractions by gel chromatography, each of which contained proteinase inhibitors. The first fraction was considered by the authors to have a single inhibitor, i.e., α_2-
macroglobulin (α_2M), with a molecular weight of 725,000-
820,000. The second fraction was comprised of several minor
inhibitors, including inter-α-trypsin and C1-inactivator.
The third fraction consisted mainly of α_1-antitrypsin (α_1A)
with a molecular weight of 52,000-54,000 [33].

Kinetic study of the inhibition of terrilytin by nonfrac-
tionated serum and $\alpha_1 A$ interaction showed the absence of an
inhibitory effect for 4 h [34]. Consequently, only $\alpha_2 M$ can
be regarded as inhibiting terrilytin.

Kozlova and colleagues [35] studied the inhibitory effect of a series of compounds (including salts of Ba, Ca,
Mg, Zn, Mn, Cu, Ni, Co, Cd, and Fe; ε-aminocaproic acid; L-
cysteine; and EDTA) on the activity of individual proteinase
obtained by gel chromatography on biogel P-150 and on a non-
purified terrilytin preparation. They found that Cu and Cd
ions at 10^{-2} M inhibit individual proteinase activity by 50%,
while terrilytin remained virtually intact. Salts of triva-
lent Fe at 10^{-4} M have the highest inhibitory effect on terrilytin and on individual proteinase. Both ascorbic acid and
EDTA slightly affect enzyme activity, and L-cysteine (at 10^{-2}
M) fully inhibits caseinolytic activity of the enzyme.

INFLUENCE OF TERRILYTIN ON PLATELET HEMOSTASIS

Terrilytin was recently found to affect both plasmic
and platelet hemostasis [36, 37]. Experiments on dogs and
rabbits with various vascular thromboses demonstrated that
administration of 10-100 PU/kg of terrilytin reduces platelet
adhesiveness. The amount of platelets decreases by 15% and
the adhesion index is reduced by 18.3% (Figure 16.4). Platelet aggregation is inhibited simultaneously [36].

Provided that 10-50 PU/kg of terrilytin are injected,
the onset of aggregation is postponed by 60%, and its duration
is shortened by 20%. The intensity of aggregation, measured
as % of optical density fall, is decreased by 12%. These
changes seem to be more pronounced when 50-100 PU/kg are administered.

Even injections of microdoses of terrilytin decreased
platelet aggregation unlike other proteolytic enzymes, such
as brinase, which caused a temporary increase in aggregation
[9]. Preincubation of platelet-rich plasma from healthy animals with terrilytin led to the inhibition of platelet aggregation caused by various inducers. Terrilytin greatly hindered platelet aggregation induced by adenosine diphosphate
(ADP) and epinephrine, and affected a process induced by thrombin and serotonin to a lesser extent (Table 16.1).

Table 16.1. Effects of Terrilytin on Induced Aggregation of
Platelets

Aggregating agents	Control	Terrilytin (final concentration)		
		0.1 µg	10 µg	100 µg
ADP ($1 \cdot 10^{-5}$ M)	53*	30	19	10.6
Collagen (50 mg/ml)	66*	48	30	22.3
Thrombin (0.5 U/ml)	78*	50	34	20.5

*Optical density

An extra series of experiments was undertaken to study
the mechanisms of antiaggregating activity of terrilytin.
During incubation of human fibrinogen at 37°C for 30 min with
concentrations increasing from 5 to 50 PU/ml, we observed
the appearance of a number of fibrinogen degradation products
in the mixture. With subsequent addition of equal amounts of
platelet-rich plasma to the mixture, platelet aggregation induced by ADP and serotonin was strongly inhibited. This finding allowed us to assume that terrilytin can inhibit platelet
aggregation due to the accumulation of fibrinogen degradation
products in plasma.

THROMBOLYTIC PROPERTIES OF TERRILYTIN

Studies in Rabbits

The thrombolytic effect of terrilytin was studied in 150
chinchilla rabbits with pulmonary thrombosis of an immunological genesis [38-41]. To induce experimental thrombosis,
the rabbits were injected with microdoses of thrombin in combination with atropine for several days and, when fibrinolytic
activity in the blood decreased by 25-30%, antipulmonary immunoglobulins were injected intravenously. Two to three hours
after the injection, diffusive aggregation of platelets was
observed in the capillary lumens of the rabbit lungs, followed
by development of complete thrombosis of the small, medium,
and large vessels over the next 5-6 h. On day 5-6, the rabbits
died from increasing heart failure (cardiac insufficiency).

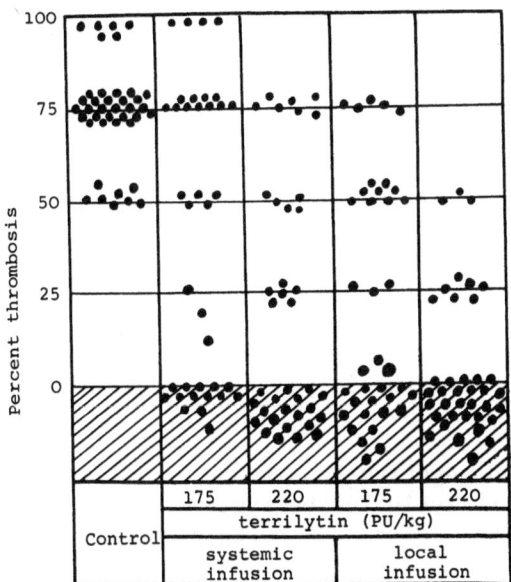

Fig. 16.5. Thrombolytic effect of terrilytin in rabbits with experimental thrombosis of the pulmonary vessels.

    The progress of thrombosis was monitored by hemodynamic study of the pulmonary circulation, determination of the contractility of the right ventricle, and morphometric assay of autopsy findings. On the second day of thrombosis, 5 ml of 175-220 PU/kg of terrilytin in polyvinylpyrrolidine solution were injected once into the experimental rabbits in either the systemic circulation (into the lateral veins of the ear) or locally (into the pulmonary artery) through a polyethylene catheter. The test animals received polyvinylpyrrolidine and/or physiologic solution in equal volumes. Twenty-four hours after injection of the terrilytin, the rabbits were killed and the pulmonary vessels were assayed for visual and macromicroscopic examination. The degree of thrombolysis was calculated as a % according to the thrombotic index value [14].

    The results of the experiments showed that pulmonary thrombosis developed in all (40) rabbits of the control group. In most of the rabbits (26), thrombotic masses obturated on the second day up to 75% of the pulmonary vessel bed with vast lobar pulmonary infarctions. In eight rabbits, the occlusion area did not exceed 50% of the whole vascular bed, and in six rabbits it covered both branches of the pulmonary

artery. Visual study showed solid, massive thrombi, and
there were no cases of spontaneous lysis during the entire
period of observation.

With systemic injections of terrilytin in the general
blood flow, the pulmonary vessels lacked thrombotic masses
in 34 of 75 rabbits (14 rabbits were injected with 175 PU/kg
and 20 rabbits with 220 PU/kg).

Despite the complete permeability of large vessels in
six rabbits given 175 PU/kg, histologic control demonstrated
fibrinous deposits in the pulmonary capillaries (less than
800 µM). Small hemorrhagic pulmonary infarctions were observed in four animals.

We obtained negative results in 23 rabbits which received
175 PU/kg of terrilytin in the systemic circulation and in
18 rabbits which received 220 PU/kg of terrilytin.

Even in the latter group of animals, however, the general area of thrombotic occlusion was morphometrically less
than in the control group (Figure 16.5). Of the 18 rabbits
given 220 PU/kg of terrilytin, thrombi were found in 6 animals
in not more than 25% of the pulmonary vessels; 50% of the
pulmonary vessels were injured in 5 rabbits, and 75% were injured in 7 rabbits.

With the smaller dose (175 PU/kg), the (100%) intensity
of thrombosis was somewhat higher: Complete thrombosis of
the main pulmonary artery was revealed in 4 rabbits, 75% in
11 rabbits, 50% in 5 rabbits, and 25% in 3 rabbits.

The response to all doses of terrilytin was good. Only
in eight rabbits given 220 PU/kg of terrilytin did we observe
a transitory (3-5 min) excitement, followed by several spasms
of the lower extremities.

With local infusion of the preparation (75 rabbits), a
positive lytic effect was seen in 48 rabbits, most of which
(28) were given 220 PU/kg of terrilytin. Complete lysis was
seen in most rabbits; in only 7 rabbits were small scraps of
thrombi found in the lumen of the pulmonary artery which, however, did not affect the circulation of blood in these vessels.
Thrombolysis occurred in 20 of 37 rabbits given 175 PU/kg of
terrilytin. The pulmonary vessels remained obturated by 75%
in 5 rabbits, by 50% in 9 rabbits, and by 25% in 3 rabbits
(Figure 16.5).

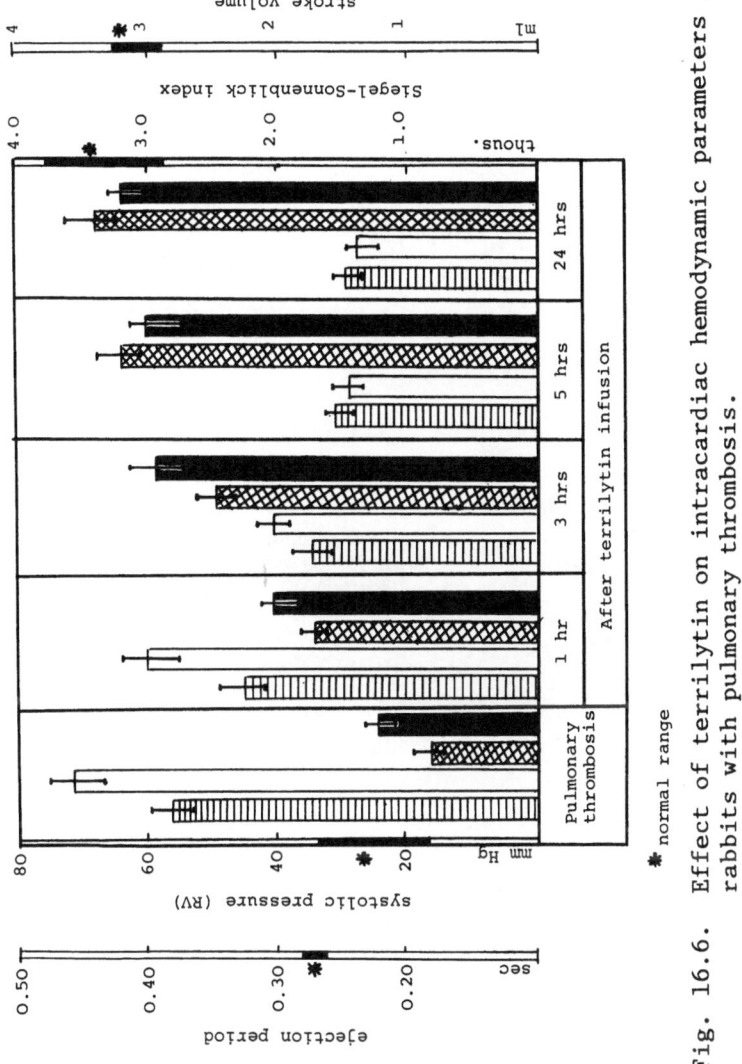

Fig. 16.6. Effect of terrilytin on intracardiac hemodynamic parameters in rabbits with pulmonary thrombosis.

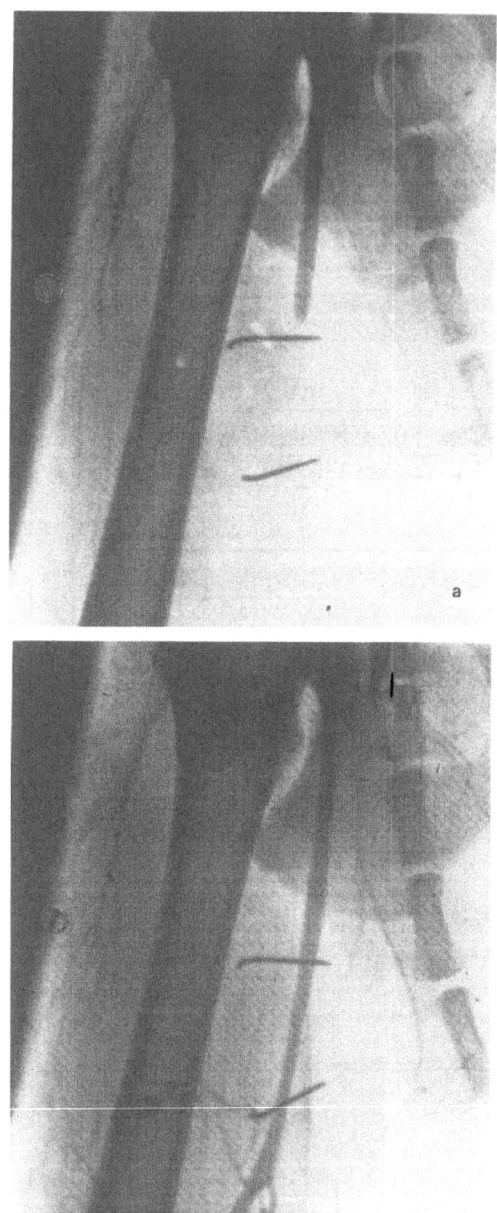

No toxic effect was detected. In addition, upon administration of 220 PU/kg of terrilytin, negligible perivascular
hemorrhages were observed which occurred less frequently in
animals receiving the lower dose.

In three rabbits which had a negative response to a dose
of 220 PU/kg, thrombosis covered over 50% of the total pulmonary artery area, and in seven rabbits only 25% of one of its
branches. With reinjection of the preparation in these animals within 24 h, a well-detected lytic effect was obtained
in four more rabbits (in three at 220 PU/kg and in one at
175 PU/kg).

Study of the hemodynamic parameters of the pulmonary
circulation during treatment with terrilytin showed that the
first signs of improvement in the pulmonary circulation associated with thrombolysis are observed 3-4 h after injection
of the preparation. Within 6-8 h, the intrapulmonary hemodynamic parameters are normalized. Rehabilitation of right
ventricular contractility is simultaneously observed (Figure
16.6).

These experimental results illustrate that the administration of terrilytin to animals with pulmonary thrombosis leads
to pronounced thrombolysis and rehabilitation of the disturbed
circulation. This effect depends strongly on the dose of
preparation used and the method of application. Analysis of
the results shows that local infusion of terrilytin to an injured vessel site is the most effective measure against thrombosis. With local infusion, the signs of normalized pulmonary
hemodynamic parameters occur much earlier than with systemic
injection. This observation definitely indicates earlier
thrombolysis in the injured pulmonary vessels.

It should be noted that, unlike previous studies [20],
in which both lytic and toxic properties of the preparation

Fig. 16.7.   Restitution of blood flow in thrombosed canine
            femoral artery after administration of terrilytin.
            The site of thrombosis is marked by the staples.
            a) Right femoral artery before intra-arterial in-
            fusion of terrilytin; b) same artery 2 h after
            infusion of terrilytin (10 PU/kg).

were detected within its effective therapeutic doses, no complications were observed in the present study despite the
use of even larger doses of terrilytin. In our opinion [38],
this result is associated with the use of polyvinylpyrrolidine solution as a terrilytin solvent instead of physiologic
solution. Polyvinylpyrrolidine solution produced a marked
detoxifying effect, favorably affected systemic and regional
hemodynamics, and rehabilitated the disturbed acid–alkaline
balance.

Studies in Dogs

This series of experiments was accomplished in 25 mongrel dogs which had old (24 h) thrombosis of the femoral
arteries and veins [43]. A model of thrombosis was developed
according to the method described previously [28].

The dogs were anesthetized with raush (producing narco-
sis). A 5- to 6-cm-long site of the appropriate vessel was
exposed and ligated with two ligatures 3 cm apart. The vessel chamber was annulated. The internal endothelial layer
was scarified by circular motions and 0.5 ml of thrombin (in
20 s) was then injected. A blood clot was formed with 2 to
3 min in the ligated chamber; this was transformed by the
sixth hour into a typical dense thrombus. After the thrombus had formed, the distal ligature was removed and the proximal ligature was slightly loosened.

Control vasography, conducted for 3 days after thrombosis
was established, showed massive occlusions of the vessel lumen
by thrombi which were not spontaneously lysing. On the second
day after thrombi formation 10 PU/kg of terrilytin was administered once, as follows: in the systemic circulation (intravenously in 14 experiments, intra-arterially in 3 experiments),
locally (into the site of thrombus localization in 3 experiments), and through regional perfusion using an artificial
circulation unit (5 experiments).

With intra-arterial administration of terrilytin to dogs
with thrombosis of the femoral vein, no toxic response was
seen; complete lysis of the thrombus was observed in 2 of
the 3 dogs (Figure 16.7). The vasogram of one dog showed
an improvement in collateral circulation. Intravenous infusion appeared to be effective in 6 of the 14 experiments,
leading to complete lysis in 5 experiments and to partial
lysis in 1 experiment.

With local injection directly to the site of the thrombi, complete lysis was observed in the three dogs within 40-55 min from the beginning of treatment. Regional terrilytin perfusion produced a positive effect in three of the five cases, leading to complete lysis in one case and partial lysis in two cases.

Thus, comparative study of the efficacy of terrilytin with different methods of administration indicates that local infusion of the drug is advantageous. The mean time of complete thrombolysis of dogs with systemic infusion was 8.5 ± 4.2 h, whereas the mean time with local infusion was 50 ± 5 min.

In a special series of studies in 12 dogs with thrombosis induced in isolated vessel chambers, thrombolysis was observed still earlier with local administration of terrilytin [44]. In these experiments, the dogs were premedicated with morphine and atropine. A 10-cm-long site of the jugular vein was then separated and ligated by four ligatures 3 cm apart to form three isolated chambers in which thrombosis was induced according to the procedure described above.

At 6 h, when thick thrombi adhering to the vessel wall in all three chambers were palpable, 10 PU of terrilytin in a polyvinylpyrrolidine solution were administered to one (test) chamber. The same amount of physiologic solution was administered to the second and third (control) chambers, respectively. In the test chamber, the thrombus began to soften by the 20th min and complete lysis was shown at the 30th to 40th min by cytologic analysis of the chamber contents. Small scraps of thrombi, which were slightly contrasted in the vasograms, remained in the test chambers at the 40th min in four cases. No thrombolysis was observed in the control chambers.

CONCLUSIONS

The above data show that proteinase isolated from the culture fluid of the fungus _Aspergillus_ _terricola_, later termed terrilytin, possesses both proteolytic and fibrinolytic activity. In contrast to other proteinases of a fungal origin, including brinase, ochrase, and tricholyase, which are characterized by a rather narrow spectrum of therapeutic effect, terrilytin affects all of the components of hemostasis. For

this reason, we recommend terrilytin for the treatment of
both venous and arterial thromboses when functional disturbances of vascular-platelet hemostasis play the leading pathogenic role.

Still, it should be remembered that the lytic effect of
terrilytin depends greatly on time, dose, and method of application. A maximum therapeutic effect, leading to complete
lysis of thrombi, is achieved in young (6 h) thromboses. With
enhanced thrombotic occlusion, the lytic ability of terrilytin decreases gradually. Nevertheless, a high thromobolytic
rate can be achieved 24 h after the development of a thrombotic process by effective use of different doses and methods
of administration.

Best results are obtained with local infusion of terrilytin to the site of thrombosis, which leads to complete
lysis of the thrombotic masses in 70-100% of the cases.

When infusion is accompanied by the administration of
plasminogen anticoagulants and activators, the rate of thrombolysis can be increased significantly by both local and systemic use of the preparation.

REFERENCES

1. O. K. Gavrilov, Problems and Hypotheses in the Theory
    of Hemocoagulation, Meditsina, Moscow (1980).
2. E. I. Chazov and K. M. Lakin, Anticoagulants and Fibrin-
    olytic Methods, Meditsina, Moscow (1977).
3. G. N. Barlow, "Pharmacology of fibrinolytic agents,"
    Prog. Cardiovasc. Dis., 21, 315-326 (1975).
4. D. Deykind, "Antithrombotic therapy: rationale and ap-
    plication," Postgrad. Med., 65, 135-140 (1979).
5. M. Verstraete, "Theoretical basis of thrombolytic treat-
    ment," Pathol. Biol., 23, 239-245 (1979).
6. M. I. Kuzin, "Modern aspects of prevention and develop-
    ment of thromboses in surgical practice," in: Current
    Problems of Hemostasiology, B. V. Petrovsky, E. I. Chazov,
    and S. V. Andreev (eds.), Nauka, Moscow (1979), pp. 219-
    225.
7. A. Sion and V. Lopez-Majano, "Incidence of pulmonary em-
    bolism," Respiration, 35, 181-185 (1978).
8. E. I. Chazov, A. V. Mazaev, V. P. Torchilin, and V. N.
    Smirnov, "Application of biocompatible preparations of

immobilized enzymes in therapy of thromboses," in: Current Problems of Hemostasiology, B. V. Petrovsky, E. I. Chazov, and S. V. Andreev (eds.), Nauka, Moscow (1979), pp. 191-200.

9. W. H. E. Roschlaw and A. M. Fisher, "Thrombolytic therapy with local perfusions of CA-7 (fibrinolytic enzyme from Aspergillus oryzae) in the dog," Angiology, 17, 670-682 (1966).

10. H. P. Klocking and F. Markwardt, "Uber die fibrinolytische Wirkung einer aus Aspergillus ochraceus isolierten Protiase," Acta Biol. Med. Ger., 26, 35-40 (1971).

11. G. V. Andreenko, T. N. Serebryakova, R. A. Maximova, and A. B. Silaev, "Fibrinolytische und thrombolytische Wirkung von Proteasen einiger Pilzkulturen," Folia Haematol., 101, 14-21 (1974).

12. N. S. Egorov, V. I. Ushakova, and L. M. Nikolskii, "On the capacity of some microorganisms for production of fibrinolytic compounds," Dokl. Akad. Nauk SSSR, 165, 217-220 (1965).

13. N. S. Egorov and V. I. Ushakova, "On fibrinolytic and thrombolytic activities of some bacteria in vitro," Biol. Nauk, 6, 93-98 (1968).

14. B. A. Kudrjashov, G. V. Andreenko, and N. S. Egorov, "Fibrinolytic agents isolated from cultures of some saprophytic fungi," Dokl. Akad. Nauk SSSR, 153, 939-942 (1963).

15. A. A. Imshenetsky, S. Z. Brozkaya, and V. V. Korshunov, "Effect of some fungal proteinases on blood thrombi," Dokl. Akad. Nauk SSSR, 163, 737-740 (1965).

16. A. A. Imshenetsky and S. Z. Brozkaya, "Selection of microorganisms with thrombolytic activity," Mikrobiologiia, 38, 1043-1049 (1979).

17. A. A. Imshenetsky, I. D. Kasatkina, S. Z. Brozkaya, and V. V. Korshunov, "Fibrin-lysing microbial enzymes," in: Problems of Medical Enzymology, Meditsina, Moscow (1970), pp. 270-277.

18. A. A. Imshenetsky, I. D. Kasatkina, S. Z. Brozkaya, and E. T. Zheltova, "Selection of active races of Aspergillus terricola — products of terrilytin," in: Enzymes for Medical Use, Meditsina, Leningrad (1975), pp. 16-25.

19. V. V. Korshunov, "Isolation and properties of proteinase of Aspergillus terricola," Mikrobiologiya, 38, 238-244 (1969).

20. G. E. Grinberg, "Terrilytin — a proteolytic enzymatic drug," Khim. Pharm. Zh., 10, 145-154 (1976).

21. L. K. Shatayeva, O. V. Orlievskaya, and G. V. Samsonov, "Effect of terrilytin, an enzymic preparation, on some

424   A. A. KUBATIEV ET AL.

components of the blood coagulation system," in: Problems of Medical Enzymology, Meditsina, Moscow (1970),
pp. 278-287.
22. I. D. Kasatkina, A. A. Imshenetsky, S. Z. Brozkaya, and
E. T. Zheltova, "The mutants of Aspergillus terricola
producing proteases with fibrinolytic activity," Mikrobiologiya, 38, 766-774 (1969).
23. O. V. Orlievskaya, L. K. Shatayeva, and G. V. Samsonov,
"Examination of the thrombolytic preparation terrilytin
and isolation of its individual enzymes," Prikl. Biokhim.
Mikrobiol., 7, 355-359 (1971).
24. G. V. Samsonov, L. K. Shatayeva, and O. V. Orlievskaya,
"Physicochemical and enzymatic properties of protease
from Aspergillus terricola," Dokl. Akad. Nauk SSSR, 206,
497-499 (1972).
25. A. A.Selezneva, T. A. Kozlova, and G. V. Samsonov, "A
comparative study of enzymatic activity and inhibition
of terrilytin and some proteases with thrombolytic action,"
Prikl. Biokhim. Mikrobiol., 8, 526-531 (1972).
26. A. A.Selezneva, G. A. Babenko, M. D. Bolshakova, T. I.
Rozhanskaya, and N. A. Margolina, "Preparative isolation
of terrilytin components and examination of their properties," Prikl. Biokhim. Mikrobiol., 12, 416-420 (1976).
27. Ya. D. Mamedov, R. G. Safarov, O. A. Narimanbekov, G. A.
Guseinov, and A. V. Reish, "Changes in physicochemical
properties of the blood coagulating system components
under the effect of terrilytin," in: Current Problems
of Hemostasiology, B. V. Petrovsky, E. I. Chazov, and
S. V. Andreev (eds.), Nauka, Moscow (1979), pp. 95-99.
28. S. V. Andreev, A. A. Kubatiev, V. A. Yurkiv, and N. L.
Koltsova, "On thrombolytic activity of terrilytin, a new
enzymatic preparation from Aspergillus terricola," Byull.
Eksp. Biol. Med., 8, 936-938 (1976).
29. Ya. B. Mamedov, G. A. Guseinov, and A. V. Reish, "Changes
in aggregation abilities of thrombocytes in application
of terrilytin and heparin," in: Current Problems of Hemostasiology, B. V. Petrovsky, E. I. Chazov, and S. V.
Andreev (eds.), Nauka, Moscow (1979), pp. 290-292 (1979).
30. Ya. D. Mamedov, G. A. Guseinov, and A. V. Reish, "Changes
in hemocoagulation process under the effect of terrilytin-
sodium salicylate complex," in: Current Problems of Hemostasiology, B. V. Petrovsky, E. I. Chazov, and S. V.
Andreev (eds.), Nauka, Moscow (1979), pp. 292-294.
31. Ya. D. Mamedov, G. A. Guseinov, and A. V. Reish, "Anticoagulation properties of terrilytin—nicotinic acid complex," in: Current Problems of Hemostasiology, B. V.

Petrovsky, E. I. Chazov, and S. V. Andreev (eds.), Nauka, Moscow (1979), pp. 255-258.

32. A. G. Beattie, D. Ogston, B. Bennet, and A. S. Douglas, "Inhibitors of plasminogen activation in human blood," Br. J. Haematol., 32, 135-143 (1976).

33. H. Rinderknecht and M. C. Geokas, "On the physiological role of α_2-macroglobulin," Biochim. Biophys. Acta, 295, 23-24 (1973).

34. G. M. Lindenbaum and I. M. Tereshin, "Interactions between human blood serum inhibitors and native and modified dextran proteinases — terrilytin and trypsin," Biokhimiya, 43, 2143-2149 (1980).

35. T. A. Kozlova, A. A. Selezneva, and G. E. Grinberg, "Study of inhibitory capacities of some compounds and rabbit blood serum on terrilytin activity," in: Proceedings of the Leningrad Scientific-Research Institute of Antibiotics, Vol. IX (1972), pp. 55-57.

36. S. V. Andreev and A. A. Kubatiev, "On the experimental application of terrilytin, a new thrombolytic preparation," in: Modern Problems of Thrombosis and Emboli, Moscow (1978), pp. 87-88.

37. S. V. Andreev, A. A. Kubatiev, I. D. Kobkova, and V. A. Yurkiv, "Terrilytin — a new thrombolytic preparation," in: Current Problems of Hemostasiology, B. V. Petrovsky, E. I. Chazov, and S. V. Andreev (eds.), Nauka, Moscow (1979), pp. 249-254.

38. A. A. Kubatiev, "Experimental efficacy of terrilytin in pulmonary vascular thrombosis," Byull. Eksp. Biol. Med., 3, 214-217 (1979).

39. A. A. Kubatiev and S. V. Andreev, "Immune thrombosis of the pulmonary vessels," Thromb. Haemostas., 38, 245 (1977).

40. A. A. Kubatiev and S. V. Andreev, "The possible use of terrilytin, an Aspergillus terricola proteinase, in the treatment of pulmonary thrombosis," Thromb. Haemostas., 42, 389 (1979).

41. S. V. Andreev and A. A. Kubatiev, "On thrombolytic activity of terrilytin, a new proteolytic enzymatic preparation from Aspergillus terricola," in: Abstracts of Eighth World Contress of Cardiology, Tokyo (1978), p. 18.

42. K. Boruach, R. N. Chakravarti, and P. L. Wahi, "Evaluation of xanthinol nicotinate (complamin) in experimental pulmonary fibrin embolism," Indian J. Med. Res., 62, 923-929 (1974).

43. S. V. Andreev and A. A. Kubatiev, "Experimental study of thrombolytic activity of a new proteolytic enzyme terrilytin," in: Abstracts of Thirteenth World Congress

of the International Cardiovascular Society, Tokyo (1977), C-4-9.

44. S. V.Andreev, A. A. Kubatiev, I. D. Kobkova, and V. A. Yurkiv, "On thrombolytic activity of terrilytin," in: Progress in the Production and Study of Antibiotics, Vol. 8, Nauka, Moscow (1980), pp. 84-94.

45. M. Kunitz, "Crystalline soybean trypsin inhibitor. II. General properties," J. Gen. Physiol., 30, 291-293 (1947).

INDEX